THE COMMON BUZZARD

THE COMMON BUZZARD

SEAN WALLS and ROBERT KENWARD

T & AD POYSER
LONDON · OXFORD · NEW YORK · NEW DELHI · SYDNEY

T & AD POYSER
Bloomsbury Publishing Plc
50 Bedford Square, London, WC1B 3DP, UK

BLOOMSBURY, T & AD POYSER and the T & AD Poyser logo are trademarks of
Bloomsbury Publishing Plc

First published in the United Kingdom 2020

Copyright © Sean Walls and Robert Kenward, 2020
Illustrations © Alan Harris, 2020
Diagrams © Julian Baker, 2020, pages 12, 13, 101, 172, 176, 239, 245

Copyright © photographer named in plate caption, with the exception of:
1a Menno Schaefer, www.shutterstock.com; 1b Bildagentur Zoonar GmbH, www.shutterstock.com;
1c Marcin Perkowski, www.shutterstock.com; 2a grusgrus444, www.shutterstock.com; 3 Giorgios Alexandris,
www.shutterstock.com; 4 (left) Ger Bosma, www.shutterstock.com; 4 (right) Chris Humphries, www.shutterstock.com;
6 KOO, www.shutterstock.com; 11 Terry Brooks, www.shutterstock.com; 16 Piotr Krzeslak, www.shutterstock.com;
17 Bernd Wolter, www.shutterstock.com; 21 Vishnevskiy Vasily, www.shutterstock.com; 24 Marcin Perkowski,
www.shutterstock.com.

Sean Walls and Robert Kenward have asserted their right under the Copyright,
Designs and Patents Act, 1988, to be identified as Authors of this work.

All rights reserved. No part of this publication may be reproduced or transmitted in any form or
by any means, electronic or mechanical, including photocopying, recording, or any information
storage or retrieval system, without prior permission in writing from the publishers.

Bloomsbury Publishing Plc does not have any control over, or responsibility for, any third-party websites
referred to in this book. All internet addresses given in this book were correct at the time of going to press.
The authors and publisher regret any inconvenience caused if addresses have changed or sites have
ceased to exist, but can accept no responsibility for any such changes.

A catalogue record for this book is available from the British Library.
Library of Congress Cataloguing-in-Publication data has been applied for.

ISBN: HB: 978-1-4081-2525-0
PB: 978-1-4729-7208-8
ePDF: 978-1-4729-7002-2
ePub: 978-1-4729-7001-5

2 4 6 8 10 9 7 5 3 1

Design by Mark Heslington Ltd, Scarborough, North Yorkshire
Printed and bound in India by Replika Press Pvt. Ltd.

To find out more about our authors and books visit
www.bloomsbury.com and sign up for our newsletters.

Contents

Preface 7

1 A Common Buzzard 11
2 Prey 31
3 Hunting..58
4 Habitat use 76
5 Territoriality and nest defence 98
6 Courtship and nesting...124
7 Incubating and chick-rearing 140
8 Dispersal and migration 167
9 Longevity and survival 196
10 Common Buzzard populations...................................221
11 Our relationship with the Common Buzzard 243

Appendix 1 Scientific names of species mentioned in the text 265
Appendix 2 References for figures with many sources.............................267
Appendix 3 Map of UK study sites and flightpaths 269
Appendix 4 Abbreviations..270
References..271
Index 296

Dedicated to Ibbie, Rosie and Ella, Bridget, Ben and James

Preface

We have undertaken fieldwork together on Common Buzzards since 1989. Before that, we had enjoyed separate experiences, which led us to our study. Sean Walls (SW) grew up in suburbia but loved any opportunity for getting into the countryside to be captivated by nature, especially raptors. As a young boy, the Common Buzzard was the most exciting bird he could reasonably expect to find in England. Birds of prey had the 'Wow!' factor for a young mind. They looked so beautiful, and yet menacing, with their curved talons, hooked bill and penetrating stare. Of course, Common Kestrels *Falco tinnunculus* were more common around his home on the south coast of England, but the Buzzard's size and its great, lamenting cry was truly enthralling. Buzzards were just common enough to enjoy regularly when visiting the New Forest or on holiday in Wales and the Lake District. It felt as though this was the nearest one could get to seeing eagles, because real eagles only lived far to the north. So, when SW found *The Buzzard* by Colin Tubbs (1974) in the school library, it was rapidly read from cover to cover. Little did he realise that more than 30 years later an opportunity would present itself to write another Buzzard monograph, to bring together all the more recent work and update a book written in the early 1970s, a book that fascinated and still charms him whenever he sees its cover. Indeed, Tubbs' book almost seems a precursor to the Poyser bird monographs that followed. His enthusiasm for wildlife led SW to do a zoology degree at the University of Oxford, as a result of which he enjoyed conservation-based expeditions to Kenya and Madagascar.

Robert Kenward (RK) had grown up on a farm, become a falconer and then also studied at Oxford, where he learnt to view life objectively as well as with the patience and passion of a naturalist. In his childhood in the 1950s there was little chance of seeing Buzzards in east Hampshire. However, during the 1960s he spent long summer holidays in the Scottish Highlands, where there were Buzzards and rarer raptor species to be seen and heard. While at Oxford, RK learnt to build radio-tags and ran projects tracking hundreds of Northern Goshawks *Accipiter gentilis* in Sweden during 1976–1987; those years informed how we were to study Buzzards and enabled a manuscript, started in 1979, to be published in the Poyser series as *The Goshawk* in 2006.

Our interests converged at Biotrack Ltd, another product of RK's early years. Biotrack underpins research on birds and other animals, both in terms of the tracking technology and by instilling an ethos to help biologists make successful science careers in ecology. We came together because we had both been working nearby, SW with the Game Conservancy (now the Game and Wildlife Conservation Trust) in Hampshire and RK with the Institute of Terrestrial Ecology (ITE), now the Centre for Ecology and Hydrology (CEH), just to the west in Dorset. SW joined RK to work as a research assistant in a post funded by Biotrack, which was run at that time by Bridget Kenward and Brian Cresswell.

We greatly enjoyed our time studying Buzzards and published much of our work as scientific papers. The contract to write this book spurred us to bring more of that work together. We could not publish all our material earlier because we had both reached managerial positions that did not support such work. However, that management experience gave us a greater understanding of the need for developments to be socially and economically sustainable, as well as ecologically sustainable, a major consideration in our Buzzard research.

We feel tremendous gratitude to all those who have made the book possible, and we would like to thank many people and organisations. To start with, projects like our research on the Common Buzzard cannot run without funding. Our primary source was the Natural Environment Research Council (NERC), through CEH. However, there was also important start-up funding from Natural England, plus shared costs, for example for flights, with the Royal Society for the Protection of Birds (RSPB), not to mention the considerable value that Biotrack provided with the supply and maintenance of the Land Rover used to track down Buzzards and finance SW's time. The British Atmospheric Data Centre (BADC) supplied the weather data, and the magnificent British Trust for Ornithology (BTO) allowed SW to extract data on all Common Buzzards ringed in the UK and provided data on the recoveries of those marked birds.

Locally, we were particularly lucky to have so many estates and farms that were not only interested and highly cooperative, but also provided a habitat diversity that proved very important. We would like to thank The Morden Estates Company, English China Clays (now Imerys), and the Bloxworth, Creech, Encombe, Keysworth, Rempstone, Smedmore, Trigon and Lees estates, the Forestry Commission, Farmer Palmer, the Blue Pool, the Baggs, Barnes, Holes, Goldsack, Kerley, Randall, Rideout, and Tory families, the RSPB, and the administration of Lulworth and Povington Ranges. Likewise, when conducting our Buzzard release experiments, or checking breeding of dispersed birds outside our study area, we were grateful to Clarendon, Crichel, Lulworth, Midhurst, Pippingford, Shaftsbury, Sherringham and Woolbeding estates, Pat and Patricia Coles, the National Trust and Natural England for the necessary licences.

Many agents, gamekeepers, farmers and other enthusiasts on those estates and in association with research at Langholm were very accommodating and supportive, especially Andrew Bachell, Tom Barnard, Ron Barnes, David Baines, Bill Beaumont, Martin Bond, Will Bond, Alex Booth, Jason Bowerman, Philip Broadbent-Yale, Ray Brooks, Damian Bubb, Rob Burncombe, Brian Burrows, Richard Caines, Robert Chadwick, Oliver Chamberlain, Harry Clark, Mick Crawley, Adrian Cullinane, Graeme Dalby, Susan Davies, Teresa Dent, Sue and David Dorrell, Charles Dutton, Eddie Cruikshank, Walter Drax, Steven Fry, Colin Galbraith, Tony Gaston, Jim and Venn Goldsack, Jeremy Greenwood, Mathew Harley, Walter Harrison, John Harvey, Norman Hayward, Guy Hole, Bridget Hooper, James House, Richard House, Alex Jameson, Glyn Jones, Peter Lardner, Stefan Leiner, Simon Lester, Oliver Lucas, Julia MacDonald-Smith, Major John Mansel, Philip Mansel, Micheal O'Briain, Mark Oddy, Duncan Orr-Ewing, Kevin Pearce, Roy Perks, Robin Perry, Mary and Norman Randal, James Ryder, David Scott, David Sekers, James Selby Bennett, Lord Shaftsbury, Gary Smith, Nick Sotherton, John Stenn, Giles Sturdy, Philida Sturdy, Rick Sturdy, Simon Thorp, Ron Thorpe, Stephen Venables, Lindsay Waddell, Merlin Waterson, Wilfred Weld and Keith Zealand.

We would not have managed all the fieldwork without dedicated time and effort from students and volunteers who had to endure carrying ladders through brambles and rhododendrons, and up steep hills, tracking during inclement weather or patiently feeding and looking after our release birds in the nest before then tracking them as they started to disperse. We enjoyed the time we had with Arjun Amar, Libby Biott, Amber Budden, Luis Cadahia Lorenzo, Hannah Dalton-Brewer, Mike Gould, Dave Hall, Jonathon Hardcastle,

Philip Harvell, Simon Holloway, Adam Kelly, James Kenward, Ginny Lindley, Santi Mañosa, Stuart Marshall, Alan Morriss, Maarit Pahkala, Renata Plattenberg, Caroline Raby, Mikhael Romanov, Jason Taylor and Tony Tyack. We are grateful to the British Falconers Club for donating some captive-reared Buzzards for releasing from artificial nests in the east of England. Torgier Nygård deserves special mention, as he arrived for our first field season to help with mounting the backpack radio-tags on raptors that led to us being able to track some birds for four years. Likewise, we would not have found Buzzards that dispersed a long way without the flying skills of Alan Morriss, once the pilot with the second highest number of commercial pilot hours in Great Britain, who flew us around southern England looking for radio-tagged raptors. Then there were those whose names we don't necessarily know who have returned Buzzard rings to provide important data, and the farmers who pulled the Land Rover out of awkward situations.

Collaborations were another important component to the research, whether getting help with post-mortem analyses from Victor Simpson and John Cooper, or discussing and publishing our results with Eduardo Arraut, Stephen Freeman, David Macdonald, Santi Mañosa, Byron Morgan, Steve Rushton and Andy South. Being part of ITE/CEH meant that we gained greatly from our collaborations, from James Bullock, Richard Caldow, Ralph Clarke, Jack Dempster, Robin Fuller, John Jeffers, David Jenkins, John Goss-Custard, Mike Morris, Ian Newton, Mike Roberts and Richard Stillman. At this point we have to give a very special thank you for the time that Kathy Hodder worked with us, dedicated to the Buzzard project and completing her PhD on *The Common Buzzard In Lowland UK: Relationships Between Food Availability, Habitat Use and Demography*, at the University of Southampton in 2001. SW is grateful for the opportunity, encouragement and assistance from Graham Holloway at the University of Reading for supervising SW's PhD on *How Sociality, Weather and Other Factors Affect the Leaving, Transition and Settling Phases of Dispersal in The Buzzard*.

Our work has also been inspired and encouraged by many raptor researchers, including Arjun Amar, Walter Bednarek, Marc Bechard, Merlin Becker, David Bird, Heinz Brüll, Tom Cade, Ian Carter, Bill Clark, Roy Dennis, Brian Etheridge, Miguel Ferrer, Nick Fox, Mark Fuller, Anita Gamauf, Rhys Green, Fran and Fred Hamerstrom, Mats Karlbom, Todd Katzner, Adam Kelly, Mike Kochert, Beatriz López Arroyo, Sonya Ludwig, Vidar Marcström, Mike McGrady, Theodor Mebs, Bernd Meyberg, Norman Moore, Stephen Murphy, Ian Newton, Mike Nicholls, Malcolm Nicoll, Torgeir Nygård, Debbie Paine, Jemima Parry-Jones, Ralf Pfeffer, Mátyas Prommer, Steve Redpath, Staffan Roos, Tasie Russel, Christian Rutz, Viktor Šegrt, Fabrizio Sergio, Janusz Sielicki, Mohammed Shobrak, Karen Steenhof, Peter Sunde, Simon Thirgood, Des Thompson, Risto Tornberg, Colin Tubbs, Petr Vorisek, Per Widén, Jeremy Wilson, Nick Williams, Reuven Yosef and Fridtjof Ziesemer, with whom we have enjoyed discussions about Buzzards at conferences, especially of the Raptor Research Foundation, and during other meetings. Nick Picozzi and Doug Weir taught RK about using mirrors to check Buzzard nests and the importance of rehydration at lunchtime, while Nicholas Aebischer and Julie Ewald ensured robust analyses. SW is also incredibly relieved that Cheryl Dykstra allowed him some time for a sabbatical from the role of Assistant Editor for the *Journal of Raptor Research*; otherwise the book would not have been completed.

To write this book we are indebted to many who have entered into discussions, sent articles and helped in so many ways. In particular, Eduardo Arraut, Peter Dare, Richard

Francksen, Anita Gamauf, Todd Katzner, Oliver Krone, Santi Mañosa, Ian Newton, Charles Nodder, Nick Picozzi, Robin Prytherch, Adam Smith and Michael Wink gave their time to review chapters, and Jason Fathers not only reviewed a chapter but also provided webcam footage of a local Buzzard nest.

The book has benefited from a great deal of artwork, many photographs and several illustrations from other authors. Alan Harris has worked his magic on the line drawings and cover illustrations, while photographs were very kindly provided to us by Chris Ashurst, Richard Clarke, Jason Fathers, Gerhard Kornelis, Antonio Mazzei, Filiep Tjollyn and Chris Wilson (with thanks also to Nick Dixon). Photographs sourced from www.shutterstock.com came from Barsan Attila, Ger Bosma, Terry Brooks, Neil Burton, David Dohnal, 'grusgrus44', Chris Humphries, Em Jott, 'KOO', Piotr Krzeslak, Marcin Perkowski, Vladimir Prokaev, Menno Schaefer, Thorsten Spoerlein, Ruth Swan, Vasily Vishnevskiy, Bernd Wolter and Bildagentur Zoonar GmbH. For permission to use other illustrations we thank Eduardo Arraut, Stuart Butchart at BirdLife International, Kathy Hodder, Lennart Karlsson at Falsterbo, Oliver Krone, Oliver Krüger, Angela Langford at the BTO, Erika Newton at the British Ecological Society, Marek Panek, Neil Paprocki, Nick Picozzi, Robin Prytherch, Roger Riddington at *British Birds*, Innes Sim, Gretchen Stillings at the Public Library of Science, Peter Sunde, Mike Toms at the BTO, Angela Turner at the Association for the Study of Animal Behaviour, Krzysztof Wiackowski regarding *Acta Protozoologica* and Popko Wiersma at the Netherlands Ornithologists' Union. The Poyser team at Bloomsbury Publishing tolerated our delays and helped greatly with advice in the later stages, for which we owe thanks in particular to Katy Roper.

Lastly, we have some very special people to whom we owe a great deal. At Biotrack, we are indebted to Brian Cresswell, who has been so supportive and encouraging that it's difficult to imagine getting to this point without him. He, together with others from Biotrack, has provided understanding and no end of technical backup and goodwill. Biotrack has been a wonderful place to work because of all the people there.

And, of course, our biggest thanks go to our families. Our wives, Ibbie Moy and Bridget Kenward, knew what we were like before making a choice, whereas the children, Rosie and Ella Walls, Ben and James Kenward, had no choice. However, they all appear to have enjoyed the quirky obsessiveness we have for raptors and that we needed to dedicate to our research, especially over the last decade while writing this book.

CHAPTER 1
A Common Buzzard

Our book about the Common Buzzard *Buteo buteo* is based mainly on our knowledge of the nominate subspecies *B. b. buteo* (Latin binomials of other species are found in Appendix 1). We have been helped by this being the buzzard species about which most has been written in scientific journals. When its scientific name was given in Latin by Linnaeus in 1758, the perception was of a species abundant throughout Europe, without immediate recognition of a subsequent need to split other subspecies from the nominate one. More recently, BirdLife International (as well as the *Handbook of Birds of the World*, Lynx Editions) uses the name Eurasian Buzzard, in an attempt to achieve a globally consistent taxonomic standard that is relevant to establishing conservation priorities. Having something that helps describe the distribution of a genetic stock seems more helpful globally than a reference appropriate to local perceptions. However, as we shall see, it is not likely we will achieve a consistent taxonomic group any time soon, even with modern genetics, and as a species it is unlikely to need conservation measures in the foreseeable future. Therefore, in keeping with local tradition in English, we will continue to use the name Common Buzzard, and thus be consistent with much of the existing scientific literature.

In other European countries, the species was often named for its feeding habits. Thus, throughout Finland, Norway, Denmark, Germany, Hungary, Poland, Bulgaria and Spain it is an 'eater of mice'. In contrast, the Swedish name denoted foraging for snakes. For the Dutch, it is simply a buzzard, whereas for the French it is one with variable plumage and in Portugal an 'eagle with round wings'. In Czech, it was a mouse-eater but is now the 'forest buzzard', while in Russian and some other Slavic tongues its root may associate it with travel. The Romanians again consider it 'common', while the Ukrainians associate it with the steppes. For our book we will include the word 'Common' to differentiate from the other buzzard species that are inevitably part of this and other chapters, rather than just shortening the name to Buzzard at all times. When we write 'Buzzards' then we mean individuals of the species last mentioned, whereas 'buzzards' will be used to denote the group of *Buteo* species, just as we use the word 'birds' to mean all species belonging to the taxonomic class Aves. So, where do we find this common, variably coloured eater of mice and snakes?

Distribution

The Common Buzzard breeds over much of Eurasia, from the Atlantic coasts of Portugal, Spain, France, the British Isles and Scandinavia in the west to northern Turkey, Azerbaijan and the Caspian Sea in the south-east of its range, curling north over the top of Kazakhstan through Russia to the north-west of Mongolia, with a thin 'finger' running from east Kazakhstan to the south-west to Kyrgyzstan (Figure 1.1). For the migratory Buzzards of Russia and northern Europe, the non-breeding distribution has patches in Kyrgyzstan, Turkey and a few small areas of West Africa, but the main non-breeding distribution covers East Africa and the southern half of the African continent (BirdLife International 2017). Currently, we are not sure which populations migrate to the different areas, but hopefully the remote-tracking equipment that is now available can shed light on that in the future. What is clear is that this distribution covers many diverse environments, immediately demonstrating the adaptability of the species to different conditions.

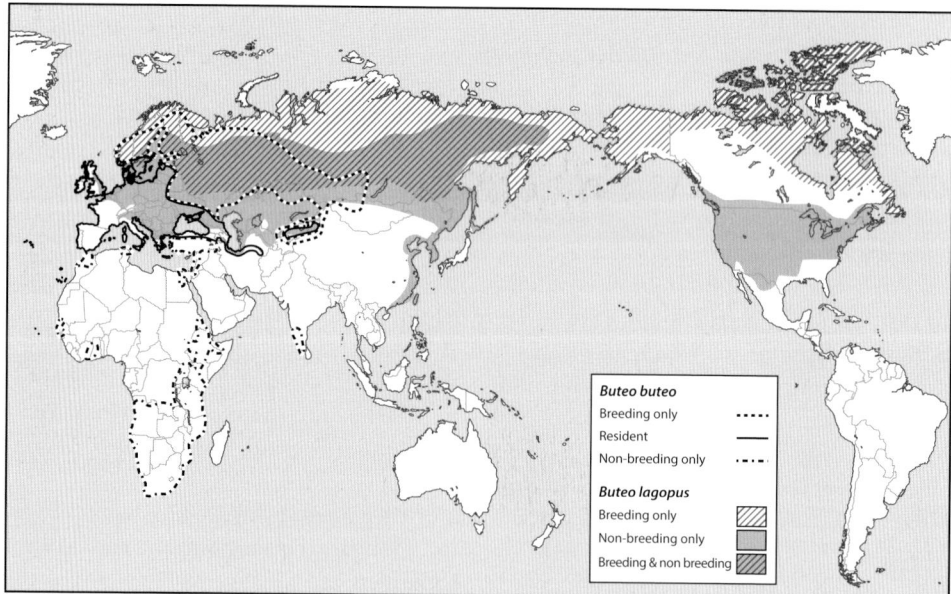

Figure 1.1. *The global distribution of Common Buzzard in comparison with Rough-legged Buzzard. The Common Buzzard breeds south of the tundra in Eurasia, whereas the Rough-legged Buzzard breeds in tundra and taiga habitats across Eurasia and North America. In winter, Common Buzzards (primarily the Steppe Buzzard lineage) migrate south of Rough-legged Buzzards, but there is overlap with Common Buzzards remaining in Europe. Plotted from www.datazone.birdlife.org with kind permission from BirdLife International.*

Several other buzzard species overlap in breeding distribution with the Common Buzzard in Eurasia. In particular, although the breeding area of the migratory Rough-legged Buzzard *Buteo lagopus* (Pontoppidan 1763) is mainly to the north (and is circumpolar, including North America), its wintering area in Europe overlaps extensively with the Common Buzzards resident there (Figure 1.1). The range of the Upland Buzzard *Buteo hemilasius* (Temminck & Schlegel 1884) in the steppes and uplands of Mongolia and China occurs to the east of the Common Buzzard's range, and the Long-legged *Buteo rufinus* (Cretzschmar 1829) inhabits the hotter regions to the south-east (Figure 1.2). The Japanese or Eastern Buzzard *Buteo japonicus* (Temminck & Schlegel 1884) breeds in Japan, northern China and Siberia east of the Baikal region, wintering widely in South-East Asia. The Common Buzzard's breeding distribution is well separated from the other Old World buzzard species found in Africa: Madagascar Buzzard *Buteo brachypterus* (Hartlaub 1860) is confined to Madagascar; the more widespread Augur Buzzard *Buteo augur* (Rüppell 1836) and Mountain Buzzard *Buteo oreophilus* (Hartert & Neumann 1914) in East Africa; and Forest Buzzard *Buteo trizonatus* (Rudebeck 1957) and Jackal Buzzard *Buteo rufofuscus* (Forster, JR 1798) in southern Africa. For these other buzzard species we use the latest nomenclature from the International Ornithological Congress in 2017 (Gill & Donsker 2017).

Looking in more detail at the distribution maps for closely related species, Common Buzzard's breeding range is predominantly south of Rough-legged Buzzard's arctic tundra

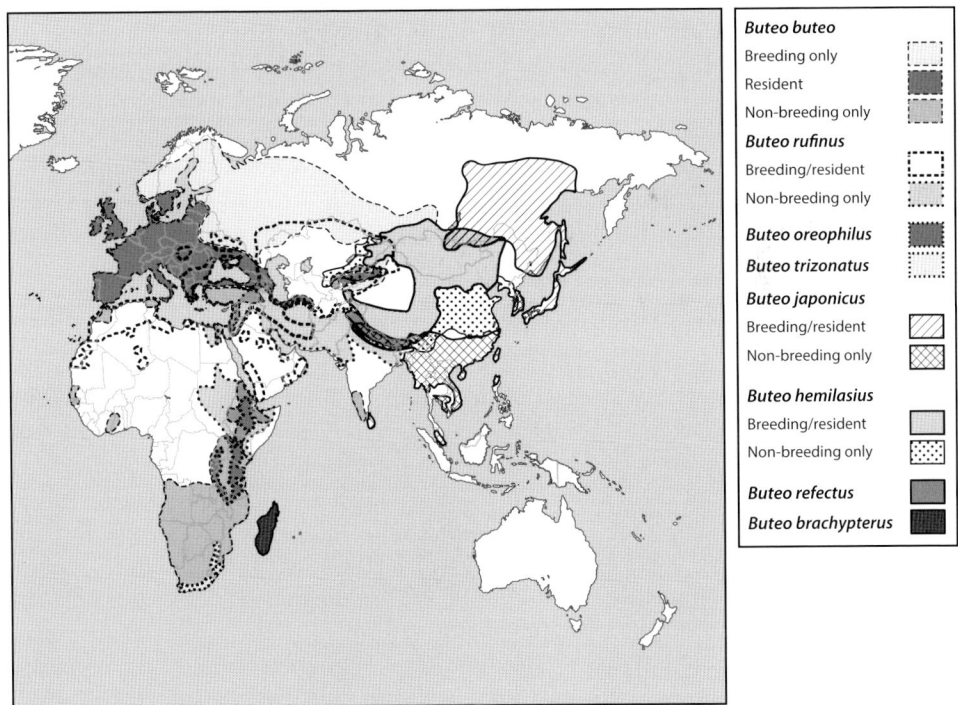

Figure 1.2. *The distribution of the closest relatives of Common Buzzard. Whereas the closely related* rufinus, oreophilus *and* trizonatus *taxa breed south of* Buteo buteo, *from Mongolia eastwards the species is replaced by three distinct species,* B. japonicus, B. hemilasius *and* B. refectus. *Plotted from www.datazone.birdlife.org with permission of BirdLife International.*

distribution, although there is some overlap in the breeding season and even more extensive overlap in the winter when Rough-legged Buzzards migrate south to avoid inclement weather on their breeding grounds. It is worth noting here, for later discussions on the origins and taxonomy, that the Rough-legged Buzzard has very separate breeding and wintering grounds throughout its range. North American Rough-legged Buzzards breed in northern Canada and Alaska, while they winter south from southern Canada to the southern United States. Likewise, in Eurasia the Rough-legged Buzzard's breeding distribution is predominantly very distinct from its non-breeding range, except in the very east of Siberia and the far north-west of Europe.

The Common Buzzard, as well as being confined to the Old World, shows much less distinction between breeding and wintering grounds in western Europe than does the Rough-legged Buzzard. Many Common Buzzards in those regions remain resident all year around, with some populations showing partial migration. Interestingly, whereas in many bird species the most northerly breeders migrate furthest south, the migratory Common Buzzards both breed and winter south of the Rough-legged Buzzards, which winter from the Amur River to North Korea in eastern Eurasia and tend to remain north of the Gobi Desert, Himalayas, Caucasus and Balkans as one moves west. Wintering ranges of Rough-legged Buzzards overlap extensively with Common Buzzards resident in large areas of eastern Europe and to the north and east of the Black Sea, but apparently not in Turkey.

On the macro scale, the southern limits of the Common Buzzard's breeding distribution are the uplands above 5,000m in Kazakhstan, Tibet and Mongolia, where the Upland Buzzard appears to be the better adapted species and presumably out competes the Common Buzzard. The Long-legged Buzzard *Buteo rufinus* inhabits the very arid lands stretching eastwards from Turkey and the Middle East to Kazakhstan, broadly as far as the Upland Buzzard's distribution. So, while Common Buzzards are able to tolerate many climatic conditions, it seems that they are not well adapted to (or are out competed in) higher altitudes or drier conditions. Its altitudinal limit varies across its range. In the UK, it becomes less common above relatively low altitudes. In Scotland, Picozzi & Weir (1974) set the higher boundary at 380m because Common Buzzards were known to nest above that altitude only in exceptional circumstances, whereas further south in Italy they nest up to 870m (Sergio *et al.* 2002) on mountains rising to 1,125m. Higher altitudes are relatively much warmer at the more southerly latitude of Italy.

There are several recognised lineages of the Common Buzzard, mostly geographically defined because they belong to islands. Hence, there is *Buteo buteo rothschildi* (Swann 1919) on the Azores, *B. b. insularum* (Floericke 1903) on the Canary Islands, *B. b. bannermani* (Swann 1919) on the Cape Verde Islands, *B. b. harterti* (Swann 1919) from Madeira and *B. b. arrigonii* (Picchi 1903) from Corsica and Sardinia. However, the Steppe Buzzard subspecies *B. b. vulpinus* (Gloger 1833), which breeds to the north and east of the Common Buzzard's range, is not completely isolated during the breeding season, as far as people can tell given the difficulty in visually separating these two forms. The distinction between them is that Steppe Buzzards are believed to migrate further south, 'leapfrogging' the nominate subspecies, to have a completely separate wintering distribution. For migration behaviour to define the subspecies seems quite a weak differentiator. For example, a recent study of migratory movements of buzzards in Estonia found that the birds exhibited short-distance migrations typical of Common Buzzards, despite being thought of as Steppe Buzzards in the past (Väli & Vainu 2015). What does this tell us? Are they Common Buzzards behaving as we expect? Or are they Steppe Buzzards making shorter than usual migrations? We still don't know.

Similarly, it has been difficult to establish accurately the distribution of the Himalayan Buzzard *Buteo burmanicus*, found to the west, south and east of this huge mountain range and migrating south in winter. Ongoing discussion concerns whether this is a subspecies *B. b. refectus* (Portenko 1935), or a separate species, originally classified as *Buteo burmanicus* (Hume 1875, Penhallurick & Dickinson 2008). Genetic analyses indicate that *refectus* is indeed a species, and confirms that status for *B. japonicus* (Krukenhauser 2004). Further west, there seems to be a consensus that *B. b. menetriesi* (Bogdanov 1879), which breeds from the Caspian Sea to the Caucasus, belongs with *B. b. vulpinus*, and that *B. b. pojana* (Savi 1831) from Italy is part of the *B. b. arrigonii* lineage.

The lumping or splitting of subspecies is a common debate in many bird groups, but in the case of the Eurasian forms of *Buteo* there is just as much debate at the species level, let alone more difficult-to-distinguish subspecies. So, let's now look at the taxonomy of the *Buteo* group as a whole.

Buteo taxonomy

True buzzards, *Buteo* spp., belong to the family Accipitridae, which includes nearly 250 species of broad-winged diurnal birds of prey. These include kites, Old World vultures, harriers, goshawks, sparrowhawks and the great assembly of eagle groups, which don't appear to have had a common origin, but rather to be a collection of species that have ended up looking similar through convergent evolution. The Accipitridae species have evolved to live on other animals, and have therefore developed speed and agility, together with sharp talons for holding on to their prey, and a hooked bill that can get through the skin, cut through flesh and pick bones clean. The Accipitridae also includes European Honey-buzzards *Pernis apivorus*, which are not *Buteo* species at all but belong to the smaller genus *Pernis*; they probably only have the name buzzard (in a few languages) through a superficially similar appearance. Considering how some of the species in this group are named, it seems that the Common Buzzard was the starting point for 'folk taxonomists' trying to find an orderly way of classifying the mind-boggling array of raptors, and therefore other species were described as Buzzard-like because 'everyone knew what a Buzzard looks like'. In some countries, including Finland and North America, all the medium-sized raptors together with the unrelated falcons were called 'hawks', while New World vultures became 'buzzards'.

There are approximately ten buzzard species in the Old World, including the Holarctic Rough-legged Buzzard. The number is approximate because there is still some debate about what constitutes a species. An old definition is that mating between separate species produces infertile or otherwise unviable offspring. Unfortunately, lack of fertility or other fitness to breed of hybrids is easier to show for plants and captive animals than for very similar-looking species over a vast geographic range. Raptors in general are hard to classify in this way. Falcon species are known to hybridise in the wild and captive falcon hybrids can be at least partially fertile. The *Buteo* species, too, are prone to natural hybridisation. Examples exist between Common and Rough-legged Buzzards (Gjershaug *et al.* 2006), Common and Long-legged Buzzards (Dudas & Janos-Toth 1999, Elorriaga & Muñoz 2010, 2013), and even very distinct raptor species can hybridise naturally. For example,

Common Buzzards have been recorded breeding with Black Kites *Milvus migrans* in Italy (Corso & Gildi 1998, Corso 2009). Unfortunately, we don't know if *Buteo* hybrids are fertile.

There have long been debates on the many proposed buzzard species, based on location and morphological differences, because the evidence for splitting species has previously been based on highly variable characters such as plumage. If a species was just very variable, it would hardly be surprising to find that certain characters were more common in different parts of its range; they might be an adaptation to the environment, as we see with island gigantism or local mate preference. However, that does not define them as separate species, or even subspecies. Therefore, the Museum of Natural History in Vienna investigated differences between some Old World buzzard species using two methods: the traditional morphological measurements of colour and dimension, alongside a more modern molecular classification (Kruckenhauser *et al.* 2004). For comparison, they used six taxa that were at that time presumed to be species (*B. buteo*, *B. rufinus*, *B. oreophilus*, *B. hemilasius*, *B. brachypterus* and *B. auguralis*) and also 13 accepted subspecies that were effectively geographical lineages.

The Austrian team found that morphological comparisons did indeed confirm that there were physical differences between the full species, largely explained by body length, flight characters such as wing length, tarsus length and bill breadth. The morphological analysis split birds in the two relatively heavy taxa, *hemilasius* and *rufinus*, into separate species, as well as separating the diminutive *brachypterus* from Madagascar. Forest Buzzard, initially considered a subspecies (*B. b. trizonatus*), was very close to all the others but nevertheless did not overlap with them morphologically, and is still widely considered to be a full species (*B. trizonatus*). The 15 remaining taxa (all 'Common Buzzards') formed one group with a lot of overlap, although certain lineages could be separated into groups within it. For example, the nominate *B. b. buteo*, was different from Steppe Buzzard (*B. b. vulpinus*), *B. b. 'hispaniae'* from Spain and the Himalayan and Japanese lineages. The Austrian team had a problem with *B. rufinus* because only the nominate subspecies (*B. r. rufinus*) had distinct characteristics, whereas *B. rufinus cirtensis*, a North African lineage from the Morocco region, was too similar to *B. buteo*. Intriguingly, although there was difficulty in separating the groups, those with adjacent distributions were more distinct than those inhabiting physically separate areas. In other words, if an area had two 'species' present, then birds in the hand could be separated on the basis of appearance, but if two birds from very distant areas were put together, it was difficult to tell them apart. Unfortunately for birdwatchers, plumage patterns proved to be a poor differentiator; those taxa with wider distributions had more variation, whereas individuals within isolated small island populations were more similar to each other. The differentiation was mainly based on size, a notoriously difficult characteristic to assess through binoculars in the field.

Results differed when molecular techniques were used to try to separate species. To find out how similar or different the taxa were, the same Austrian geneticists assessed variations between genes in the pseudo-control region, being the most variable section of the mitochondrial genome, and therefore the best for investigating the smaller intraspecific differences (Haring *et al.* 1999). Minor differences in the nucleic acid chains, and hence the protein that is coded by them, accumulate with time, provided that they do not adversely affect the operation of the protein, which is important for energy production in cells. The process of change of the protein is, in effect, a slow molecular clock, compared to changes in deoxyribonucleic acid (DNA), because the function of important proteins in the mitochondrial powerhouse of cells needs to be conserved.

Only the Eastern Palearctic taxa were clearly separable using this molecular clock: *B. hemilasius* and *B. brachypterus* were again separated as species, as were two buzzards previously considered subspecies of *B. buteo*, namely *B. b. japonicus* and *B. b. refectus*. There appeared to be another cluster, although they were not unequivocally differentiated genetically, containing *B. oreophilus oreophilus*, *B rufinus rufinus* and the island lineages *B. b. bannermani* and *B. b. socotrae* (again considered a subspecies at the time). Undisputed species, for example, *B. buteo* and *B. hemilasius*, were 1.0–1.6 per cent genetically different, compared with some similar-looking species from the New World that had been geographically separated for millenia, such as *Buteo jamaicensis* and *Buteo galapagoensis*, which were 3.8–4.4 per cent different. The remaining 'species' based on their morphological features could not be genetically separated by analysis of the pseudo-control region.

Despite lack of much help with fine detail, these molecular results are useful for indicating that all the Western Palearctic species evolved within a short time frame and that the morphological radiation developed comparatively quickly compared to the genetic change in mitochondria. The authors hypothesise that this *buteo–vulpinus* complex was a result of reticulated evolution; in other words, where populations were periodically split for long enough to become different, but not long enough to become separate species before the populations joined up and bred together again. A potential mechanism for this was intermittent glaciations temporarily separating populations and then retreating to allow re-mixing. There is evidence for this from fossil remains of *Buteo*s in the British Isles dating from around the Pleistocene (Moore 1957), a time when humans were thought to have emerged in their modern form, in the company of woolly mammoths, sabre-toothed tigers and many birds as we know them today. It was also a period when repeated glaciations separated populations.

With evidence of so much genetic similarity between *B. buteo*, *B. rufinus* and *B. oreophilus*, contrasting with some unexpected differences within *B. buteo*, Krukenhauser *et al.* (2004) proposed treating *Buteo buteo* as a superspecies. This would contain almost all the current *Buteo buteo* subspecies and Forest Buzzard *trizonatus* within *Buteo* [*buteo*] *buteo*, separate from *Buteo* [*buteo*] *rufinus* (including current *Buteo rufinus* lineages and possibly also *B. b. socotraensis* and *B. b. bannermani*) and *Buteo* [*buteo*] *oreophilus*. However, *Buteo japonicus* and *Buteo refectus* were considered sufficiently distinct genetically to be separate species, along with *Buteo hemilasius* and *Buteo brachypterus*. Based on the extensive genetic analyses, we favour this view (Figure 1.3).

Investigating further the origins of this putative superspecies, the mitochondrial analysis corroborated earlier proposals by Dean Amadon (1982) that the *Buteo* genus probably originated in the Neotropics where there are many more Buteonine groups (Riesing *et al.* 2003). It also proposed that the likely origin of the ten Old World buzzards was an isolated population of Ferruginous or Rough-legged Hawks (Buzzards) from the Americas in Beringia, the area around the Bering Strait, which intermittently linked Alaska and Russia and provided a land bridge between the New World and the Old World. The geneticists considered the different Old World species to have emerged from pulsed invasions between periods of glaciation. In contrast, an analysis from New York investigating the genetic phylogeny of the Accipitridae as a whole (Griffiths *et al.* 2007) has suggested that the single *Buteo* clade was a relatively recent and single dispersal event into the Old World, which

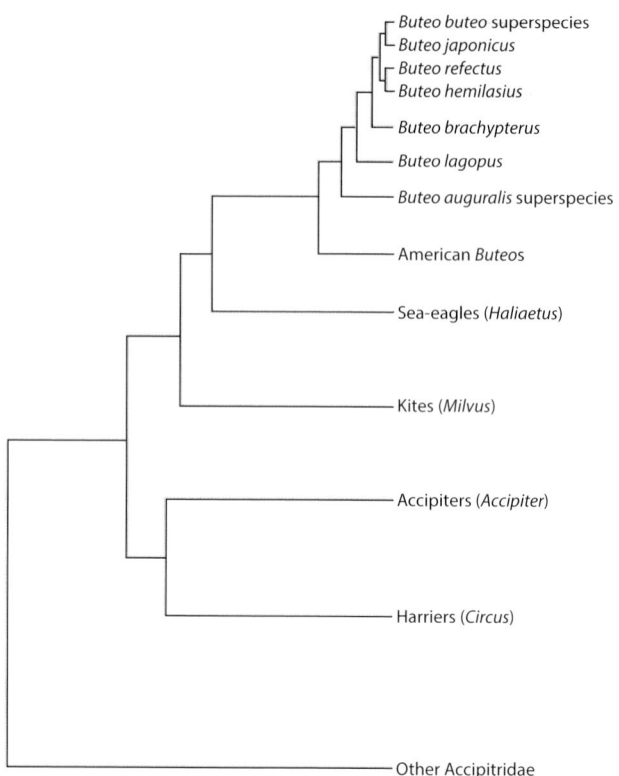

Figure 1.3. *Tentative classification of* Buteo buteo *in relation to its near-relatives and other Accipitridae, based on the analyses of Krukenhauser et al. (2004), in which* B. buteo *includes as subspecies* B. b. buteo, B. b. hispaniae, B. b. harterti, B. b. insularum, B. b. arrigonii, B. b. rothschildi, B. b. menetriesi, B. b. vulpinus, B. b. trizonatus, B. b. socotrae *and* B. b. bannermani, *and also* B. b. oreophilus, B. b. rufinus *and* B. b. cirtensis.

would be a good explanation for the lack of genetic diversity found by Krukenhauser and her colleagues. They agreed with Kruckenhauser *et al.* (2004) that, once the founder population had spread through the Old World, periodic separations of populations within that could then have created the different species we know today by reticulated evolution. Since Carole Griffiths' study, Raposo do Amaral *et al.* (2009) conducted a further, more specific genetic analysis of the Buteonine hawks, corroborating the idea that their ancestors arrived from the Americas. It showed that the Old World taxa (*B. buteo, B. rufinus, B. refectus, B. hemilasius, B. japonicus, B. auguralis, B. augur* and *B. rufofuscus*) clustered next to (and were therefore closely related to) the circumpolar Rough-legged Buzzard and New World Ferruginous Hawk *Buteo regalis* (Riesing *et al.* 2003). The fact that Rough-legged Buzzard breeds in both the New and Old worlds makes the species a good candidate as the origin of the Old World buzzards. Alternatively, perhaps a dispersal event produced a common ancestor of both Rough-legged Buzzard and Ferruginous Hawk. As yet another alternative, there could have been several arrivals, the first being pushed down into Africa during a glaciation and producing the Augur, Jackal and Madagascar Buzzards; then, after the glaciation had finished, a second species could have arrived to radiate into the more northerly buzzard species, with Forest and Mountain Buzzards perhaps being settlers from migrants into southern Africa. Given that Raposo do Amaral *et al* (2009) found these species so closely related to today's very similar Rough-legged Buzzard, and that the most parsimonious route would have been from the north through Beringia, it seems likely that the common ancestor, even if not Rough-legged Buzzard, was a very similar species or subspecies adapted to the northern tundra. As with the modern Rough-legged Buzzard, its main prey would have been small mammals, the prey for which Common Buzzards are so often named today.

New field evidence supports Krukenhauser *et al.*'s inability to separate the Common Buzzards from African Long-legged Buzzards of the *cirtensis* type. These Long-legged Buzzards have been found breeding north of the Mediterranean near Gibraltar (Elorriaga & Muñoz 2010), overlapping more with Common Buzzard's distribution. Moreover, there have been cases of crossbreeding between Long-legged Buzzards and Common Buzzards in Hungary (Dudas & Janos-Toth 1999) and southern Spain, producing offspring with characteristics from both species (Elorriaga & Muñoz 2013).

As recently as 2010, a separate Socotra Buzzard species has been proposed (Porter & Kirwan 2010). Although the proponents acknowledged that there is a lot of morphological and behavioural overlap with other buzzard forms in this superspecies complex, they felt there was sufficient differentiation and isolation to separate Socotra Buzzard, living on the island of Socotra in the Arabian Sea. Porter & Kirwan (2010) even proposed a new scientific name *Buteo socotraensis* to reflect the provenance of a new taxon, rejecting the previous subspecies nomenclature (*B. b. socotrae*). They regard this species to be closest to Cape Verde Buzzard (*B. b. bannermani*), although geographically completely separated from it. There is also the possibility that Socotra Buzzard originated as a hybrid between Long-legged and Steppe Buzzard, as Socotra Island lies on the border of both species' migration and non-breeding quarters. However, it seems that few individuals of these other species ever get to the island; with over 15 years of intensive fieldwork on Socotra, Porter & Kirwan only had one record of Steppe Buzzard, and that bird would have had to reject its migratory urges and stay for the breeding season if it were to contribute genetically to the population. Over time, the isolation may well tend towards a new species, but maybe the occasional Long-legged Buzzard could arrive and diversify the gene pool in a form of reticulated evolution.

So, as we can see from the ongoing debate, a tight definition of Common Buzzard is virtually impossible. Nevertheless, a species (or superspecies) like Common Buzzard can become a helpful package for conservation. Work on DNA is becoming more rapid and inexpensive these days, thanks to firms such as Oxford Nanopore, which can read the letters of the code as DNA passes through artificial pores with equipment that will soon be hand-held. Future conservationists may consider more carefully the importance of maintaining the diversity of genetic traits within a species, as genetic diversity is one of the pillars of the Convention on Biological Diversity.

Fortunately, our Common Buzzard is a generalist raptor, which by virtue of being both abundant and common to many countries is likely to have great diversity in its nuclear genome, even if its mitochondrial genome has not had long to diverge. We cannot hope to capture the diversities of behavioural and morphological traits that may be present in the species and its near-relatives, with which it still hybridises and may or may not produce offspring able to reproduce. However, most of the research that has been published in accessible peer-reviewed journals has been undertaken in western Europe, where the most common form is the nominate Common Buzzard species, so it makes sense for us to focus on the birds of this lineage, which we know best.

Characteristics

Buzzards are raptors. The word 'raptor' comes from a Latin root inferring violent seizure or capture. It conjures up an image of birds hunting with strong feet, equipped with long, sharp talons. Those characteristics apply to the species in several groups: hawks (Accipitriformes), falcons (Falconiformes), owls (Strigiformes) and often the New World vultures (Cathartidae). Common Buzzards look like classic hawks, unremarkable in appearance in that they haven't got baza-like crests, vulturine bare heads, kite-like tails, the long legs of a Crane Hawk *Geranospiza caerulescens*, or stunning colours, patterns or even size to differentiate sexes and ages. Instead, they are one of the better-camouflaged birds, with classic proportions that are very practical, looking like a base-line model hawk from which other raptors could have diverged. In fact, we now know from the molecular analyses that this is a misperception in evolutionary terms, because Old World buzzards are some of the more recently evolved birds of prey.

However, even if Common Buzzards may at first sight seem drab as individuals, closer examination of whole populations within some parts of their distribution tends to show a remarkable variety of markings. Indeed, as a testament to this, the French use the name *Buse Variable*. Although the back of the wings and body are a uniform brown of some shade, the head and underside can vary from very pale to really quite dark chocolate-brown. There are many different ventral patterns, from very few brown splotches on a creamy background, to almost entirely brown with a paler chest-band. The variations of the underside plumage can be used to categorise different 'morphs'. Most commonly, three of these morphs are recognised (Plate 1 a, b, c; see colour section): light, intermediate and dark (Glutz von Blotzheim *et al.* 1971, Melde 1983, Krüger *et al.* 2001). Dittrich (1985) in northern Bavaria (Germany) defined five morphs; some authors use a seven-morph categorisation (Kappers *et al.* 2017) and Staffan Ulfstrand (1970) even defined 16 morphs. The reason why it is possible to have so many classifications is that Buzzard polymorphism is continuous, with any number of intermediates possible, rather than discrete, easily identifiable plumage patterns. The continuity of variation has been demonstrated more conclusively with a quantitative assessment of Buzzards in the Netherlands, which showed a continuous variation in colour index from computer-aided image analysis of digital photographs (Kappers *et al.* 2017). Continuous polymorphism is also expressed in an American *Buteo*, Swainson's Hawk *Buteo swainsoni* (Briggs *et al.* 2011), as compared with the discrete polymorphism that can be seen in another American *Buteo*, Ferruginous Hawk (Schmutz & Schmutz 1981) and an *Accipiter*, Black Sparrowhawk *A. melanoleucus* (Amar *et al.* 2013). Note that these morphs are completely different to the very rare aberrations that produce white buzzards in several of the species/lineages (Robb & Pop 2012). White buzzards are often leucistic and not albino, judging by eye colour that still retains some pigment.

In common with other birds of prey, the raptorial feet of Common Buzzards have other adaptations for catching prey apart from their sharp talons. Once a raptor has managed to grab an animal, it will hold it securely, partly by reflex and partly by a ratchet system. The latter is described more fully in the chapter on hunting, where we will discover how the feeding habits of these birds do not always live up to the expectations set by their long talons and hooked bill. Nevertheless, they have another of the typical raptor attributes, a large fleshy cere that surrounds the nostrils at the base of the bill. Perhaps the most surprising discovery about raptor ceres was made by James Dwyer (2014), who noticed that the colour of Crested Caracara *Caracara cheriway* ceres can change instantaneously to threaten other birds when feeding communally, not something that has been observed in other species. No such short-term colour change has been shown for Common Buzzard ceres, which are usually a waxy rich yellow that matches the bright yellow legs and feet, although that is not always the case even if the bird is in good health. We have seen well-fed captive-hatched Common Buzzards that have been raised on laboratory mice that presumably did not contain the right pigments, and the ceres of those birds were a washed-out blue-yellow (Plate 2, a, b; see colour section). There are even wild Common Buzzards that show washed-out blue ceres, so the yellow cere is not a reliable trait.

Behind the cere are the typically large, forward-facing raptor eyes. The eye colour is not as piercing as some of the incredibly bright eyes found in other raptor species. Instead, adults have dark brown, rather inconspicuous eyes, camouflaged against the brown plumage that surrounds the eye. So, while they can look a little grumpy, with a heavy-set eyebrow, a Common Buzzard doesn't look as frightening as a Northern Goshawk, for example, which can have stunning bright yellow or fiery orange eyes, the effect being exacerbated by pale eyebrows that look quite intimidating. Juvenile Common Buzzard eyes are rather lighter brown or greyer compared with the rich chestnut of an adult. The forward-facing eyes allow binocular vision, which are better for depth perception, well adapted for spotting camouflaged prey, focusing on them and moving quickly and accurately to catch them. What is a little less usual, not found in non-avian vertebrates but in around 54 per cent of bird species, is that the eyes have two fovea (high focus areas), one for central and the other for peripheral vision. Birds have large eyes relative to the size of the skull. For this reason, they do not need much muscle to move the eye, compared with humans, for example. Nevertheless, they need acuity of vision for seeing prey. So, having acuity in two areas reduces the need to move their eyes about as humans do. This obviously suits their hunting needs when in chase, but it doesn't help them focus very close; they appear not to be able to see something just in front of their beak and will lunge at food they know is there from having seen it further away. Raptors also need steady vision for hunting, and so they have very complex and fast-acting musculature in their neck to keep their head level and vision steady as their body moves up and down in flight, or twists and turns at a stupendous rate when stooping or pursuing prey.

For anyone intrigued to see the bone structure and musculature inside a Common Buzzard's skull, Stephan Lautenschlager *et al.* (2013) at the University of Bristol have made a 'digital dissection', using contrast-enhanced X-ray-computed tomographic (CT) scanning. They were interested in the specialised muscle structure of raptors in comparison with other bird species, their focus being for the reconstruction of damaged muscle on traumatised birds of prey. As well as showing the relatively slight eye musculature, the scan also reveals how big the brain is, taking up a significant proportion of the skull.

Unlike many raptors, Buzzards are not strongly sexually dimorphic in either size or plumage pattern. Males tend to weigh about 800g and females more like 1,000g, but there is great variability and overlap (550–1,364g, Demongin 2016). It is therefore difficult to tell what sex a Buzzard is when it is in the hand, let alone at a distance in the field. Working on Common Buzzards trapped and rehabilitated in southern Europe, Zuberogoitia *et al.* (2005) found the only significantly different characters between the sexes were minimum tarsus width, bodyweight and wing length. However, although different on a group basis, actually there was considerable overlap between the sexes, except in tarsus width; those birds with a tarsus width less than 7.0mm were male, and those more than 7.9mm female. Some with a tarsus width between 7.0–7.9mm could not be categorised. Our assessment with growing nestlings was that males had a tarsus width less than 6.0mm, and this was proved to be the case in four post-mortems, although we thought there was a possibility of incorrect classification based on such a small sample size (Walls & Kenward 1995). Despite being a large bird, the feet at the end of those tarsi are proportionally very small for a raptor. Falconers have known for centuries that having small feet reduces the ability to bind to larger prey (seize them securely), thus limiting potential prey species for a Common Buzzard.

Identification

We often see Buzzards as they soar on thermals and updraughts, hardly flapping their broad wings, or perching on conspicuous high points, calling with their loud, plaintive territorial mewing. In cooler, flatter areas where there are fewer updraughts, Common Buzzards will also hover, rowing their wings back and forth like a Common Kestrel, but this is not apparent in most areas. Buzzards have distinctive, very broad wings and a short tail, just shorter than the wing breadth. The comparatively large ratio of wing area to body-mass (low wing-loading) means they are well adapted to catching any rising thermals or air forced up by the landscape and using it to their advantage, often hardly seeming to make any effort to rise hundreds of metres in the air. They are at their most exciting during the breeding season, when they will make frequent territorial stoops and fly up to challenge any intruding Buzzard or other raptor. The low wing-loading also allows them to soar at slow speeds or hover with less headwind; this has advantages when searching in flight for prey on the ground.

Common Buzzards can be distinguished from similar raptors at a distance by shape and behaviour. When flying, they often hold their wings pushed forward and slightly raised in a distinctive, shallow V-shaped dihedral (Figure 1.4). This dihedral may appear to contradict the idea of having the biggest surface area to trap rising air when soaring, because a V-shape will reduce the downward-facing area slightly. What's more, it can make the birds look more unstable, rocking from side to side, as anyone who has watched Turkey Vultures *Cathartes aura* – which hold their wings even higher – will confirm. However, the appearance of instability is deceptive. With wings like that, they can take advantage of less stable updraughts more typical of the outer reaches of a thermal or gusty wind that is buffeted upwards when it hits a hillside. The way it works is that if the updraught is uneven and forces one of the wings up, the other wing becomes more horizontal, increasing the surface area towards the airflow and so being pushed back up relative to the other wing, whose downward area is temporarily decreased. So it is self-righting. If they had held their wings flat and one wing had gone up relative to the other wing, they would be forced into taking corrective action with a flapping movement that would consume energy. Instead, soaring birds with a dihedral wing profile can wait for the auto-correct and continue to soar without exerting effort, making for more efficient flight. Some believe that the juveniles do not have such a sharp dihedral and soar on flatter wings. Maybe it takes experience to become confident that the rocking will eventually self-right before the need to take corrective flaps.

Another auto-adjust comes from the flexible thinner primary tips that curl up at the end with stronger airstream, and straighten if the wind force reduces, dampening the effects of sudden gusts. The ends of the primaries are more likely to bend than the base of the feathers, because they are thinner than the rest of the primary, and they do not overlap other feathers. This gives the classic 'big fingers' at the far end of the wing. The wide air slots between the ends of the primaries also reduce turbulence, facilitating the Buzzard's smooth flow through the air, again making for more efficient gliding. When soaring, the tail with its straight sides and sharp corners is often fanned, to the extent it almost touches the back of the wings at times. When gliding from one thermal to the next, the wings are flat or slightly lowered, and if on a fast glide they tuck their primaries back to reduce head-on wind resistance, looking quite angular. Usually when flying, they require a few deep wingbeats to get going, but will soon progress into a glide or make quick, rather stiff, shallow wingbeats.

Figure 1.4. *The spread wings of soaring Common Buzzards can have a quite strong V-shaped dihedral (© Em-Jott, www.shutterstock.com).*

Buzzards have a broad, short neck, so the head does not protrude much, a characteristic of true *Buteo*s. They do not have the sharp head and long, rounded tail of the completely unrelated European Honey-buzzard, and compared with kites they are far more compact and have a much steadier flight, usually without any W-shape to their wings. Harriers can look similar, but their wings are much narrower and their tail much longer, plus harriers often quarter low to the ground and look even more unsteady in flight. The marginally smaller Steppe Buzzard subspecies has slightly narrower and longer wings (not noticeable to the untrained eye), but this subspecies is still quite distinct from harriers.

Feathers need replacing as they get sun-bleached, weaker and tatty over time (Plate 3; see colour section). The distinctive primary fingers are very dark brown on top and black below, probably because dark-coloured feathers are more robust and can withstand the wear that comes from being pushed through the air, at considerable speed when stooping. Often, the primary tips are broken by the time they are moulted. Flight feathers are moulted symmetrically, primaries from the innermost out and tail feathers from the centre outwards. The secondaries have a more complex moult, starting at different places, from the outermost inwards, the innermost outwards and another point between the two that moults inwards. Moulting of the secondaries starts a little later than the primaries and tail feathers. Because the Common Buzzard is such a large bird, and feathers have a limited growth rate, there is not sufficient time in the four- to six-month moulting season to completely replace all the secondary feathers. Therefore, adults moult irregular feathers, still moving in sequence, replacing secondary 1, 3 and 6 in one season, for example, and the following year moulting 2, 4 and 8.

The number of feathers replaced can depend on a bird's nutritional state, so more feathers can be replaced if the bird is well fed at the beginning of the moulting season and can start sooner. Moult is energetically costly, primarily due to the growing of the feathers, but also as a consequence of flying without a full set of feathers. Using Magnus Sylvén's 1982 estimates of the energy required to grow feathers (57kcal per gram), Peter Dare (2015) equated feather growth to requiring one or two shrews per day. It's easy to imagine how, if food is not plentiful, feather growth can be arrested, so feathers are grown in the warmer half of the year when more food is available. The female can start moulting during incubation, as she is not so active and can use an appreciable proportion of the food brought to her by the male to convert into feather growth and warmth for the eggs. Males tend to start much later, when the youngsters are fledging and the female is also hunting. Adult males that lose clutches can start moulting earlier. Moulting can continue until autumn, depending on food availability.

Juveniles have to grow all their flight feathers at the same time, so that they are ready to fly from the nest as soon as the time comes to leave. As with all birds, the first year's feathers are consequently not as strong as in later moults. All Buzzards will still have some juvenile feathers in their second year, and there are likely some juvenile feathers left after their second moult, heading into their third year. With all juvenile feathers being grown at the same time, all the wing feathers will have the rich, dark chocolate-brown in their first autumn. On the other hand, Buzzards more than a year old will have some feathers from the previous year that will be bleached and tatty, and some will have feathers grown two years previously; these will look even paler. This is also true of the contour feathers on the back. There is a tendency for juvenile greater coverts (the line of feathers directly above the secondaries) to have pale tips that contrast with the dark brown secondaries, giving a continuous line across the wing (Plate 4; see colour section). Adults will have variously bleached secondaries and less of a consistent line. So, a good view of a perching or feeding Buzzard (or more often a good digital photo these days) can help age it by judging the degree of colour contrast between different-aged feathers on the back, and how neat the line of the greater coverts is. Earlier in the annual moult, the head, nape and back of an adult can form a rich brown cape-like pattern that contrasts with the sandier light brown of the old feathers. This can even be spotted with the naked eye when driving past a perched Buzzard by the roadside. On the front, juveniles tend to have a streaky throat and a paler chest and belly, whereas adults have a more solid dark throat and barring on the belly, divided by a pale cream horseshoe across the chest (Prytherch 2009). Unfortunately, that is not an easy, or indeed reliable guide for aging, because of the great plumage variability that also makes Common Buzzards difficult to distinguish from other buzzard species.

Common Buzzards are often seen from below when in flight, or the back is obscured by flapping wings. This makes it harder to see the variation of feather wear and bleaching that reveal the age of a perched Buzzard. However, through the first winter it is still possible to distinguish juvenile Buzzards hatched in the current year from older birds, using characteristics that apply to other raptors; this is especially true in autumn and early winter. The reason for this is that juveniles will not moult their primaries and secondaries until the following spring (often starting before breeding adults), so there are no missing feathers in

Figure 1.5. *Flight silhouettes of Common Buzzards, showing how the secondaries of a juvenile (left, © Ruth Swan; www.shutterstock.com) create a bulge in the trailing edge of the wing, which is relatively straight in adults (right, © Thorsten Spoerlein, www.shutterstock.com).*

the autumn. Even later in the autumn, when sub-adults' and adults' feathers are nearly fully grown, the hind edge of the wing will be less tidy than the smooth, even curve of a juvenile's, with feathers that were fully grown by August. In the first year, the hind edge of the wing has a distinctive S-shape, where the middle secondaries bulge out further than those close to the body or next to the primaries (Figure 1.5). Subsequent secondaries grow to form a straight line from the primaries to the body, and this feature can sometimes be used to age birds in winter. During the transition years, where some of the juvenile feathers are retained, the central secondaries protrude further than the newer feathers, creating an untidy hind edge. Once all of the juvenile secondaries have been moulted, the wing takes on the straighter edge of an adult. The juvenile S-shape can be distinguished from an untidy or straight line with an almost silhouetted bird. In contrast, good light is needed to see the pale panels towards the tip of the underwing (Figure 1.5). If they are visible, the white flash towards the end of the wing is more strongly contrasted against the surrounding bars and wing-tips in adults (Dare 2015), and we will explore more of how they use that signal when discussing territorial display.

Plumage markings are the feature most often used for distinguishing between different bird species in the wild. Unfortunately, Common Buzzards are difficult to positively identify where there are similar species, because their variety of morphs gives them the most variable plumage of all Palaearctic raptors (Porter 1974). Unfortunately, even the categorisation of morphs remains open to discussion. A long recognised issue has been that the characteristic plumage of juveniles (more streaky on the throat and chest, with a poorly defined chest-band and less colouring on the legs) looks lighter, and so some have questioned whether age would affect their morph categorisation. Elena Kappers *et al.* (2017) investigated that particular problem by getting 13 independent people to qualitatively assess the morph of 10 Common Buzzards caught at different ages and photographed with a wing held out to reveal all its characteristics. They showed that Buzzards do darken with age, more so on the front than on the underwing. In six cases, the morph category remained the same, but for four individuals the later pictures were assessed as the next darkest morph on their seven-morph scale: three from intermediate to dark-intermediate, and one from light-intermediate to intermediate. Whether they would have darkened sufficiently to change morph type within a five-morph or three-morph scale was unclear.

Plumage variation is dependent on melanistic pigments that are both genetically determined and likely to be dependent on body condition (Roulin *et al.* 2008). Melanins can be used by the body's immune system, and could therefore be in short supply for feather pigmentation of an immuno-compromised bird. Nevertheless, despite this potential link to health, the morphs continue through generations, and evolutionary biologists get excited as to why different morphs should persist. Oliver Krüger's laboratory has devoted much effort to investigating the causes and consequences for a population of Common Buzzards in Westphalia, Germany. Later in the book, we will see how he and his colleagues have shown that Buzzards appear to choose mates of a similar morph to their mother (Krüger *et al.* 2001) and display varying morph-dependent degrees of aggression to predators around the nest (Boerner & Krüger 2009), and also that intermediate morphs produce more offspring over their lifetime than dark and light morphs (Krüger *et al.* 2001).

While distinct morphs may be apparent in some areas, it can be difficult to assign individuals to a particular morph category in other regions, due to the continuous nature of the variation. In the past, morphs tend only to have been recorded when researchers were studying morphs, so we do not have a good idea of how prevalent distinct morphs are throughout the whole distribution. Around us, in southern England, there is no obvious categorisation: most birds are intermediate. Before the 1970s, the official Swedish bird list (*Förteckning över Sveriges fåglar* 1970) among others, considered that Steppe Buzzards were breeding in the north of Sweden, and the nominate subspecies in the south. However, there was some debate because Leslie Brown & Dean Amadon (1968) had set the western limit of Steppe Buzzard's range in Finland (east of the Baltic Sea), and Gustav Rudebeck (1957) thought that the birds in northern Sweden were too dark and less rufous than the Steppe Buzzards that migrated to Africa. Therefore, Ulfstrand (1977) made a concerted morphological study of museum specimens from the whole of Sweden. He could not geographically divide Swedish Buzzards into morphs or subspecies. Although birds in the south tended to be smaller and paler, he could not define any geographical demarcation line, and so agreed with Brown & Amadon (1968) that Steppe Buzzards were not common in northern Sweden. From what we now know of adjacent populations being more different than distant populations, if there really was a difference then we would expect to see genetic evidence (Kruckenhauser *et al.* 2004). As we write the book, it's exciting to see a Buteo Morph citizen science project being conducted by Max-Planck institutes, using social media to look at morph types throughout Europe. They had nearly 7,000 images in September 2016.

In darker Common Buzzards, the carpal patch on the underwing is not as distinct from the dark underwing-coverts that spread back towards the body, although the coverts can vary in colour from very dark brown to pale brown or ginger. The chest and belly are dark, but often there is still a pale horseshoe on the chest. In paler Common Buzzards, the chest and belly can be white, cream or slightly yellow with only a few dark feathers. In this case, the dark carpal patch under the wing stands out more, becoming a comma shape in the palest birds. These pale versions are the birds that cause most confusion with inexperienced birdwatchers, who are convinced that they must be something rare. When seen against a brown backdrop, such as a ploughed field, the brown wings and back fade into the background, and at a glance they can appear almost white, like the gulls with which they sometimes forage.

The most difficult bird to distinguish from nominate Common Buzzard is the very closely related Steppe Buzzard subspecies, which also has different colour morphs, including a fox-red phase that shows particularly red-tinged underwing-coverts. At least this rufous phase is much easier to distinguish from Common Buzzards due to its uniform reddish body and wing-coverts, but that makes it more difficult to differentiate from Long-legged Buzzards. In turn, while there is overlap in plumage between Steppe and Long-legged Buzzards, the latter tend to have paler heads and darker bellies and often an unbarred tail. This makes them easier to distinguish from nominate Common Buzzard, with its darker head, pale breast-band, often pale belly and always regularly barred tail. Some dark-phase Long-legged Buzzards have a wide black sub-terminal tail-band, similar to the easiest identifier for adult Rough-legged Buzzards, which may also have one or two other narrower bands. Adult Common Buzzards can have a dark sub-terminal band, but it is much narrower than those of Rough-legged and Long-legged Buzzards. If seen from below, then Rough-legged Buzzards mainly have much paler underwings. If the tail does not have the wide, dark band, then it is worth checking the carpal patches, which are solid rectangular blocks in Rough-legged Buzzards, rather than the crescents of a Common Buzzard. Some Common

The Common Buzzard

Buzzards have very pale underwings, but then the carpal patch is a much thinner dark crescent, not a solid block. Unfortunately, all three species have morphs where the front of the underwing is solid dark from the carpal ends right to the body, and that can make the underwing very difficult to use as a distinguishing feature.

When attempting to confirm a Common Buzzard's identification, a good feature to search for is the tail, which consistently has regular barring all the way down, unlike Rough-legged Buzzards, honey-buzzards, kites and harriers. Dark-morph Common Buzzards may be confused with dark-phase Booted Eagles *Hieraaetus pennatus*, but can be distinguished from the underwing plumage, particularly by the Eagle's shoulder 'lights', absence of lighter inner primaries, and all-dark secondaries, rather than just the trailing edge band of a Common Buzzard. The very lightest Buzzard morphs could be confused with much larger Short-toed Snake Eagles *Circaetus gallicus* at a distance, but when closer the speckled underwing, together with the more pronounced head and squarer-cut tail of the eagles, are apparent (Porter 1974).

Lastly, another distinguishing feature, which often alerts people to the presence of a soaring Common Buzzard, is its vocal repertoire. The flight calls have a trailing downward pitch and volume from a harsh start, approximately one second long. The calls are quite variable and are often described as mewing, with reference to them sounding a bit like a cat's mew, sometimes written starting with the letter 'P', for example, *pee-ee-ya* or *pee-yaaa*, or

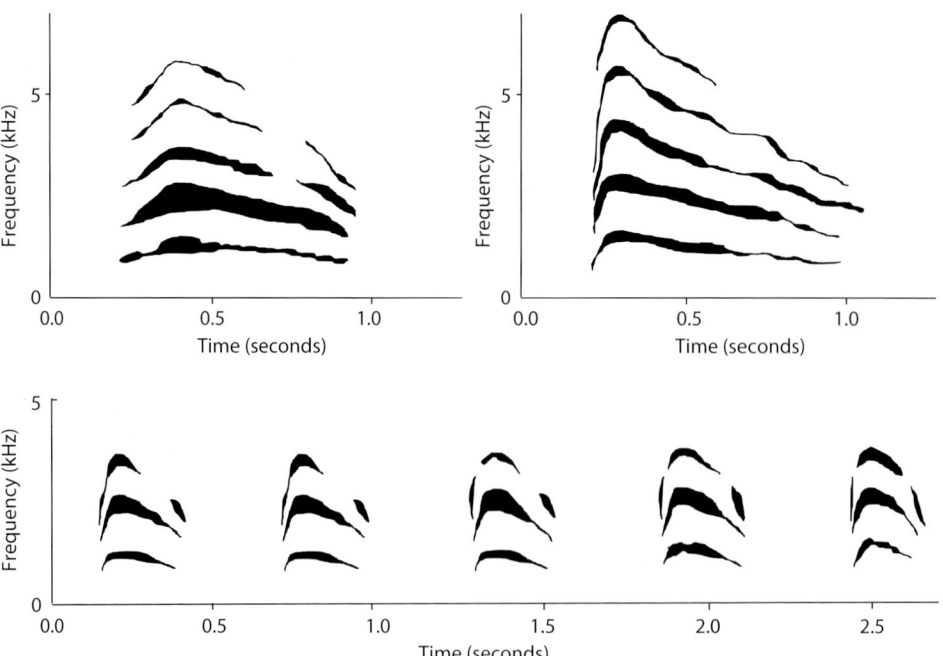

Figure 1.6. *Common Buzzard sonograms (Jeserich 1970) show that calls occur on several frequency harmonics, albeit differing in volume on each, as shown by the thickness of the lines. Thus, the most common call of soaring birds (top left) is strongest on its second harmonic, whereas the alarm call (top right) rises more sharply on all harmonics and tends to waver; a male faced with an aggressor (below) repeats a shorter territorial call with fewer high-frequency harmonics.*

even *peeyooo*. But Prytherch's (2009) *ca-au* represents the start better, because the call sounds as though it is coming from the back of the throat, with a sharp start. The diminuendo has the result of making it sound softer, but listen carefully and the sudden start is harsh. If distressed, the trailing end warbles down, rising up and down quickly but still diminishing in volume and decreasing in pitch, rather like how our voices wobble when we are stressed. The territorial call is a sharper and quicker *eyoo*, around a quarter of a second long, rising and then falling evenly, but with the tone falling lower than the starting pitch. A juvenile Common Buzzard gives a call like the adult's flight call but it has multiple frequencies so can sound very discordant, far harsher than the pure adult note. In late summer and early autumn, juveniles spend a lot of time begging with a yipping noise, described by Dare (2015) as *pi-ya–pi-ya–pi-ya*, with the pitch and volume rising fast, followed by a dramatic fall, in less than half a second. If you are not able to hear these calls in the wild, it is worth visiting the excellent Xeno-Canto website (http://www.xeno-canto.org/species/Buteo-buteo), where you can hear various vocalisations, and see their sonograms, from all around Europe (Figure 1.6). As a comparison with other species, the flight call of Long-legged Buzzard sounds similar, whereas the Rough-legged Buzzard flight call maintains the higher pitch and volume, so sonograms look more level and for longer, about one and a half seconds. Gjershaug *et al.* (2006) have described, and obtained sonograms of, the juvenile begging call of a hybrid between a female Rough-legged Buzzard and a male Common Buzzard. Those calls were more like those of a Rough-legged Buzzard but had elements of Common Buzzard and incomplete calls.

When writing about the Common Buzzard we are dealing with a polymorphic species, which is not well differentiated from other species and is incredibly widespread. In the following chapters we are going to explore details of Common Buzzard life and how these fit in with the changing environment and how they relate to humans. This chapter has attempted to characterise the Common Buzzard and consider the constraints it has inherited from its origins, in order to understand how natural selection has worked over time to define the very successful species we enjoy today.

Conclusions

1. Common Buzzard is a widely distributed species, ranging east across Eurasia from the British Isles and the Azores until it is replaced by the Japanese Buzzard in eastern Mongolia. To the north, the Rough-legged Buzzard dominates the high tundra, while to the south, from the Mediterranean and Turkey to eastern Kazakhstan, a variety of closely related species blend into the nominate Common Buzzard.
2. Its wide distribution demonstrates a key characteristic of this species: the adaptability of this raptor to different environments.
3. In the field, it is difficult to differentiate the nominate Common Buzzard from subspecies and other Old World buzzard species. Morphologically, its plumage is very variable, as is that of other similar species.
4. Even using the modern techniques of genetic cladistics, it is difficult to unambiguously categorise some individuals into a particular species. It has therefore been suggested that a *Buteo buteo* superspecies would be more appropriate, although it would have to be carefully constructed to meet conservation objectives.

5. Given the lack of genetic differentiation between the Old World buzzard species, their origins appear to have been relatively recent (during the last glaciation), from either Rough-legged Hawk (Buzzard) or Ferruginous Hawk, or a common ancestor of those species that originated in the Neotropics.
6. Common Buzzards have very modest features, camouflage-brown plumage, small feet, a short tail and wings that are somewhat broader than many other raptors. While evolution has moulded these features to suit their hunting of small mammals in open country, none of their features are extreme, suiting them perfectly to be adaptable generalists.

CHAPTER 2
Prey

Common Buzzards are relatively easy birds to train to return to a falconer because they tolerate humans very well. Rather than relishing the chase like a Peregrine Falcon *Falco peregrinus* or Northern Goshawk, they are very happy to take the easiest food they can find. We will see later in the chapter how the Common Buzzard adapts to anthropogenic change by exploiting the food that change reveals. For a falconer, it means that the inert meat placed on a gauntlet can have a greater appeal than a kicking young Rabbit *Oryctolagus cuniculus* running for cover 50m away. Unfortunately, speaking from experience, that also means a trained Buzzard can be an embarrassment to display. Rather than making an elegant chase, or showing obedience by promptly thumping the fist as soon as meat is displayed, a Common Buzzard may merely find the nearest bit of ploughed field in order to pull up a few worms and look like an inelegant cross between a thrush and a waddling duck (Plate 5; see colour section) – hardly what the spectators are yearning to see. Yet, is this something the trainer should feel embarrassed about? Isn't it this adaptability of exploiting different food sources, after its plausible origin from a species evolved to feed on fluctuating populations of small mammals, that makes the Common Buzzard a highly successful species? If the Common Buzzard were a business we'd be applauding the flexibility that makes it so widespread and common. So, let's explore the dietary adaptability that has made the Common Buzzard such an evolutionary success story.

Food requirements

Food provides the fuel required for a bird's thermoregulation, growth, dispersal, hunting, breeding and defence of itself and its chicks. We are lucky to have good information from tame Common Buzzards on their calorific requirements. As with many aspects of Common Buzzard behaviour, we can start by looking back to Peter Dare's PhD thesis (Dare 1961), in which he assessed intake rates from young birds, and also conducted some feeding experiments to get a more precise estimate of what was being eaten, and what effect that had on body mass. Dare initially fed two Buzzards of 6–8 weeks old as much Rabbit as they could eat. They consumed on average approximately 150g per day, or 17 per cent of their body mass. There were days when they ate up to 30 per cent of their body mass, in more than one meal, but these days were infrequent. That's a considerable amount of food, and far more than humans eat comfortably as a proportion of body mass.

Later in the autumn, and again the following spring, Dare provided the same Buzzards with an excess of Grey Squirrels *Sciurus carolinensis*, a relatively rich meat. In autumn, the birds tended to gorge one day and then eat less the following day. One ate about 125g per day (12 per cent of body mass) while the other ate 150g per day (14 per cent body mass). In spring, they both ate approximately 140g per day, again sometimes gorging over 30 per cent of their body mass in one day. These were tame birds, perched waiting for food to be delivered, not wild Buzzards that must burn more calories pursuing prey. To get an impression of Buzzard diet in the wild, Dare also spent time analysing pellets to estimate the daily food intake in their natural habitat. It seemed that in autumn and winter (September–February) they ate approximately 100g per day (150kcal or 3–4 voles). If Rabbits were sufficiently abundant to make up a larger proportion of the diet, then the estimated intake was higher, presumably because once the Buzzard had a Rabbit there was a lot of flesh available to be devoured easily. These estimates of food requirements fit well with what others have found of generalist raptors, such as buzzards, kites and eagles, needing 10–15 per cent of their body mass of wet food per day, compared with more specialised avian predators, such as falcons or small *Accipiters*, which require more like 20–25 per cent of their body mass daily (Craighead & Craighead 1956, Brown 1978, Kirkwood 1979).

When raising a brood of chicks, a great deal more food is needed than for two adults. Dare (2015) made an estimate for the different stages of rearing a chick. He concluded that nearly 7kg of prey (350 adult Short-tailed Voles *Microtus agrestis*, as in Plate 6; see colour section) are needed to get each chick to fledging. In the first few days after hatching the additional food the parents must provide is quite modest, at only a vole or two per chick per day (30g), but it rises quickly. In the second week after hatching they require between three and five voles (75–100g), and in the third week it's between five and seven voles (125–150g) per chick per day. The highest intake is in the last weeks of nesting and the early branching (days 22–42), when there is still a lot of feather growth and the birds are becoming more energetic. A nestling requires between six and nine voles (150–175g) per day to survive that period. After that, the demand reduces to around six voles (150g) per day.

These estimates agree with the observed 14kg of food brought to a brood of two on Dartmoor, from when they hatched to 53 days of age. The intake is considerably higher than Sylvén's (1982) estimate of energetic requirements (calorific, rather than mass) of a growing chick. He estimated a peak at 165kcal per day, but that is only just over 100g of natural prey, only two-thirds of Dare's estimate during the maximum growth period. However, these energetic estimates were made on flesh alone, whereas Dare's field and experimental observations were based on whole animals. So, if we are to think in terms of the complete prey being caught, Dare's estimates are more appropriate.

The male provides all the food during the first three weeks of chick-rearing, so at 20 days he is likely providing food for himself (120–150g), slightly less for the female as she is not being so energetic (100g) and 150g for each chick. With a brood of three that could be nearly 600g of food per day, four to five times what he would be catching in winter. Food is more abundant during the breeding period, and the daylight hours provide longer hunting periods, but this is still a significant increase. It's not so bad if there are sufficient Rabbits, as that is about two young Rabbits of 350g, but if only voles are available that would be more than 20 to catch in a day – quite a challenge. Not all nests raise three nestlings, and so it is evident that not all males in all years can escalate their food-gathering ability to provide all that is needed, especially if Rabbits are not abundant. After the third week, the female will start hunting and the labour can be shared, although by then some of the food supply may have been depleted by the male's previous hunting, so the prey encounter rate may be reduced.

Dare also looked to see what happened to the mass of full-grown Buzzards that weren't fed. In cold weather they could lose about 2 per cent of their body mass per day, or 150g over seven days. After a week, they were fed again until they regained their healthy mass. It was interesting to know that when the experiment was repeated in Spain (García-Rodríguez *et al.* 1987), the body mass again declined at 2 per cent per day, even though the climate was a lot warmer. It is thought that Common Buzzards can withstand 8–10 days of total starvation even in cold conditions such as a continental winter. Considering the mass that emaciated Buzzards can be reduced to before dying, Dare (2015) estimated that it would take more than two weeks without food before a Buzzard that started in reasonable condition would die of starvation.

Falconers must manage the mass of their birds for flying; that is, they need sufficient food to have the energy to fly with ease and power, but not so much food that they are lethargic and uninterested in returning to the falconer for food. Therefore, falconers are very sensitive to how particular foods change the mass and attentiveness of their birds. If fed beef, most birds of prey will increase their mass, but if fed whiter meat such as Rabbit then there is less mass gain, or possibly the bird will lose mass. For example, In *A Manual for Falconry*, Michael Woodford (1960) considered Woodpigeon *Columba palumbus* to have twice the energetic value of Rabbit. However, Buzzards are renowned for 'living on air', putting on mass no matter what is fed to them.

In a controlled experiment at Glasgow University, Scotland, Nigel Barton & David Houston (1993) fed some captive Peregrines and Common Buzzards equal amounts of Domestic Pigeon *Columba livia domestica* breast, and then Rabbit meat, each for a period of eight days. The quantity was such that the Peregrine maintained its mass with Domestic Pigeon meat, but lost mass with the Rabbit meat. On the other hand, Common Buzzards increased mass with the same amount of Domestic Pigeon meat, and managed to maintain their mass with the diet of just Rabbit. Buzzards can absorb more nutrients from the less nutritious meat than Peregrines. This apparent greater digestive efficiency of Buzzards is most likely to do with the length of the small intestine. Barton & Houston showed a clear correlation between digestive efficiency and residual intestine length (taking account of body size) for several raptors. Buzzards had one of the longest guts relative to size, although another generalist and scavenger, the Red Kite *Milvus milvus*, had an even greater relative gut length and was even more efficient at absorbing nutrition from food. Such is the efficiency of their guts that Buzzards and Kites can thrive on carrion, but can also maintain themselves by eating earthworms. Of course, energy requirements are not only affected by the bird and its morphology, but also by the environment, which for much of the Buzzard's distribution changes seasonally. Barton & Houston also indicated that for captive-fed birds the digestive efficiency was greater at temperatures over 20°C compared with those fed at 0°C. Thus, a need for more food during cold weather may result not only from the greater heat loss, but also because the gut is not working as well.

Twenty years later, while conducting her PhD on Common Buzzards in Ireland, Eimear Rooney (2013) obtained the energy content of typical Common Buzzard food using a bomb calorimeter. Small mammals, which are normally swallowed intact, were tested whole, whereas only the muscle of larger items such as Rabbits was used because that's how the Buzzard would eat it. Although earthworms had the lowest energy per gram of dry mass, at 21.7kJ/g it was not very different to a Woodpigeon at 24.3kJ/g or the highest value of 25.1kJ/g for a Wood Mouse *Apodemus sylvaticus*. With small vertebrates averaging a dry mass about 25 per cent of whole body mass, it's possible to work out how many of each prey type are needed per day during the winter, when Buzzards are attempting to maintain their condition rather than meeting the extra needs of breeding. Working on a Buzzard's body mass of 900g (somewhere between the male and female average mass) and using Sylvén's calorific estimate of 630kJ per day, a Common Buzzard needs three or four 35g voles, or half of a typical 350g Rabbit or around 50 2g earthworms per day. This agrees with Dare's work on inactive Buzzards back in the 1960s. On the other hand, Rooney used the estimate from Kenneth Nagy *et al.* (1999) of field metabolic rate in kJ per day as $10.5 \times mass^{0.681}$. That formula gives an energetic requirement of about 980kJ/day (male) or 1160kJ/day (female) during the non-breeding period, which equates to 50 per cent more than Dare's estimate and may be more appropriate for an active free-living bird. This would estimate the daily intake to be more like 80–100 worms, most of a small Rabbit or between five and six voles per day. As Buzzards spend rather little time actively hunting, their demand for food outside the breeding season probably lies between these two estimates.

As well as energy, food provides the vitamins and minerals necessary for health. When selecting mates, many birds have specific signals that indicate their health. Examples are the classic tail of a male Indian Peafowl *Pavo cristatus*, or the evidence that female Great Tits *Parus major* appear to choose males on the width of the black stripe that runs down their breast (Norris 1990). We will explore mate choice more in the chapter on courtship, but there are various ways in which food can affect the appearance of Buzzards to indicate health – something that may influence potential mates. The first is true for all bird species: if there is insufficient food when feathers are being grown, then 'stress bars' develop (Newton 1968). These have been called 'hunger traces' or similar by falconers for centuries and are visible weaknesses across the feathers. In juveniles, which grow all their flight feathers at the same time, a weak line across all feathers may mean their whole tail breaking at a particular point, or several feathers from the same wing breaking, creating a potentially more serious flight imbalance than likely from the regular moulting process. If food supply is good, then feathers still have 'growth bars', which show as alternating light and dark bands, much more subtle than 'stress bars'; these are likely due to the daily timetable of food delivery and sleeping physiology (Riddle 1908, Wood 1950). Ptilochronology, the measurement of those growth bars, can also be an indicator of condition when growing the feather. Gombobaatar Sundev *et al.* (2009) have shown that those Upland Buzzards in areas with better vole populations, which produce more fledglings, have chicks with wider growth bars because they have more available nutrients and so can grow more feather in a day.

Another mechanism that will indicate health is the use of melanin in feathers. Barn Owls *Tyto alba* with elevated levels of corticosterone (a hormone associated with stress) have less colourful feathers, due to the lack of the melanin pigment (Roulin *et al.* 2008). So, a bird that has been suffering stress during the feather growth period will look more washed out. It is likely this is true for other birds; for example, anyone who has handled a lot of birds will have seen tails on juveniles change abruptly from normal colouration to a washed-out appearance, perhaps with a stress bar at the start of the change in colour (Figure 2.1). This represents a period when there was a sudden and extended lack of food, in which the nestling would have continued to grow using reserves within the body, and the tail feathers would have kept growing, too, even if they didn't have sufficient quantities of pigment. Stress bars and growth bars indicate the health (and likely food intake) while the feathers are being grown, and will remain on the bird until the next moult, nearly a year later.

A more current indication of a Buzzard's health can be seen in the cere. The normally yellow colour comes from carotenoid pigments that depend on both diet and condition (Chakarov *et al.* 2008). Buzzard legs are also yellow, and Jésus Martínez-Padilla *et al.* (2013) have shown that the coloration of both the cere and legs in Common Buzzards can be correlated with corticosterone, one of the most studied hormones associated with stress. Although the exact mechanism has not been established, carotenoids are thought to be needed for the body's defence in times of stress, whereas if not stressed they can be stored in the cere and legs, which results in the fleshy areas appearing brighter and more colourful (Møller *et al.* 2000). Hence, they are good indicators of the stress and nutritional conditions of a Buzzard, and so are useful signals of health for sexual selection and chick-rearing. Plate 2 (see colour section) contrasts the yellow cere of a wild Buzzard with a captive-bred Buzzard

Figure 2.1. *From the tip of this tail feather towards its base, five fault-bars can be counted, followed by a complete change in pigmentation (© Sean Walls).*

that we collected as part of our release project. While the captive-raised chick (Plate 2b) appeared healthy, it had been fed on nothing but laboratory mice, and the cere had a blue hue. The lack of yellow suggested that it had not got the carotenoids available in a wild diet. François Mougeot and Beatriz Arroyo (2006) have shown correlations between the ultraviolet (UV) reflection of ceres with male body mass and the timing of broods in Montagu's Harriers *Circus pygargus*. More intriguingly, Deseada Parejo *et al.* (2010) have found that the UV reflectance peak of the ceres of juvenile Eurasian Scops Owls *Otus scops* was related to relative mass, and that parents preferentially fed those with the UV reflectance that indicated underweight chicks. However, the extent to which UV is seen or used as a cue by raptors is not fully understood (Lind *et al.* 2013).

Data collection and biases in diet studies

Having established how much food is needed by Common Buzzards, and the consequences of not receiving sufficient, we are ready to look at what prey they eat to obtain that food. First of all, let's consider the ways in which Buzzard diet data are collected, and their inherent biases. Whilst thinking about methods and biases may not be as exciting as watching birds, without such thoughts the observations we make may mislead us. There are several diet studies where more than one method has been used on the same set of Common Buzzards, resulting in very different impressions of what they are eating, so let's look at each method in turn.

Chance observations provided the earliest data on raptor diet, and are still recorded today. These are the events all naturalists happen to see, such as a flying Buzzard with an especially long snake dangling limply below it or a dramatic stoop on a Rabbit. There was no forethought to go and record what Buzzards were eating, but having seen something exciting the observer may want to record it in a local bird report or online. Why should a modern scientist bother with such unsystematic data when trying to make an assessment of Buzzard diet? With the Common Buzzard it is particularly worthwhile, because there is a long history of opportunistic observations of the 'unexpected', and Common Buzzards have many 'unexpecteds' due to their generalist nature. So, while such chance observations can't be used on their own to understand Common Buzzard behaviour, they do supplement the more rigorous data collection and help to complete the picture of diet diversity. Older observations also help to assess whether there have been changes through time, though not in a quantitative manner.

Planned field observations still involve sightings of Buzzards and recording their behaviour, but they differ from chance observations in that someone has designed a systematic way of collecting the data. They are more robust for comparisons between sites or quantifying what is happening over time at one site, because researchers collect the data in a defined, repeatable manner. However, such observations are still susceptible to bias, primarily because researchers can only record events when birds are visible, or where the observer is expecting to see them. Common Buzzards spend a lot of time in woodlands where it is difficult to get more than a glimpse of them, so much behaviour will be missed. Likewise, due to the effort of making observations, researchers will go to where they hope to get the most data, and perhaps unwittingly leave the area where a few Buzzards are doing something different.

Prey remains at nests have been the easiest way of recording what Common Buzzards (and other raptors) are eating. Sometimes, it is not the remains but the entire prey bodies that are found if food is in sufficient surplus (Plate 7; see colour section). A study in Poland reported an impressive case where 23 uneaten prey items (18 were female voles) were found on one visit to a nest (Olech & Pruszynski 2000). With this method, data can be collected consistently and comparisons can be made between areas. Unfortunately, the method is fraught with issues that have been openly discussed by researchers. The problems revolve around the misrepresentation of prey items. Smaller items may be swallowed whole, leaving no remains, and many have found that small prey are underestimated (Selås, Tveiten and Aanonsen 2007; Tornberg and Reif 2007; Francksen, Whittingham and Baines 2016). There is a risk of overestimating large prey, too, unless bones from one particular limb or part of the thorax are used to count each individual.

Cameras at Buzzard nests can reduce the bias that occurs when some remains are more likely to be found than others, because theoretically all the prey fed to the youngsters can be seen. Risto Tornberg and Vitali Reif (2007) compared the results of video and prey remains from the same nests. They found that cameras at Buzzard nests captured four times as many prey deliveries as items identified in prey remains. Large birds were very likely to be recorded whereas small mammals and herpetofauna were under-represented among prey remains. Similar results, confirming the over-representation of bird feathers and Rabbit fur in remains collected on nests (Figure 2.2) were shown more recently by two studies in Scotland (Swan 2011, Francksen *et al.* 2016b). Selås *et al.* (2007) combined data from pellets and nest remains, and also found that large birds were more likely to be recorded than smaller prey, but in terms of the main taxonomic groups, only amphibians differed significantly between cameras and nest remains: if these were taken out then the two techniques were comparable.

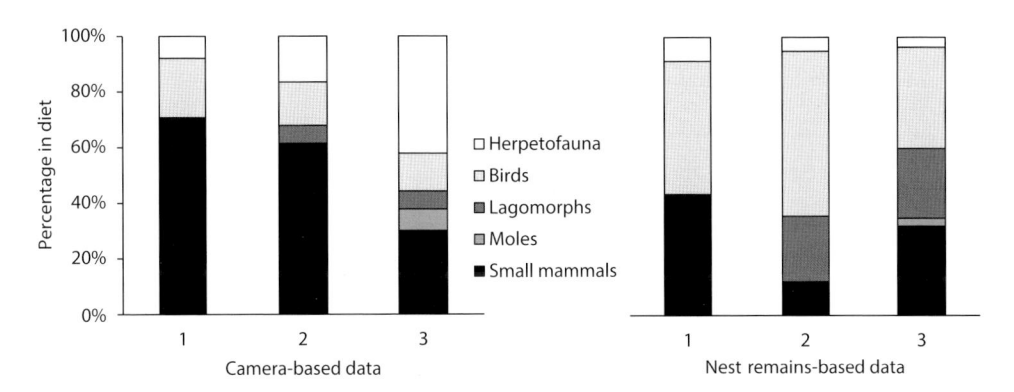

Figure 2.2. *Three studies that estimated Buzzard diet showed that cameras at nests recorded more small mammal prey, whereas collections of prey remains recorded more birds and lagomorphs (Rabbits). Data from 1) Tornberg & Reif (2007), 2) Francksen* et al. *(2016a) and 3) Swann & Etheridge (1995).*

Cameras at other bird nests have had surprising success at capturing Buzzards extracting nestlings for food. These cameras have usually been mounted because a bird species has been declining and researchers are interested if anything happening at the nest, such as predation, could be causing the decline. Squirrels, Magpies *Pica pica*, Jays *Garrulus glandarius* and other corvids are often suspected, especially for effects on small passerines. Surprisingly, however, Buzzards are more common as predators than many would imagine. Just like the opportunistic observations, this adds rather little to our knowledge of Buzzard diet as a whole, but can surprise us, and could help those trying to conserve the prey species.

Pellet analysis is as traditional as collecting and counting nest remains (Plate 8; see colour section). However, in common with field observations it has the potential to identify prey species during the non-breeding period, too. If there is sufficient food then Common Buzzards generally regurgitate one pellet per day. As with the analysis of prey remains at nests, this presents a problem with quantifying how many individual items are involved, but the added issue that some bones (for example, of amphibians) and other food items (for example, earthworm chaetae) are under-represented. At least for small mammals, the enamelled teeth appear to survive well. However, when counting bones or teeth in pellets, it is tricky to decide how many individual animals they belong to. If two femurs of similar size but opposite legs are found, they could be one each from two animals or both from the same. Or does the jaw bone come from the same individual as the pelvic bone? The usual assumption is conservative, to attribute the remains to as few individuals as possible. Furthermore, quite often results are given as the number of individuals of particular species per so many pellets from so many nests. However, it can be unclear whether equal numbers of pellets have been taken from each nest, or whether the majority came from a single nest.

Bones of larger animals, such as Rabbits, may also not feature in pellets because they are too large for Buzzards to swallow whole. Therefore, such prey can be missed if estimates are to be based on identifying bones. However, Rabbits can be counted in several pellets following meals on the same kill if hairs in the pellet are also being used to identify prey. In Poland, Goszczyński & Pilatowski (1986) compared prey size classes found at the nest with those in pellets. For Common Buzzards, 67.5 per cent of pellets had prey up to 20g (vole size), while the nest remains had only 21.6 per cent of this small mass class. On the other hand, only 4.9 per cent of items in pellets were from prey over 300g, compared to 31.7 per cent of prey remains at nests.

When conducting his PhD, a long time before all the wonderful technological tools we have now, Dare wished to establish what Common Buzzards ate in winter and considered that pellets were the only way to get a reasonable understanding. Unfortunately, pellets were not often dropped in the same place, because the Buzzards periodically moved roost sites. Where they were dropped they were often scattered in inaccessible places or under vegetation where they were difficult to find, and they were liable to disintegration in wet weather. Despite the effort, he believed he found less than 15 per cent of them. To deal with this problem, Dare (1961, 1998) fed two Buzzards to establish a correction for the differences between what is eaten and what appears in the pellet. After finding the fraction of each prey type that appeared in the pellets, he then multiplied up the remains found in pellets from wild birds. Richard Francksen *et al.* (2016) found that pellet contents over-estimated invertebrates, compared with the examination of nest contents and nest camera observations. They therefore also used feeding trials to estimate that, because Buzzards tend not to ingest feathers from large birds, feathers were likely to be detected only once for every two meals on the carcass of a Red Grouse *Lagopus lagopus scotica* (Francksen 2016). Unfortunately, most studies do not try to correct the values and just accept the biases.

Stomach content analysis is another instance of opportunistic observation. While it is true that someone has made the effort to extract the contents and analyse them, nonetheless it relies on the luck of acquiring Buzzard carcasses that are fresh enough for the stomach contents to be identifiable. Without systematically killing Buzzards (which would be both unethical and illegal nowadays), the samples are necessarily chance events and therefore are likely to be biased. For example, Buzzards eating certain foods may have been more likely to be killed because their foraging was dangerous, or those being hit by wind turbines might be migrants rather than residents.

Tagging and tracking has been used more recently to investigate Common Buzzard diet, a technique that was still in its infancy when Colin Tubbs wrote *The Buzzard*. We and others have radio-tracked Buzzards to attempt to observe what they are eating, and tracked prey species to see how many of them are eaten by Common Buzzards (Figure 2.3). Not surprisingly, this also has its biases. Theoretically, you can follow particular birds continuously to determine exactly what they are eating, therefore reducing biases caused by visibility. In

Figure 2.3. *A young Buzzard marked with a backpack radio-tag shortly before fledging; tags are used to monitor dispersal, survival and foraging behaviour in early life (© Robert Kenward).*

addition, there is the potential to get to the fresh kill early enough to see the predator still eating the carcass, and even take samples to investigate factors, such as disease, which may have contributed to the predation (Kenward 2006). However, in practice, this is not easy when studying Buzzards. Trying to stay close to a sharp-eyed Buzzard in open country without affecting its behaviour is very difficult, and inevitably you are not going to see everything that happens, especially when small prey are swallowed whole. The financial resources to keep trackers close to Buzzards or check radio-tagged prey very frequently are not usually available to most research projects. So, tracking is liable to produce samples of Buzzard behaviour that rarely include a successful kill, unless you count the pulling of earthworms, which was our particular interest.

Combination techniques include a recent study that used pellets to study diet away from the nest combined with radio-tracking and global positioning system (GPS) tags to facilitate finding roost sites with pellets on Scottish grouse moors (Francksen *et al.* 2016a). The tags concerned transmitted the GPS locations to mobile phone masts, which often delayed reception of locations, and which were in any case not taken frequently enough (due to battery limitations) to have tracked a tagged bird's trajectory in detail. However, the locations did reveal roosts, and a relative lack of trees for the Buzzards to roost in meant that birds were more faithful to roosts for Francksen than they had been for Dare 50 years earlier in Devon. Knowing which Buzzards were attending the roosts also made a comparison between ages possible, and gave useful information about predation on grouse. Pellets from adults tended to include more voles and birds, whereas youngsters had more lagomorphs and invertebrates. However, there was great variation from individual to individual, and so the differences were not significant.

Another combination approach, which has yet to be used for raptors, is the attachment of cameras as well as GPS to the birds. This technique has already been used to investigate tool use by New Caledonian Crows *Corvus moneduloides* (Rutz *et al.* 2007), and these birds are smaller than Buzzards. In the future, the use of computers-on-chips for doing mundane tasks, like timing GPS samples to be only one at night but frequent when hunting, will be used in much more sophisticated ways, such as for detecting focal behaviours. A camera attached on either predator or prey that switched on only when a kill occurred could be very useful for detecting predation impacts. However, it requires a lot of intensive work calibrating accelerometer readouts to determine the likely combination of movements associated with hunting or being hunted, and the event is most easily recognised after it has occurred, which doesn't help turn on the appropriate recorders during the event. Running cameras and accelerometers all the time is power intensive and quickly uses up the relatively small power cells appropriate for tags on birds.

Many papers recognise that for a species with a very varied diet, from small creatures like earthworms and shrews to larger species such as Hares *Lepus europaeus* and even deer carrion, it is not really the number of individuals being eaten that matters to the predator, but the biomass consumed. In mass terms, one small Rabbit is equivalent to many voles. However, there will be a lot more skin and fur on the voles, so it is still not the same nutritional input, even if the meat were of the same calorific value. Moreover, the richness of the food is also important for understanding what the minimum requirements are and how much more is needed to raise young. An individual specialising on Woodpigeons or Ring-necked Pheasant *Phasianus colchicus* poults may only need half the biomass of prey compared with a pair that can only capture young Rabbits. This needs to be kept in mind while we discuss the various prey groups.

Prey

Due to shortcomings in the techniques used to study Common Buzzard diet, it is difficult to estimate exactly what they are eating. However, the diet does tend to change according to location and time of year, as can be seen when a single technique is used to study diet in different locations and years, thereby holding a similar bias throughout. To make the story easier to navigate, we have divided prey into their broad taxonomic groups.

Mammals

Whether the language is Latin, Germanic, Uralo-Altaic or Slavic, the Common Buzzard tends to be named for eating mice and other small rodents. Thus its German name is *Mäusebussard*, and its Spanish name *Ratonero Commun*. The scientific literature bears out the accuracy of the country lore that is implied. All studies investigating the whole diet (rather than specific elements) record small mammals in the diet. When records are based on the recording of prey delivery by cameras or by pellet analysis, which are the more accurate approaches for small prey liable to be swallowed whole (Figure 2.2), small mammals tend to be the primary prey numerically (Figure 2.4).

There has been a lot of evidence collected on the importance for breeding success of voles in Common Buzzard diet (see Chapter 7). Although it has been presumed that small mammals continue to be the most important prey group outside the breeding season, there is much less evidence for that. Stefan Schindler *et al.* (2012) found that the density of wintering Common Buzzards in northern Germany was higher where they counted more vole holes, which would suggest the Buzzards were feeding on the voles. However, although

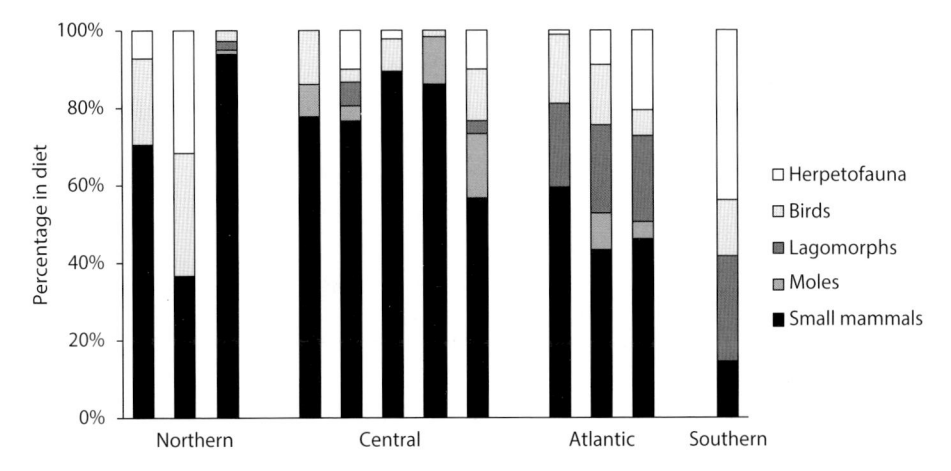

Figure 2.4. *Prey frequency recorded by cameras or pellet analyses at nests in northern Europe (Fennoscandia), and by pellet analyses in central Europe, Atlantic (British Isles) and southern Europe (Spain), showing the predominance of small mammal prey except in southern Europe. Data sources in Appendix 2.*

there was some arable land, the study area was dominated by grassland, which is worth remembering when we discuss the importance of invertebrates in arable areas in southern England, below. On Scottish grouse moors, too, winter pellet analysis showed small mammals to be responsible for around two-thirds of the diet, and lagomorphs another quarter, indicating that mammals were a very dominant food source, unless invertebrates and amphibians were seriously underrepresented in the pellets (Francksen *et al.* 2016c).

The most frequent small mammals identified as prey are Short-tailed Voles in the UK and Poland, and Common Voles *Microtus arvalis* in Norway, Finland and Spain. Other vole species, such as Bank Voles *Myodes glareolus*, are commonly eaten, but Common and Short-tailed Voles (and others from genus *Microtus*) appear to dominate the small mammal pickings. The reason why these species are more commonly recorded than others is thought to be simply that voles are more diurnal than mice or other small mammals, and so most likely to be active when Buzzards are hunting. Karol Šotnár & Ján Obuch (2009) showed different species were eaten at different altitudes. During the rodent population peak in 2007, Buzzard pairs nesting in a valley preyed mainly on Common Voles, whereas pairs living at the rim of the valley fed on Bank Voles and Moles *Talpa europaea*, and those higher in the Vtáčnik mountains fed mainly on Yellow-necked Mice *Apodemus flavicollis*; again they were probably just taking what was most available to them. Reif *et al.* (2001) recorded Water Voles *Arvicola amphibius* being taken almost as commonly (15 per cent) as *Microtus* species (17 per cent) in Finland, which is unusual.

We often found that voles laid on the nest were headless, presumably because the head had been eaten by one of the parents, and this has been noted by others, too. When wild raptors were fed experimentally, they typically ate the heads of smaller mammals first, a possible explanation being the brain's high fat content giving it a higher nutritional value (Slagsvold *et al.* 2010). Eating the head is also the best way to make sure that a caught animal is killed and unlikely to bite the predator or escape.

Other rodents are included in the diet to a much smaller degree than voles and mice, including Red Squirrels *Sciurus vulgaris* (Swann & Etheridge 1995) and insectivores such as shrews and West European Hedgehogs *Erinaceus europaeus*, and even the occasional carnivore such as Least Weasel *Mustela nivalis* (Jędrzejewski & Jędrzejewska 1993). Moles often appear in the literature. For example, Jan Pinowski & Lech Ryszkowski (1962) found them to be the most numerous species after voles (see Figure 2.4, column 8), and an old description from Morris (1862) observed that the Buzzard 'destroys numberless moles, of which it seems particularly fond'. A camera on a nest in Hungary recently showed Moles to be a major part of one Buzzard pair's diet (Mátyás Prommer pers. comm.). More usually, Moles appear as a minor but consistent part of prey remains and hunting observations. Actually, when feeding Moles to a captive Common Buzzard, it didn't seem very keen on the taste, so maybe Morris was watching particularly hungry Buzzards, or perhaps where they are fed to nestlings due to their abundance compared to other species, then individuals will acquire the taste over time and find them more palatable.

Although in many areas small mammals are the main prey, they can be only secondary prey when Rabbits or lizards are more available. Many British publications demonstrate how important young Rabbits are as a major food (Graham *et al.* 1995, Dare 1998, Sim *et al.* 2000, Hodder 2001), and this becomes especially clear when diet is considered as biomass across four sets of data, including Dare's from Devon (Figure 2.5). Dare (1957) goes as far as to say 'Rabbits are still caught at every opportunity, and I have several instances of Buzzards capturing young rabbits in localities where neither the farmers nor myself had suspected their presence. There is no doubt in my mind that Buzzards can winkle out the survivors of myxomatosis far more effectively than can the Ministry of Agriculture and, for this reason alone, they deserve the fullest protection.' Common Buzzard nests in southern Scotland were recorded closer to each other if lagomorphs (including Brown Hares and Rabbits) were more abundant, an indication that they were a key food source (Graham *et al.* 1995). In southern England, again we found that Buzzard breeding density increased with more Rabbit burrows, and also that nests with more Rabbits nearby produced more young (Hodder 2001).

However, the importance of Rabbits during the breeding season appears to be primarily a British phenomenon, and the quantity of papers on this prey is a product of the number of nest-based UK studies published recently compared with other countries. If one goes to the earlier studies recorded on the European mainland, as summarised by Glutz von Blotzheim *et al.* (1971), the emphasis is very much on small rodents. Even within the British Isles there can be considerable geographical variation. Swann & Etheridge (1995) found that two Scottish study sites (Appendix 3) only 65km apart had markedly different prey remains in nests. Rabbits appeared in 99 per cent of nests in Moray, but only 31 per cent of nests in Glen Urquhart contained lagomorphs, predominantly Brown Hares rather than Rabbits in that case. Rooney & Montgomery (2013) found that only Rabbit numbers (not other prey types) correlated with breeding success in Northern Ireland, where historically *Microtis* voles have been absent (Harris & Yalden 2008). In Spain, Rabbits can also be important; for example, Mañosa & Cordero (1992) found young Rabbits to be the most frequent nest remains, although pellets suggested that lizards were eaten more frequently.

Where Rabbits are not so common, or even absent, young Brown and Mountain Hares *Lepus timidus* sometimes feature, as noted in Finland (Tornberg & Reif 2007), Norway (Selås *et al.* 2007) and Italy (Sergio *et al.* 2002), as well as Scotland, albeit always in much smaller quantities than Rabbits. A questionnaire of farmers in England and Wales indicated that there was no relationship between sightings of Brown Hares and Buzzard sightings,

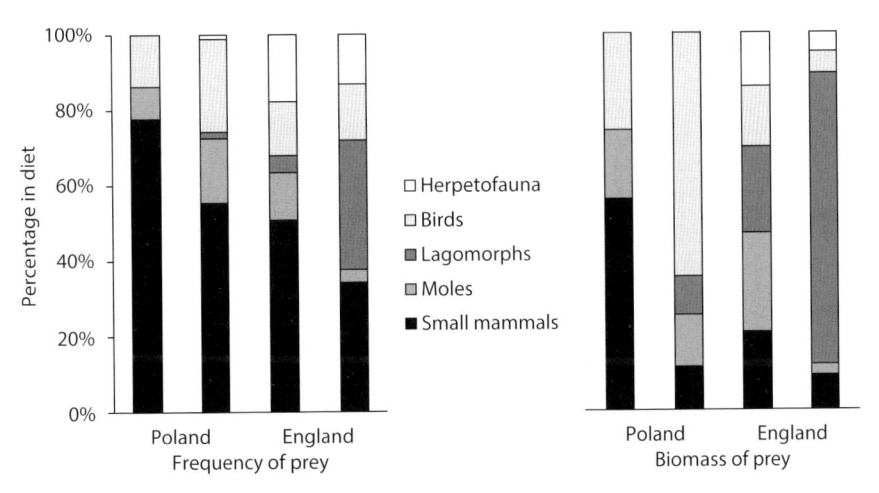

Figure 2.5. *Although small mammals tend to dominate diet in terms of frequency (left), when biomass, too, was recorded in the same four studies (right) birds and lagomorphs were often of greater importance. Data from Goszczynski & Pilatowski (1986), Dare (1998) and Skierczyński (2006).*

whereas there were fewer Hares with Red Fox *Vulpes vulpes* sightings (Vaughan *et al.* 2003). So, Common Buzzards do not seem to be responsible for the decline in Hare numbers, even if they sometimes appear in their diet. Hares are far less common than Rabbits and also larger, which is probably why researchers do not record them as often. Even Rabbits found at the nest are not fully grown. Dare (1998) noted that 98 per cent of Rabbit prey were 150–350g, but larger Rabbits, up to 700g, were sometimes recorded. Likewise, in some Spanish nests, Rabbit tarsus lengths corresponded to Rabbits of less than 550g (Mañosa & Cordero 1992). This could be either a mass limit to carry the carcass back to the nest, or an inability to catch and kill larger animals.

Rabbits are of a size that means they are perhaps over-represented by prey remains at nests, because the bones of the hind limbs and fur are not eaten but left on the nest. Thus, we need to look at other forms of evidence to get a sense of proportion of Rabbits in the diet. Some studies have found strong correlations between the number of young fledged from a nest and the density of Rabbits nearby (Graham *et al.* 1995, Austin & Houston 1997, Dare 1998, Hodder 2001, Rooney & Montgomery 2013). Others have used the temporal changes in Rabbit numbers to investigate changes in diet and productivity. Dare (1957) considered that the Rabbit had gone from a major to a minor food source after myxomatosis had seriously reduced the population, although Buzzards still caught them at every opportunity. There was a sudden drop in the Skomer Common Buzzard population after myxomatosis hit the Island in 1955 (Appendix 3). However, Davis & Saunders (1965) were not convinced that the Rabbits were the only factor, because Rabbits regained their numbers quickly but Buzzards failed to increase again. It has to be remembered that even if Rabbits are important prey for breeding Buzzards, this is unlikely to be the case for ensuring Buzzard survival once Rabbits are full grown: even male Goshawks, with much larger feet than Buzzards, are less likely than females to remain in areas where full-grown Rabbits are the main winter prey (Kenward *et al.* 1993a).

In summary, small mammals are the major prey of Common Buzzards throughout Europe, but where Rabbits are available, particularly in Great Britain, in terms of biomass they can dominate the food source during the summer breeding season.

Birds

Remains on the nest and nearby plucking points show that avian prey can make an important contribution to raising young Common Buzzards (Plate 9; see colour section). There are only a few studies where birds are shown to be the predominant prey (Figure 2.6), up to as much as 75 per cent by mass (Jędrzejewski *et al.* 1994, Goszczyński *et al.* 2005), but in most cases the high bird proportion only occurs in low vole years (Spidsø & Selås 1988), or if voles are absent (Rooney & Montgomery 2013). Although birds are not often the primary food on a population scale, individual Buzzards can specialise on catching birds even if mammals are present. For example, when climbing to nests in our area it was apparent that only certain nests had Woodpigeon remains, so it seemed only a few Buzzards could manage to take such a large, strong flier.

A consistent predominance of avian prey was found in pellets of Common Buzzards within the Białowieża Forest National Park in Poland (Jędrzejewski *et al.* 1994). Pellets had a very small percentage of small mammals (2–5 per cent), compared to bird bones (42–75

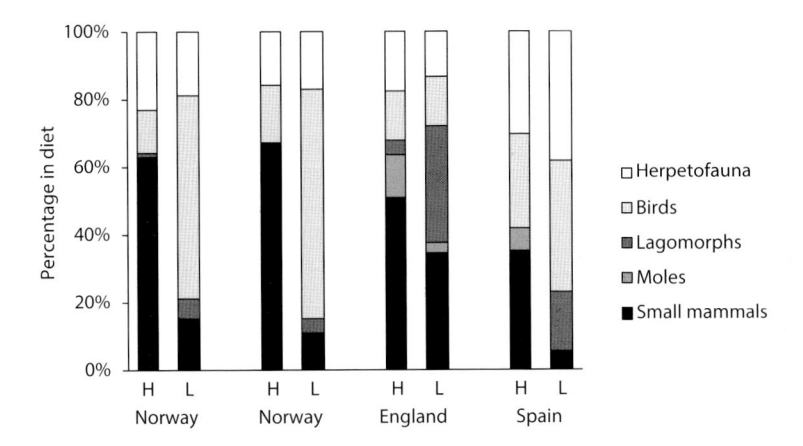

Figure 2.6. *Diet recorded at nest remains, in years or areas when small mammal counts were high (H) or low (L) relative to other prey, by the same teams in each country. When small mammal counts were low, the frequency of bird prey increased greatly in Norwegian studies, and lagomorphs increased in England and Spain. Data sources in Appendix 2.*

per cent), of which nearly half were Song Thrush *Turdus philomelos* bones. The data primarily came from 'forest nests', which were more than 1km from edge of forest. Their study focused work on the special pristine forest that makes Białowieża so famous, rather than 'ecotone nests' at the forest edges, which were only included if the researchers serendipitously came across them rather than systematically searching for them. So, as a whole, the study considered birds to be the more prominent prey. Reproductive success of forest nests was low (0.7 fledglings per pair), suggesting the birds were not as easy a prey for Common Buzzards, compared with small mammals in other studies. Their breeding performance did not correlate with any small mammal species, whereas the success of ecotone nests, at forest edge where *Microtis* voles were more available, had a strong correlation with small mammal abundance. Despite those correlations, even Buzzards at the ecotone nests mainly ate birds, including European Starlings *Sterna vulgaris* that were not available to the forest nests. Birds were also the most abundant group (46 per cent) in the diet of Common Buzzards in the Italian pre-Alps (Sergio *et al.* 2002); again, these were mainly woodland birds, and European Blackbird *Turdus merula* was the most common species recorded in food remains.

Juvenile birds

With many more species of bird available for Buzzards to hunt, compared with other vertebrate groups, the list of bird prey species is very long. What is apparent is that rather than specialising in taking certain species, Common Buzzards focus on juvenile birds (Fryer 1986, Spidsø & Selås 1988, Voříšek 2000, Tornberg & Reif 2007, Šotnár & Obuch 2009). Young birds can be easier to catch, especially if still developing in the nest when they haven't got strong enough feathers to fly away; also, being lighter they require much less effort to transport back to the nest. The majority (61 per cent) of the woodland bird prey that could be aged in Białowieża Forest were fledglings (Jedrzejewska *et al.* 1994). On the other hand, if too small, birds are not very nutritious. Selås *et al.* (2007) showed from video and nest remains that birds over 50g in mass were selected; the highest selection rank was for birds of 50–120g. Jacek Goszczynski & Tomasz Pilatowski (1986) found that the average mass of prey was 90g across mammals and birds. From nest contents, Fabrizio Sergio *et al.* (2002) found that only 8 per cent of bird remains were adult, 72 per cent fledglings and 19 per cent nestlings. Buzzards appeared to be exploiting the vulnerability of young birds rather effectively. From nest camera footage in Scotland, it was noted that if a Meadow Pipit *Anthus pratensis* fledgling was delivered, it was invariably followed by another or even several more fledglings in quick succession, suggesting that a pipit nest had been found by the parent and exploited one chick at a time (Richard Francksen pers. comm.).

Fledglings do not always outnumber adult birds in the Buzzard diet. At one Scottish site, Robert Swann & Brian Etheridge (1995) showed fledglings were only predominant for Ring-necked Pheasants, a species where the adult is too big to carry to the nest unless dismembered. They identified nest remains over 13 years from two study sites, resulting in records of nest contents for just over 100 nests at each site. They identified eaten fledglings by whether the feathers still had the waxy sheath that is present at the base when growing. If anything, this would misidentify some full-grown carcasses as fledglings because moulting adults would also have some growing feathers. In Moray, where Rabbit was the predominant prey, there was no particular indication that fledglings were being selected, except for Pheasant (50 chicks/poults, compared with one full-grown bird). Indeed, for Woodpigeon, the only other species taken in any quantity, there were 60 full grown versus 15 fledglings. However, at Glen Urquhart, where Rabbits had low abundance and small mammals were more common, there were more fledglings taken of most bird species, the most numerous being Common Chaffinch *Fringilla coelebs* (of which there were 110 fledglings compared with 38 full grown). For Meadow Pipit, the next most common species, 85 of 137 were fledglings. Meadow Pipits were the most common bird prey remains at nests in the northern English Lake District (Fryer 1986). So, the Moray site shows that where alternative prey are abundant, opportunistic prey captures may not result in fledglings predominating. However, if Buzzards have greater need of birds for feeding their own young, then it seems that they can specialise in finding young birds, because they are easy to catch and abundant during chick-rearing.

Common Buzzards have also been caught in the act while taking young from songbird nests that are being surveyed for predators with cameras. The species taken can be surprisingly small, such as Pied Flycatchers *Ficedula hypoleuca* (Bolton *et al.* 2007) or Wood Warblers *Phylloscopus sibilatrix* (Mallord *et al.* 2012). Weidinger (2009) found Buzzards to be responsible for 7 per cent of 178 nest predations recorded by cameras at songbird nests. They took mainly fledglings of Jays and thrush species. Moreover, Buzzards took nestlings in four consecutive years, indicating that nestling passerine exploitation is not always confined to years of low rodent abundance. Wolf Teunissen *et al.* (2008) investigated predators of Black-tailed Godwit *Limosa limosa* and Northern Lapwing *Vanellus vanellus* eggs and chicks, using both cameras at the nest and radio-tagging. The nest cameras primarily recorded mammalian predators eating the eggs, and they did not record Buzzards during that period. However, Buzzards accounted for over 12 per cent of chicks once they had left the nest, particularly of Lapwings (Black-tailed Godwit chicks being more vulnerable to Grey Herons *Ardea cinerea*). In one case, they recorded the Lapwing chick's

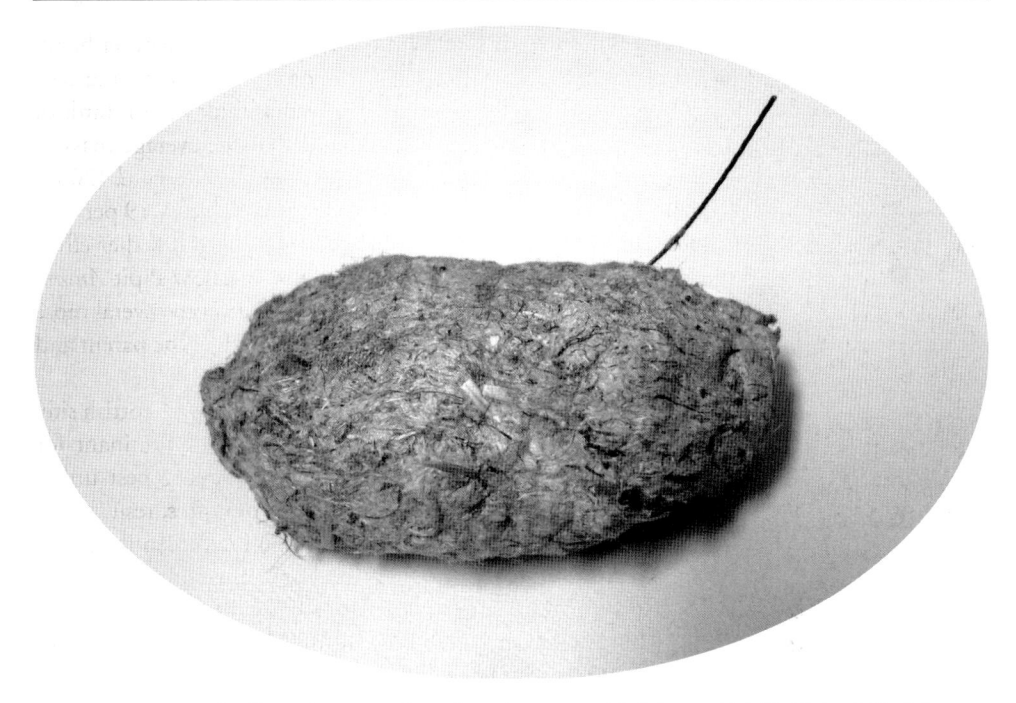

Figure 2.7. *Buzzard pellet with antenna from the radio-tag on a Pied Avocet chick (© Richard Clarke).*

radio-tag transmitting from inside a Buzzard! A Welsh study tracking 10 Pied Avocets *Recurvirostra avosetta* found two radio-tags under plucking posts thought to be used by a Buzzard, and one was found in a Buzzard pellet (Dalrymple 2014) (Figure 2.7).

Historically, Tubbs (1967) thought that the Common Buzzards in the New Forest (UK, Appendix 3) were more reliant on birds. This, he put down to the lack of suitable diurnal mammal populations, as the Rabbits there appeared to be mainly nocturnal at that time. Instead, Buzzards focused on young Western Jackdaws *Corvus monedula* and other corvids, although over time the large Jackdaw colonies in the old woods diminished to a scattering of a few pairs, which meant Buzzards had to find alternative prey (Tubbs & Tubbs 1985). Similarly, until recently Ireland has not had voles, and there are still areas where they are absent. In these regions, while Rabbits can be a large component of the diet, corvids and other young birds are as important for breeding Common Buzzards. Indeed, one estimate suggested that more than 135,000 young corvids could be taken each year in Ireland (Rooney & Montgomery 2013), which is probably regarded as useful by both those who see corvid damage to their crops, and those who want small passerines to survive their youth. Perhaps this offsets the effects Buzzards have taking other young birds, because there are fewer Buzzards and so their impact might not be as great as that of the corvids they kill; the corvids are likely to take a much larger number of juvenile passerines. However, across the Common Buzzard's world range, Ireland is probably unique in its dearth of voles, and while there is evidence that many Buzzards do eat some corvids, they are not normally such a common prey species. Like Moles, corvids are thought to be distasteful to raptors. Falconers must train Peregrines for Rook *Corvus frugilegus* hawking by tying tastier meat to lures that look like Rooks, so that the falcon associates black-winged birds with a good reward.

Nevertheless, a Buzzard will eat many things a Peregrine won't and can develop a taste for corvids should the circumstances require.

Not all birds caught are young, but quite often they are disadvantaged in some way, such as Tufted Ducks *Aythya fuligula* on a frozen pond (Bright 1955), or Atlantic Puffins *Fratercula arctica* as they emerge from their nesting burrow, as seen on Skomer after the Rabbits crashed in 1954 (Davis & Saunders 1965), or more recently a travel-weary Grey Phalarope *Phalaropus fulicarius* that had been blown off course into southern England (Dare 2015). The same is true of gamebirds injured during shooting.

Gamebirds

Gamebirds receive a lot of attention in the literature in relation to Common Buzzards because they are a source of livelihood and recreation. The issue is that gamebirds often require intensive rearing or habitat management to boost numbers to an economically viable level for managed shooting; Buzzards can hamper those efforts and so are seen as a pest species by some game managers. We will return to how that influences the Common Buzzard's relationship with humans in later chapters; here we will just focus on what they eat. While Buzzards do take gamebirds of all species in the UK, including the huge Western Capercaillie *Tetrao urogallus*, they have rarely been shown to kill large numbers of them (Park *et al.* 2008). For example, Mark Watson *et al.* (2007) found only one radio-tracked Grey Partridge *Perdix perdix* was attacked by a Common Buzzard, compared to 25 killed by Eurasian Sparrowhawks *Accipiter nisus*, despite Buzzards being twice as likely to be seen. We found Pheasant remains in 7 per cent of nest visits, and a questionnaire with gamekeepers in the local area suggested the loss of their birds to Common Buzzards was less than 5 per cent overall (Kenward *et al.* 2001). This was in an area predominated by Pheasant rearing, rather than partridge shooting, and even a released Pheasant poult is a larger bird than Buzzards routinely take. When we visited nests on an estate rearing Red-Legged Partridges *Alectoris rufa* there were occasions when the nests appeared filled with their hip and leg bones. However, compared with the numbers of birds being released, the number on the nests appeared inconsequential. Vitali Reif *et al.* (2001) found Common Buzzards in Finland regularly took gamebirds, particularly the moorland Black Grouse *Tetrao tetrix* and the woodland Hazel Grouse *Tetrastes bonasia*, which together reached 7.4 per cent of individuals eaten, or 30.3 per cent by mass in the recorded diet. These specialists took grouse even in years of high vole abundance, but overall the proportion of gamebirds in the diet increased in years when voles were not so abundant. Indeed, even in the absence of voles in Northern Ireland, where birds are the dominant prey, only 2.3 per cent of the prey recorded were gamebirds, compared with nearly 15 per cent corvids (Rooney & Montgomery 2013).

Although the proportion of grouse chicks is overestimated in prey remains at nests (where it can be four or five times higher than that recorded with cameras), gamebirds can be an important group in terms of mass, even if their numbers are low, because they are some of the largest live prey Common Buzzards take. That said, like other avian prey, Buzzards mainly take young grouse (Tornberg & Reif 2007). Compared with the number of other bird species, and the dominance of mammals, Swann & Ethridge (1993) found that gamebirds were a small proportion of the diet, especially given the bias in finding large gamebirds in nest remains. The only gamebird that Selås *et al.* (2007) detected in Norwegian nests was one Eurasian Woodcock *Scolopax rusticola*, among a great variety of other bird species.

Domestic Chickens *Gallus gallus domesticus* are not game, but their predation is worth noting as another species in which humans take a keen interest. Common Buzzards have been observed worming next to chickens, preferring the easy worm pickings rather than a fight with a chicken, which would likely damage the Buzzard. There are, however, a few reports of Buzzards taking chickens when other food is in short supply or the weather harsh (Dare 1957, Goszczynski & Pilatowski 1986).

Invertebrates

The primary evidence of particular prey species in the diet comes from pellets, prey remains and other observations at the nest, and thus during the breeding period. During the remainder of the year it is less easy to assess what is eaten. In the past, there tended to be an assumption of similar diet throughout the year. However, we do have a lot of anecdotal data about what others have seen Common Buzzards eating outside the breeding season, and these show how invertebrates are very much part of their diet. As Common Buzzards recolonised further east in England, where most people had not been used to them, we started to hear surprised reports of Buzzards pulling earthworms from ploughed fields (Plate 5; see colour section). This is not a new phenomenon; it has long been known that Common Buzzards also eat invertebrates, such as beetles and worms. Insects were present in 6 per cent of pellets analysed from adult Buzzards in Poland (Truszkowski 1976). Hayman (1970) described Buzzards worming on freshly sown pasture in southern England and he thought that worms might play a much larger part in the diet of Buzzards than was generally realised. Earlier, Dare (1957) had made an effort to watch his study Buzzards during winter and concluded that worms could potentially sustain a Common Buzzard through the winter if available in sufficient quantity. While earthworms may have become more of a staple diet now that we have so much short grassland and ploughed fields in winter, worming was observed as far back as the 17th century: 'yet for want of better food it will feed on beetles, earthworms and other insects' (Willughby 1676, in Tubbs 1974, page 37), and recorded by D'Urban & Mathew in 1892. We made a more systematic assessment, which involved radio-tracking individual Common Buzzards continuously from dawn until dusk during midwinter, to see what they were feeding on. The only vertebrate that came into the diet was from one bird that spent time on a Woodpigeon *Columba palumbus* carcass, and from the field evidence that looked like it had been killed by a Eurasian Sparrowhawk, so in essence was scavenged carrion. It was difficult to get to view Buzzards most of the time without disturbing them and changing their behaviour, but the radio-tags sent a motion-sensitive signal which made it obvious that a lot of the time they were simply perched, so seeing them would not reveal anything about their foraging. In the end, having tracked each of 35 Buzzards throughout a complete day, only 20 were observed feeding. Of these, all were feeding on earthworms at some point during the day. While it is true that seeing exactly what was eaten was not easy, the birds were primarily on ploughed fields, and where the prey were observed they were earthworms (being stretched from the soil in iconic cartoon fashion), rather than other invertebrates. At other times of year, the Buzzards were also often to be seen worming on grassland.

Worms are not suitable for raising Common Buzzard nestlings. Some songbirds can raise their young on beaks full of worms, but songbirds are considerably smaller, move shorter distances, and are unlikely to require the energy a Buzzard needs to fly back to the nest. Therefore, during the breeding season it is not surprising that the emphasis for Common Buzzards is on catching vertebrates, which are useful-sized food parcels to be carried back to the nest, containing bones for calcium and other vitamins as well as fur or feather roughage for casting pellets. There are also fewer earthworms available during the breeding season, because the vegetation is taller and the ground drier. Unlike European Blackbirds, which appear to be able to hear earthworms moving in the soil to pluck them from it, Buzzards require visual recognition of prey on the surface. Earthworms may be nutritionally poor in some ways, but they are rich in essential amino acids, and we have never seen a worming Buzzard with a dull cere. Looking to the Americas, *Buteo*s such as Swainson's Hawks are well known for exploiting sources of abundant insects outside the breeding season, while most other *Buteo*s are known to eat invertebrates opportunistically.

Besides earthworms, Buzzards eat many other invertebrates. Nore *et al.* (1992) found that the main item of their diet in a poor vole year in France was the Field Cricket *Gryllus campestris*. In the Catalan region of Spain, Mañosa & Cordero (1992) used the opportunity of a large number of dead Buzzards confiscated from hunters to check stomach contents. Only 45 of the 69 stomachs examined revealed prey items, but 73 per cent of the prey were invertebrates. This compared with only 1 per cent of avian prey (presumably what the hunters were trying to protect by shooting the Buzzards). This provided hard evidence that Buzzards eat invertebrates in large quantities. Beetles are certainly part of the mix, with Dare (1957, 1998) considering the primary victims being slow-moving dung beetles *Geotrupes* spp. and *Typhaeus typhaeus*. Briggs (1983) made an opportunistic observation of a Buzzard eating 36 beetles in 27 minutes; inspection of the ground afterwards suggested that they were dung beetles. A more unusual observation came from Tor Spidsø & Vidar Selås (1988), who found the most abundant invertebrate from prey remains and pellets were carpenter ants *Camponotus* spp., although there is a chance that these ants were feeding on remains in the nest rather than being eaten. Francksen *et al.* (2016), studying the pellets of Buzzards on grouse moors, recorded nearly a fifth of prey items as invertebrates, but primarily beetle elytra rather than earthworm chaetae, although they knew from the same study that invertebrates were overestimated in pellets around the nest (Francksen *et al.* 2016b). Grubs such as Lepidopteran caterpillars, leatherjackets (crane-fly larvae *Tipula* spp.) or horse-fly larvae (family Tabanidae) are also on the Common Buzzard menu (Dare 1998). In fact, it is likely that Buzzards will eat anything that moves, as it is likely to provide some fat and protein. More unusually, a Buzzard in Devon was seen feeding on Large White butterflies *Pieris brassicae* in summer as well as earwigs, spiders and woodlice. That can't be an easy way to maintain the nutritional requirements of a bird the size of a Common Buzzard.

Carrion

Falconers have labelled Buzzards as 'lazy', because they look for the easy meat. In his book, *Falconry for You*, Humphrey ap Evans (1960) sneeringly says of the Buzzard, 'Its cumbersome flapping, when close to the ground or taking off from a tree, makes it an object of ridicule to the haughty falconer. But worse still, it is a carrion feeder … They are alas branded as cowards too, or "slack mettled" as it can be less harshly termed.'

It's true that Common Buzzards tend to look for the easy meat. Captive birds would much rather return to the falconer's fist for feeding, rather than chase something that would require effort and, more critically, might harm them. This behaviour derives from Buzzards' behaviour in the wild, where they scavenge as much as possible. Likewise, Buzzards will be attracted to shot Rabbits left out by a keeper trying to distract them from a Pheasant pen, or carrion laid out by a photographer for an opportunity to get close, or indeed any other carcass. Perhaps this carrion-feeding behaviour led to New World vultures (and especially the Turkey Vulture) being known as 'buzzards' in North America.

For many people in the UK, their closest encounter with raptors will be when Buzzards are feeding on killed animals by the roadside. Unfortunately, that can also lead to Buzzards becoming road casualties themselves, unless they are quick to avoid cars, and often they will try to pull the carcass off the road. One of the nests in our own study area had, for more than one year, butchers' bones that had been cut by a saw. This nest was very close to a waste-disposal tip, from which the Buzzard had likely scavenged the remains. Dare (1998) also noted a butcher's meat bone at a nest, and Common Buzzards also visit abattoirs and feeding sites for reintroduced Red Kite populations.

In our study area, we found the occasional remains of a lamb *Ovis aries* at the nest, but there was no evidence that Common Buzzards ever killed a lamb. This could only be a vanishingly infrequent event given the size of lambs and the evidence that Buzzards don't even bring full-grown Rabbits to the nest. There is evidence from a study in the Welsh hills that sheep carrion may be seasonally important: pellets collected outside the breeding season

Figure 2.8. *If larger species die, Buzzards are not shy to eat them (© Neil Burton, www.shutterstock.com).*

were more likely to have sheep remains in them (59 per cent of pellets) compared to the breeding season (27 per cent) (Newton *et al.* 1982).

The shortness of this section belies the use of carrion. Observations are few in literature because it is not exciting to record. However, if a wildlife photographer were to ask how they would get close to a Buzzard, based on photographs in books and what appears on the internet these days, we would recommend putting out carrion within a Buzzard's foraging range. As for which species the carrion should be, it really doesn't matter. Dare has recorded them eating such unpalatable carrion as Least Weasel and Red Fox (Figure 2.8). Pinowski & Ryszkowski (1962) thought that the only Coypu *Myocastor coypu* a Buzzard was found eating on a Coypu farm was one that had already been killed by something else. Buzzards not only follow the rule 'if it moves and is small enough, then kill it and eat it', but also 'if it used to move and it's now dead then eat it, whatever its size!' Meat is meat.

Reptiles and amphibians

Amphibians and reptiles occur with great regularity in Buzzard diets, but rarely form the bulk of the food eaten. In the warm south of Europe, lizards and snakes can make a significant contribution to the diet. Mañosa & Cordero (1992) found that reptiles, especially Ocellated Lizards *Timon lepidus* and Large Psammodromus *Psammodromus algirus*, were the second most consumed prey (14 per cent in remains at nests, 36 per cent in pellets), while Sergio *et al.* (2002) found a quarter of remains at nests to be a variety of snakes and Common Toads *Bufo bufo*. Even in the colder north where the herpetofauna is less abundant, a study in Norway found that reptiles, and to a lesser extent amphibians, formed a third of remains from the nest, confirmed with video at the nest (Selås *et al.* 2007). This makes the Common Buzzard's Swedish name of Ormvråk, or 'snake-slayer' more understandable. Surprisingly, snakes were more frequently brought to the nest in good vole years, perhaps because they attracted Buzzards to hunt over open ground where the snakes were basking (Selås 2001), and a Buzzard carrying a snake from open ground to a nest is, of course, very conspicuous. Elsewhere in the literature, herpetofauna rarely rise above 20 per cent of the Buzzard diet (Figure 2.4). In the UK, Dare (1998) estimated that amphibians and reptiles made up almost 20 per cent by mass of adults' diet during the summer. Common Frogs *Rana temporaria* were preyed on intensively in March–April (on their migrations for spawning), while scales of Adders *Vipera berus*, Common Lizards *Lacerta/Zootoca viviparus* and Slow Worms *Anguis fragilis* were found in pellets during the spring and summer. In Hungary, a Midwife Toad *Alytes obstetricans* was brought to a nest every other day, but the juveniles did not seem to like the taste and the female nearly always ate it (Mátyás Prommer pers. comm.). In our study area we were lucky enough to have Chris Reading studying Grass Snakes *Natrix natrix* and Common Toads by putting microtransponders into them to identify them on recapture (Reading & Davies 1996). One year, we took the reader for these tags up to all the Buzzard nests in his study site and found the tags of several toads, which had provided meals for the nestlings. Considering the chances of uneaten toad remains being removed by adults, or pellets with the tiny transponders falling from the nest, these records were remarkable and suggested that it was not uncommon for Buzzards to eat toads in our area.

Oddities

It is difficult to think of any animal a Common Buzzard won't eat, and so here is a short collection of rare observations that helps characterise quite how opportunistic and versatile it is as a species. Fish is a pretty unusual food, noted by Dare (1957), and probably scavenged since Buzzards are not adapted for plunging into water. There was also the case of a European Eel *Anguila anguila* (Madge 1992), but Dare rightfully pointed out that this eel could have been on land, and hunted as any snake would be. Although others have reported Common Buzzards eating fish (Jenkins 1984, Mitchell 1984, Dravecky 2003, Temmink 2004), this was generally as carrion. Nevertheless, Buzzards have been seen to forage among seaweed along the shore, so there is a chance that they could pull a fish out of a shallow pool.

There are also some observations of Common Buzzards killing and eating unusual birds. Barnard (1978) observed a Buzzard attacking and killing a Short-eared Owl *Asio flammeus* in flight, which it then appeared to eat. It is not impossible that this was an interspecific territorial act, rather than foraging, but it ended in a meal. On the other hand, Mikkola (1976) reviewed the literature on raptors killing owls, and found 12 records of Buzzards killing Tawny Owls *Strix aluco*, 11 for Long-eared Owls *Asio otus* and two for Barn Owls. Dare also reports a Barn Owl being carried by a Buzzard before managing to escape, and Schwartz (2014) found the remains of young Tawny Owls in a Buzzard nest. However, when looking at other raptors, Common Buzzards are more likely to be killed than to kill (e.g., Sergio 2008). That said, there is a nest-cam film of a nestling Osprey *Pandion haliaetus* being snatched from the nest by a Buzzard (Webster 2012). Cameras set up on carrion have also shown storming attacks by Buzzards on adult Northern Goshawks, which fled the scene, so it can be expected that some bold Buzzards kill other raptors. Certainly, adult Common Kestrels have been killed by Buzzards, and there is a chance that the Kestrel's declining population could be at least partially influenced by the increase in Buzzards (Riddle 2011).

Prey switching

A strong theme from this chapter has been the ability of Common Buzzards to switch to different prey species. This may have been an adaptation to deal with the propensity of vole populations, as a staple of the diet, to fluctuate wildly. Obviously, Buzzards can only predate what is available, but it is possible that their predation on certain species could have an impact on what is available in future years. If too many breeders are taken from a prey population, it can take a long time to recover, or it may never recover at all. Vole populations tend to cycle dramatically, regardless of raptor predation, but what about the other prey species that Buzzards must use when their main prey is scarce?

The impact of predators on their prey has been investigated using various competing hypotheses of what predators do when their main prey is in short supply. These theories are worth exploring because they illustrate ecological principles that can help to manage wildlife populations. However, there is additional interest when it comes to a species that can hunt gamebirds, because of the financial and recreational problems this causes for stakeholders of the prey species. As we shall see in later chapters, the maintenance of habitats for game as well as crops can be an important motivation for retaining land that remains diverse and rich in nature.

Dietary switching in the north of Europe

Researchers in Finland, Norway and Sweden have for a long time been investigating the potential impact of raptors on gamebirds. Of course, Common Buzzards are not the only predators, and ignoring others can produce misleading conclusions. These studies have therefore also included the effects of the Northern Goshawk, owls and mammalian predators. A relatively low density of humans in the large countries concerned has previously created less pressure for intensive land use than in the UK, so that there is much semi-natural habitat, although even the use of forest has intensified greatly in the last half century.

The Alternative Prey Hypothesis (APH), proposed by Per Angelstam *et al.* (1984), suggests that predators switch to alternative prey during years when their primary prey are very low and this switch then reduces the populations of the alternative prey. Common Buzzards have the potential to fit this model, primarily feeding on voles but able to switch to grouse and other birds during poor vole years, as do Rough-legged Buzzards (Pokrovsky *et al.* 2014). The impact may simply be to cause the alternative grouse prey populations to cycle with a time-lag behind the primary prey, but if the predators have grown to very high density in years when the main prey are super-abundant, the switch of lots of predators to a less abundant prey could decimate grouse populations.

A different theory is the Main Prey Hypothesis (MPH), for grouse-specialised predators (Tornberg *et al.* 2012). In this case, grouse populations are not adversely affected by raptors until they have been reduced for other reasons, at which point the density of predators does not fall with declining grouse numbers as they have alternative prey to sustain them during low grouse years. This results in relatively high predator pressure on a low grouse population. Tornberg *et al.* suggested that was the likely scenario with Northern Goshawks, focusing on grouse as the main prey. Indeed, some early modelling on the first microcomputers in the 1980s showed that Goshawks could coexist with grouse populations, provided there was i) variation in the ability of the hawks to catch prey, ii) a non-breeding season for both predator and prey species and iii) hawks could breed as fast as prey (Kenward & Marcström 1988). Goshawks typically had broods of 2–4, and experimental removal of predatory mammals on islands showed that the severe impact of these mammalian predators was driving breeding success of grouse to less than four per brood (Marcström *et al.* 1988).

Theories about complex systems tend to proliferate, and another perspective was provided by the Shared Prey Hypothesis (SPH) (Norrdahl & Korpimaki 2000), which suggests that if one prey is abundant it will increase the number of predators, and that could have a big effect on rarer prey species because even though the predation is incidental, there are so many predators that a lot of the rare species get eaten. One example might be the increased numbers of reptiles taken by Buzzards in high vole years, possibly because the Buzzards ended up hunting over bare ground in clear-cut areas where lizards had chosen to bask (Selås 2001). Such theories provide interesting perspectives, and can be shown to give accurate predictions on which to base management if they can be tested. Another theory, the Predator Facilitation Hypothesis (PFH), suggests there could be situations where one predator might inadvertently facilitate the predation by other predators (Tornberg *et al.*

2012). For example, if voles are few in number, mammalian ground predators may chase grouse, flushing them or exhausting them, making them more vulnerable to being caught by bird specialists such as Peregrines or Goshawks. This theory is very difficult to test unless the right mechanism can be isolated.

Of course, the right mechanism often cannot be isolated, because the richer the diversity of predators and prey, the more complex the system becomes. For example, other animals such as hares, shrews and forest grouse species in Sweden and northern Finland have cycles of their own, which may be synchronous with the vole cycles, or lag behind as predator numbers fluctuate. It is also possible to ascribe the 'alternative' prey to the wrong species when a study is short. If starting a study of Common Buzzard predation in a low vole year you may get the impression that the main prey are birds, but the following year the Buzzards will switch to an 'alternative' prey of voles. Thankfully, these Finnish studies had long datasets (more than 17 years), and the breeding success has correlated with voles, not other species (Reif *et al.* 2004a), in the long term. Moreover, when vole numbers were low the prey species became more diverse, which has also been recorded in other studies by Likhopeck (1970), Spidsø & Selås (1988) and Jędrzejewski *et al.* (1994). It is therefore a fair assumption that voles, where they occur, are the staple prey of Common Buzzards, at least in the summer breeding season when these nest-based studies were conducted.

Reif *et al.* (2001) found that the quantity of small gamebirds brought to the nest was inversely proportional to the abundance of voles that year, and independent of the number of grouse available, which is consistent with the Alternative Prey Hypothesis; more grouse and hares (depending on specialisation at individual Buzzard nests) were captured when voles were less abundant. In fact, they concluded that grouse were the 'classic' alternative prey, and as such could be vulnerable to an abundant predator suddenly focusing on them. However, the researchers didn't think that Common Buzzards alone could affect gamebird numbers. This was especially true since sometimes when vole numbers were low, Buzzard productivity remained high, suggesting that the alternative prey were sufficiently abundant for Buzzards despite the latter not showing signs of taking large numbers of grouse. Even the combined predation of Buzzards, Goshawks and Hen Harriers *Circus cyaneus* would not have sufficient effect to significantly change grouse numbers, which were more directly impacted by mammalian ground predators. This confirmed our earlier work with computer modelling and experimental removal of predators on islands (Marcström *et al.* 1988). There was no evidence for Shared Prey Hypothesis, because the numbers of grouse taken did not increase when voles were at their peak.

Dietary switching in the British Isles

In many study areas to the north and east in Europe, Common Buzzards depart for warmer climes before the winter, leaving other predators to hunt the voles in their snow tunnels through the months of short daylight. However, in much of the UK there are Buzzards and voles available year-round. So what happens there? Most of the studies in the UK are dominated by super-abundant Rabbits. However, locally there can be sudden changes. Waves of disease, including viral haemorrhagic enteritis (VHE) as well as myxomatosis, may hit Rabbit populations. Resistance of Rabbits to both afflictions seems to be building so the proportion dying in each wave of the disease appears to be smaller and smaller, but there certainly were times in the past where the effects were very noticeable.

Unfortunately we don't have so much evidence of whether individual Buzzards have been switching, except on the island of Skomer (Davis & Saunders 1965). We can presume that they are in other places, too, based on the fact that not all Buzzards have high densities of Rabbits nearby and knowing the diversity of what Buzzards eat. There is concern about the decline of species such as Wood Warbler, and we know their young can be taken by Common Buzzards. Nevertheless, although it is quite unthinkable that Buzzards have been the primary driver in the drastic decline in Wood Warbler numbers, and predation rates do not seem to have increased over a period when Common Buzzard numbers have multiplied (Mallord *et al.* 2012), there is a chance they could contribute to the demise of species such as this Warbler when their numbers have been driven critically low by other factors.

Viewed from an annual perspective, dietary switching is a simple matter of fact. Just as voles may be obscured by snow in the north, so the breeding seasons of Rabbits and young birds come to an end, making these prey much less available for Buzzards in the UK. Intensification of agriculture probably contributed to a reduction in small rodent numbers, but British Buzzards have alternative prey. It appears that in the damp, cool winter months many of them switch to eating worms.

They are not the only generalist predators in the UK to find this fall-back food a convenient product of modern land management. They are joined in worm-eating by corvids, Red Foxes and Eurasian Badgers *Meles meles*, all of which switch to other prey in the dry summer months when worms appear less on the surface of grassland. Together with those other generalist predators, they then turn to Rabbits and young birds of many species in summer.

Generally there doesn't seem to be too much concern voiced about what Buzzards would switch to, unless it is gamebirds. However, we find it interesting that pressure from Buzzards, which are now strictly protected because they were once rare (as a result of human reaction to predation), are now combining with other factors to create pressure on other popular birds. The predation by Buzzards of nestling thrush species is interesting, because a project studying European Blackbirds and Song Thrushes also indicated substantial predation on them by another well-protected worm-eater, the Badger, in the flightless stage after fledging (Hill 1998). Buzzard predation is also starting to be noticed on wader chicks. In cases where there is proof that the combined force of predation is critical for a species that is only just clinging on to its extant status, conservationists may have to address issues of how or if we can manage predation by Common Buzzards.

Conclusions

1. Common Buzzards can eat many different prey species; they have a long and efficient digestive system that will extract all available nutrients from what they eat. This also means that they are able to withstand food shortages better than similar birds.
2. With such a diverse food supply, it is interesting to compare and contrast findings in different areas, but it is important to know the methods used, because they affect the results.

3. In the breeding season, camera traps at nests appear to give the best picture of what is eaten, although analysing pellets can give a better representation of small prey than collecting nest remains, which may omit prey small enough to be swallowed whole.
4. All studies of Common Buzzard diet record voles and other small mammals as prey, and most analyses indicate they are the most frequent prey at nests, unless Rabbits or small lizards are abundant.
5. In terms of biomass, young Rabbits have been shown to be an important component of the breeding season diet in Great Britain, often positively correlating with nest success.
6. Birds feature consistently in the Common Buzzard prey list at nests, especially when voles are scarce in northern countries. Buzzards mainly take young birds, which have poor flight skills, and raid the nests of very small birds as well as taking chicks of gamebirds and waders.
7. Earthworms on grassland and ploughed fields can form the most abundant winter prey items for Common Buzzards in the UK, and possibly elsewhere.
8. Common Buzzards often scavenge carrion such as roadkill, abattoir discards, prey remains from other predators and even meat put out for other raptors such as Red Kites.
9. Studies of prey switching in Finland appear to show small mammals as the primary prey, but in low vole years a switch can be made to young gamebirds.
10. In common with other generalist predators, which have become abundant due to favourable feeding on modern farmland, increasing numbers of Buzzards in the British Isles may become a conservation concern for some other bird species with dwindling populations.

CHAPTER 3

Hunting

Where we live in southern England, we can observe the hunting behaviour of common raptors such as Peregrines chasing pigeons, or the characteristic active hovering of a Common Kestrel, or the flip over a hedge, and off in a flash, of a Eurasian Sparrowhawk. In contrast, we do not often see Common Buzzards in hot pursuit. Instead, we see them standing on prominent perches, staring inquisitively at the ground, or possibly soaring low and slow along a hillside searching for ground prey. In fact, if you want to see Buzzards actually attacking prey, your best chance is to look for them running around on a ploughed arable field or short grassland in winter, picking up worms. Of course, they have the lightning swiftness of raptors, and can certainly be dramatic when they are keen to grasp evading quarry. They conduct all the hunting manoeuvres we would expect from a bird of prey, but rely a lot more on surprise and a very short chase, and so the probability of seeing a striking hunt is very low, compared with the likelihood of observations during the hours they stand around on perches or run around fields. Despite the relative rarity of hunting observations, people have managed to study how Common Buzzards hunt, capture and kill their prey.

The previous chapter considered the diverse diet of Common Buzzards. Correspondingly, their hunting methods must be adaptable to harvest all the different prey species. Here we start with how they hunt for their most common prey and how evolution has shaped and constrained their abilities, before moving on to see how they have adapted to the anthropogenic landscape.

Adaptations for hunting

Common Buzzards are generalist predators, but they are not 'Jacks of all trades' when it comes to catching that diversity of prey. Whereas they have morphological characteristics that sometimes constrain what they can do, their effectiveness is masterly with some prey, although with others they appear less skilful. To understand why, it is worth considering their origins and the differences with other raptors that define what they can hunt. As we saw in Chapter 1, molecular analyses indicate that Common Buzzard shares a close common ancestor with Rough-legged Buzzard. These are very similar species morphologically, difficult to tell apart in the field except for the adult tail colour and underwing pattern. Rough-legged Buzzard does have a more harrier-like appearance (Génsbøl & Bertel 2008), likely due to the 7–8 per cent longer wings and 5–7 per cent longer tail (Glutz von Blotzheim *et al.* 1971, Cramp & Simmons 1980) (Figure 3.1), as well as differences in their foraging behaviour. What would not be so obvious to an observer watching them hunt, is that Rough-legged Buzzards also have 3–8 per cent shorter legs than Common Buzzards (Figure 3.1) and 5 per cent shorter toes. Perhaps the small morphological differences are adaptations for the Rough-legged Buzzard's increased flight activity, taking smaller prey on the barren arctic tundra, whereas the Common Buzzard has relatively longer legs and toes to seize mobile larger prey, therefore enabling it to exploit a wider dietary spectrum. This would make the Common Buzzard more successful as species richness increases towards the equator, but unable to outcompete the Rough-legged Buzzard at bringing small mammals to nests further north. Let's now look at the evidence that other research has provided.

A detailed paper by Magnus Sylvén (1978) directly compared the hunting behaviour of Rough-legged Buzzards with Common Buzzards. Sylvén observed both species between October and March in the same 12km² area of southern Sweden, to which Rough-legged Buzzards had migrated south from their breeding areas whereas the Common Buzzards were resident. He estimated the two species were eating an almost identical diet. The most

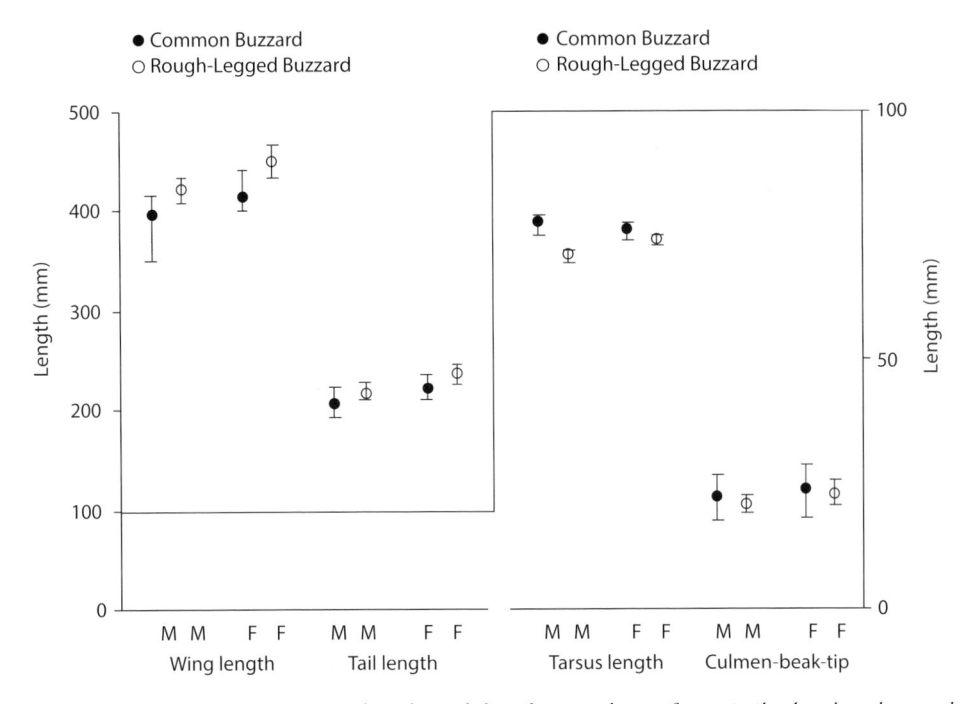

Figure 3.1. *Whereas Common Buzzards and Rough-legged Buzzards are of very similar length and mass, the latter have longer wings, a slightly longer tail, but shorter legs. Data, showing mean and range of values, are from Glutz von Blotzheim et al. (1971) and Cramp & Simmons (1980).*

significant difference between the two species was the amount of hovering during foraging observations, which amounted to as much as 33 per cent for Rough-legged Buzzards, whereas he did not observe hovering at all in Common Buzzards. It is important to differentiate active hovering from 'hanging' in the air, which is more passive and requires a headwind or updraught. In his paper, Sylvén remarks that 'hovering is a most power-demanding type of activity being, for example, much more expensive than ordinary flapping flight'. We may take these Swedish observations to be what Dare (2015) defines as 'active hovering', differentiated from the more passive 'hanging' that is used more usually by hunting Common Buzzards. The latter do hover actively, but this systematic recording showed how uncommon it was, and this is consistent with other studies that have quantified hovering time. For example, both Andrzej Wuczyński (2005) and Marta Bylicka et al. (2007) estimated it was less than 3 per cent of the time (Figure 3.2) in Poland. In the Swedish study, Common Buzzards apparently spent less than 2 per cent of their time on the ground, compared with 11 per cent for Rough-legged Buzzards, and were thus spending much more of their time on perches than were the 'Rough-legs'. In Poland, too, Common Buzzards spend more time perched than in other activities, except in autumn (Figure 3.2). Unfortunately, we have not found further studies that compare the hunting behaviour of these species, so we need to compare with species in the wider raptor assemblage, starting by further consideration of their hunting flight.

On the tundra, there are few perches higher than the surrounding low vegetation, so birds must fly to search for small mammals. Large broad wings require less pressure on the leading edge of the wing to get the same lift (Pennycuick 2008), which means that birds with broad wings do not have to move so fast through the air to maintain height efficiently. Broad wings with a pronounced alula, characteristic of buzzards, are useful attributes for birds that need to fly slowly. They give them time to inspect the ground carefully for small mammals among the vegetation, and time to react before being a long way past what they have spotted. Narrower-winged falcons travel too fast to scan the ground carefully and react,

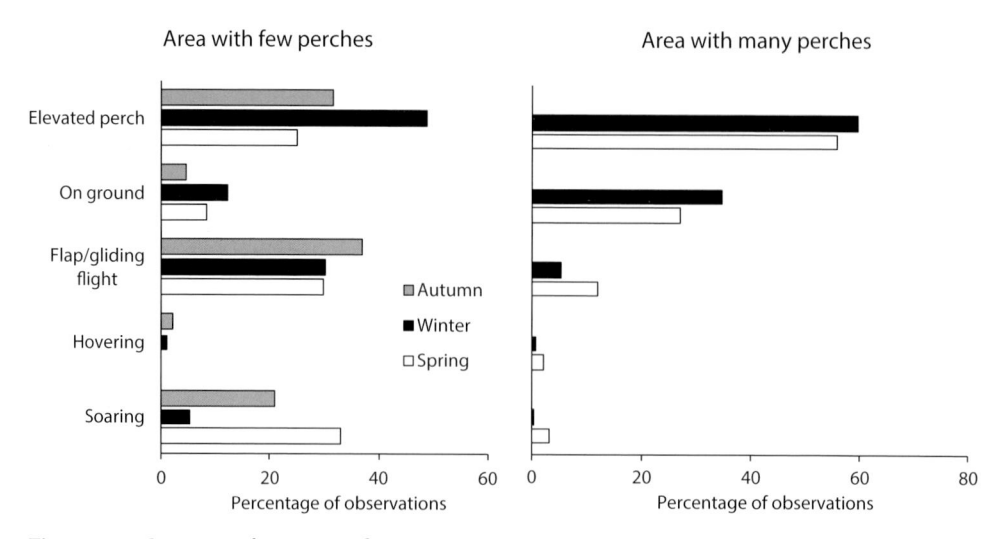

Figure 3.2. *Systematic observations of Common Buzzards in Poland recorded more flying in areas with relatively few elevated perches, but little hovering in either region. Data are combined from Wuczyński et al. (2005) and Bylicka et al. (2007).*

but some have evolved hovering behaviour to keep them stationary to search the ground below. Common Buzzards can hover with active wingbeats but, unlike Common Kestrels, which can hover on the stillest of days, buzzards need a headwind to provide some of the lift. They prefer updraughts that provide free lift for hanging on the wind without hovering. Wherever possible, Buzzards hold their wings out straight and still, so that there is little energy cost, and the lift is maximised for the body mass.

The broad wings that make slow flight possible conversely mean that buzzards cannot attain the speed of other raptors, because the increased area that generates lift also produces drag that slows the wing as it moves forward through the air. Likewise, the added effort it takes to move the wings when flapping gives buzzards very slow acceleration from the ground or low perches, and makes hovering less energy efficient. Perhaps hovering is necessary for Rough-legged Buzzards, which exploit the perchless arctic tundra, but it is a behaviour less often used by Common Buzzards, which can hunt more efficiently by using perches in the landscape where they breed. While buzzard species cannot attain the extraordinary flight speeds of narrower-winged falcons, the resistance of buzzard wings enables them to be more manoeuvrable; they can twist much more sharply than many birds of a similar size. This is important when they spot the majority of their prey at close quarters and must instantly turn onto it, possibly changing direction very rapidly if the target spots their attack and accelerates or swerves rapidly between patches of vegetation on the ground.

Noting these differences between raptor hunting techniques, Barton & Houston (1994) placed Common Buzzards in a 'searcher' category of hunter, searching and then swooping down onto prey in an ambush, rather than an 'attacker', which must pursue its prey through the air, such as a falcon. When looking at the physiological differences between the two hunting groups, they found that searchers such as Common Buzzards had less pectoral muscle and a smaller area on the sternum on which to attach the muscle. So, not only do their wings have greater resistance to moving quickly through air, but they haven't got as much muscle for rapid wingbeats, either. Another characteristic of searchers was that they tended to have longer guts for the greater extraction of nutrients from prey. Maybe this counteracts any reduction in muscle mass, but it would enhance the overall efficiency of obtaining nutrition from the prey that they do catch, thereby reducing the amount of time they have to spend hunting. Common Buzzards are not the most dynamic hunters to watch, but they are able to thrive due to the diversity of their diet and the ability to extract energy and nutrients efficiently from it.

Another morphological characteristic that differentiates Common Buzzards and their close relatives from other birds of prey is their raptorial apparatus. Buzzards have relatively small feet, which, together with their perceived 'laziness' (or efficiency, depending on perspective), makes them poor hunters for falconers who want to put larger prey on the supper table. Again, ap Evans (1960) was rather scathing of their feet, describing them as 'impotent-looking' with diminutive claws, which he considered unable to hold prey, suggesting that Buzzards were best adapted for carrion feeding. To catch large prey, raptors need long toes to hold their prey securely. If you get the chance to see different raptor species close up at a raptor centre, it is remarkable how small the feet of Common Buzzards appear, especially compared with a raptor of similar body size. For example, comparing a male Northern Goshawk that has a very similar mass, body length and tarsus length, the feet (toe plus claw from museum specimens) are 20 per cent larger in male Goshawks (Cramp & Simmons 1980) (Figure 3.3). Or, put another way, the span across the longest

The Common Buzzard

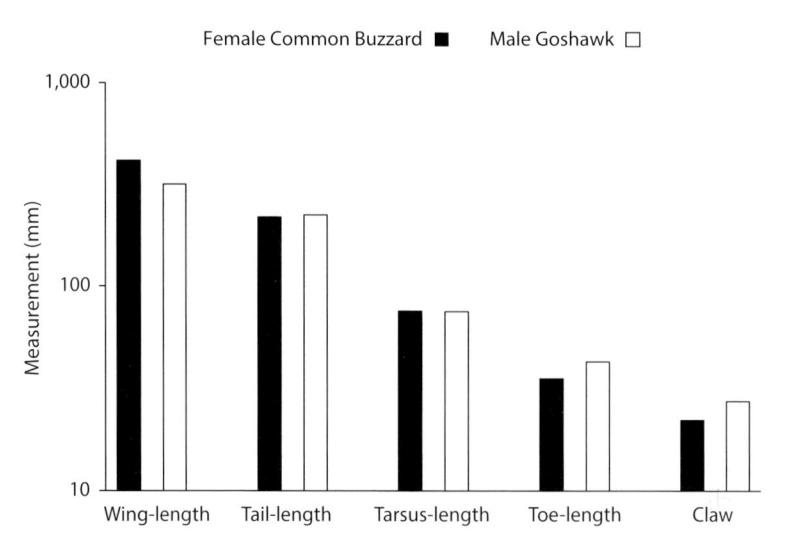

Figure 3.3. *The maximum body-mass of both female Common Buzzards and male Northern Goshawks is about 1200g. Whereas Goshawks have the shorter wings, they have appreciably longer toes and claws. Data are from Glutz von Blotzheim et al. (1971) and Cramp & Simmons (1980).*

toes on the foot is equal to the full length of the tarsus for a male Goshawk but only 77 per cent of the tarsus length of a Common Buzzard. Moreover, male Goshawks have smaller legs, relative to other linear measures, than females (Kenward 2006).

So why does a Common Buzzard have such small feet and talons? A Peregrine has an even larger span across the toes relative to the tarsus than a Northern Goshawk, and this is presumably an adaptation for capturing birds in aerial pursuit. The surface feathers of birds are slippery, perhaps for defence as well as aerodynamics, and they also come out easily if grasped in small clumps. The rump feathers of gamebirds and pigeons are especially easily shed and would be the first point of contact for an inexperienced predator approaching from behind. Having a wide toe-span is an adaptation of falcons such as the Peregrine for grasping birds securely. Within reason, the larger the feet the larger the prey that can be seized. In the past, wealthy falconers sometimes flew pairs of Peregrines (called 'casts') at geese, herons and even cranes, demonstrating how their large feet can secure relatively large prey.

Thus, for hunting larger prey, especially birds, there are advantages in having large feet. However, are there also disadvantages to having large feet? Again, it is the origins of vole hunting in tundra that suggest a possible process of natural selection. At least two disadvantages are possible, the risk of entanglement in vegetation, and the increased muscle needed to produce pressure from talons leveraged at the end of long toes. Peregrines mainly catch birds in mid-air, so the feet only need to deal with the prey and do not risk entanglement with vegetation surrounding it. On the other hand, Goshawks can take large prey (for example, Rabbits and hares) on the ground and also arboreal mammals (for example, squirrels). However, the large and arboreal prey are often clear of dense vegetation, so again the bigger the grasp size the better. In contrast, Buzzards most often have to extract their food from thick, small-scale vegetation. As a raptor grabs its prey, its toes are outstretched, all working to grab the biggest area of prey that they can, or for smaller species to maximise the capture area, just as we use nets to increase a trapping area. That big span is less optimal for grabbing small animals in vegetation. Long toes might prevent a foot from penetrating thick vegetation. If the prey was a small animal in grass, a larger foot would also 'catch' more grass around it, quite possibly disrupting a firm hold and allowing the vole to escape. Therefore, although there would have been selective pressure to evolve the grab area to be as large as possible, there would have been selection against increasing foot size too much, because that would decrease its effectiveness at catching prey in dense vegetation. There may also be an advantage of leverage; shorter toes can exert greater pressure towards the ball of the foot, and the better the grip the more effective the kill will be, especially if it requires a prolonged hold to subdue prey larger than voles.

A third consideration arises from a colleague's remark that Rough-legged Buzzards in Canada tend often to be found at the top of relatively spindly trees (Mike van den Tillaart pers. comm.). This is not something we have particularly noticed in Common Buzzards, but it may help the tundra birds take advantage of any vegetation that stands up. Grabbing spindly branches would be much easier with smaller, better-leveraged feet. In contrast, large-footed falcons tend to perch on ledges with larger footholds, where small feet are not so advantageous. These last thoughts are speculative, and the perching ability is more likely a by-product of having evolved feet appropriate for grabbing the prey that the different species are exploiting. What is not speculative is that, with both of us having experience of training Northern Goshawks as well as Common Buzzards, we have noticed the comparably great force that the short toes of an excited Buzzard can exert on a hand inside a falconry glove. Their feet need to be measured with care (Figure 3.4), and it is not surprising that Common Buzzards can subdue lagomorphs, even if training them to hunt Rabbits is not easy.

Figure 3.4. *Kathy Hodder is measuring the span across the hind and middle toes of a female Buzzard in our Dorset study (© Robert Kenward).*

Having used their large, broad wings for searching and their small feet for seizing and killing their prey, Common Buzzards have yet another feature that points to their origin as hunters of small mammal: a large gape, or bill width, that allows them to eat quickly. Tore Slagsvold & Geir Sonerud (2007) investigated prey handling across a range of raptor species. Although there was only one Common Buzzard in the study, nevertheless it was consistent with findings relating to other species. In general, raptors ate mammals faster than birds, and mammal specialists ate the mammals even faster and spent less time trying to avoid the less-nutritious fur, bones and skin. Species that ate small mammals also had a larger gape, which allowed them to swallow rodents whole, or take large chunks of meat from larger prey. Common Buzzards can simply swallow voles intact, reducing handling time to the absolute minimum. In comparison, narrow-gaped bird specialists such as Peregrines and Goshawks spend a much longer time preparing the prey, carefully plucking out the feathers and then only eating the muscle; the remains are left as carrion, which scavengers such as Buzzards can then feed on. Slagsvold & Sonerud conjectured that narrow bills would be selected when it was necessary only to take certain parts of the body, requiring better plucking and smaller mouthfuls. A wing has a lot of indigestible feathers, whereas the short hair of a vole makes up a much smaller percentage of the food parcel, and so is more manageable for binding and casting up pellets.

This also concurs with the hypothesis of Barton & Houston (1993) that hunters of birds need to be more agile and have greater acceleration, so they have only short guts to minimise body mass and therefore need high-quality food. It would be a slow task to pick the muscle off a vole carcass. Therefore, a vole-eater's digestive system must be able to deal with fur and bones. Moreover, in an open environment, with other raptors having good eyesight and hunting the same moors, it is helpful to be able to eat a meal as fast as possible, in case others come to steal it. In such an environment there could well be strong selection for a wide gape that allows the prey to be gulped down as quickly as possible, and for a long gut to maximise the energy absorption.

Before we go on to look more specifically at how Common Buzzards use these characteristics, it's also worth remembering that they have large eyes, high acuity and two high-focus foveae in each eye to cover the central and lateral peripheral vision simultaneously, so they don't have to move the eye much (Chapter 1). Vance Tucker (2000) found that raptors view distant objects by turning their head to the side, and interpreted that as the raptor needing to use the deep fovea that has its highest focus at about 40 degrees to the side of the beak. In contrast, objects closer than 8m are observed more directly because they are more in focus from the shallow fovea, which has the best line of sight at 15 degrees to the central axis. Knowing this, Tucker proposed that raptors had to take a 'spiral' flight line to their quarry. If they had to turn their head sideways, in order to see the distant prey ahead, then it would cause drag and be a strain on the neck. Instead, raptors view their prey to the side of them, and take a curved path to it. This may be more important for faster, stooping falcons, but nevertheless is a characteristic that affects how Buzzards view prey, and probably explains the head movements a perched bird of prey will make when searching for prey. Being so large, their eyes will collect whatever light there is, and Dare (2015) has observed that in winter they can go to roost approximately half an hour after sunset, by which time the light has become very dim. Indeed, Truszkowski (1976) considered they mainly hunted for vertebrate prey during the crepuscular periods. However, we have to remember that the large number of cones, which improves colour vision in good light, will not absorb light in the same way that rods do, so the eyes cannot be as precise in low light. This explains why Buzzards can often look clumsy when trying to land in trees during low light.

So, Common Buzzards have large, broad wings for slow flight and careful scanning, small feet for snatching prey from dense vegetation and a wide gape for speedy consumption of small mammals. How are these adaptations used to best effect when hunting the wide diversity of prey that Buzzards consume?

Hunting vertebrates

When hunting for small mammals, for which Common Buzzards are so well adapted, they hunt alone, either from a perch or from the air. Aloft, they use a combination of gliding and hanging on updraughts along a slope, and sometimes hovering, to observe the ground often 15–30m below, or higher if hunting for Rabbits. Typically, once a Buzzard spots its quarry, it will drop a little by closing the wings slightly, adjust its position and then drop almost vertically to gain speed before a short horizontal high speed glide. At the last moment, the feet are swung forwards to strike the prey. If you are lucky enough to be close when this happens, then you can hear a loud clap as the wings make contact with the ground and the air under them escapes. If the prey is large, the wings may be held back for a while like a 'fallen angel' (see Chapter 5), helping to stabilise the Buzzard while the feet subdue the prey. Their hunting stoops are not as long and dramatic as those performed by falcons such as Peregrines, which is just as well because the food is on hard ground rather than in the air. It is safer to gain speed by dropping fast, but then ambush the quarry horizontally, so that there is less risk of injury from hitting the ground at speed.

Buzzards have probably evolved to maximise their chances of finding small mammals, and will use all the clues they can that indicate the presence of potential prey. One proposed indicator is small mammals' urine, because an interesting study showed experimentally that Common Kestrels and Rough-legged Buzzards spent more time over areas that have vole scent marks than those without (Viitala *et al.* 1995, Koivula & Viitala 1999). The urine and faeces that voles use to scent-mark their territories absorb UV light and therefore they are seen darker in the UV spectrum (Viitala *et al.* 1995). All diurnal birds that have been studied (at least 35 species) have an extra cone type that allows them to perceive UV light (Honkavaara *et al.* 2002). However, while this is a plausible mechanism, more recent experiments looking at what a Common Buzzard eye can detect, and which wavelengths the urine absorbs relative to the vegetation and water in the environment, cast doubt on whether they can actually use such a process (Lind *et al.* 2013). Buzzards also employ many other cues, including those provided by humans. Bernard King (1986) reported Buzzards hovering 4–5m above and just behind a combine harvester. To maintain their position, their wings were held well above the body and rotated, as in active hovering. Dare (1998) also observed Buzzards 'waiting-on' to catch animals flushed from hay-cutting. Presumably this strategy presents both mice running from the harvester, and 'ready-chopped' small mammals to feed on. Hunting Buzzards may be able to use other raptors as indicators of prey availability. Serge Daan *et al.* (1989) observed a few Rough-legged Buzzards, and a greater number of Hen Harriers and Common Kestrels, flying over moorland most often at times of day when voles were active, which was assessed by checking Longworth traps every 20 minutes.

Temporary local conditions, such as weather, can affect how raptors hunt. When studying raptors that specialise in catching small mammals in Norwegian forests, Geir Sonerud (1986) recorded that Common Buzzards and other diurnal species did not use clear-cut areas when snow cover was complete, preferring to use the snow-free older stands or to leave the area completely. He presumed this was because they were visual hunters, unlike owls, which could maintain their capture rate by using auditory cues. Similarly in Poland, while walking regular transects over several years, Dominik Wikar *et al.* (2008) saw that Common Buzzards usually preferred open fields during the non-breeding periods, but would move to the woodland and use higher perches if there was heavy snow. They are also opportunists when the environment offers a temporary abundance of vulnerable prey. For example, Dare (2015) reported that 15 Buzzards were seen hanging in the air around the edge of some heathland that was being burnt to rejuvenate the heather (Dartmoor swaling), in order to capture anything fleeing the burn, probably reptiles and small mammals. Other raptors will also do this where animals are being smoked out around natural fires.

Where there are perches, Common Buzzards tend to use them most of the time (Figure 3.5). Presumably, they use less energy perching compared even with fixed-wing gliding in an updraught, whereas transferring from perch to perch requires energy-intensive flapping, even when an initial boost can be obtained by dropping sharply off a high perch to build speed before their wings are spread to gain lift. Of course, there will be many places where there is no updraught, or the headwind is too erratic, such as a clearing in a forest, and then perching is definitely the most efficient way to watch for prey. Provided there is no snow, Buzzards will also hunt on the ground. For example, a winter roadside survey in south-west Poland in the 1990s found Common Buzzards on the ground approximately one-third of

Figure 3.5. *Telegraph poles are favoured perches for Common Buzzards (© Barsan Attila, www.shutterstock.com).*

the time, and quoted other studies recording between 11–66 per cent of wintering Buzzard sightings on the ground (Wuczyński 2005). Hunting from the ground was deemed necessary in areas with few elevated perches (Figure 3.2), but also considered more energy efficient than flying back to a perch all the time. The focus was still on hunting small mammals, although Wuczyński showed that during winter Common Buzzards were more likely to be on the ground in arable stubble or plough, so those on the ground could have been feeding on invertebrates.

Dare (2015) considered perches to be a prime requisite of a good territory. Perches can be used for hunting, for taking a rest and for conspicuous advertising of preparedness to defend a territory. Although hunting behaviour is readily identifiable, most of the studies were only recorded 'perching' (for example, Wuczyński 2005, who made observations from a bus travelling at 50–70km/h) rather than hunting while perching. There is good experimental evidence for the importance of perches for hunting, from the temporary construction of artificial perches in some clear-cut areas in the boreal forest of southern Sweden. Those clear-cuts with artificial perches were used significantly more by foraging Common Buzzards than the control clearings in April and May. What's more, if the perches were then moved to the control regions lacking perches, the Buzzards switched their foraging to those sites, too (Widen 1994).

When on a perch, Common Buzzards will, like birds in general, face into the wind so that the air smooths the feathers over the body. If inexperienced Buzzards try turning the other way, the wind catches under the feathers, ruffling them up, and – if strong enough – exposing the long, fluffy white undertail feathers as it tries to tip the bird forward. When a Buzzard is known to be hunting, it frequently looks at the ground and bobs its head; this looks somewhat comical, especially when the neck is extended at an odd angle to get the best view (remember that the eyes are relatively fixed in the skull, so the bird needs to move its head). If no prey are spotted, then the Buzzard will move to another perch, often working along a fence or hedgerow sequentially (Dare 1998). The Buzzard needs to stay at a perch long enough for any potential prey animals to move and make themselves visible. Sonerud estimated that birds moved on within eight minutes if they did not see potential prey (in Widen 1994), whereas Dare (1998) considered that they could scan a favoured patch for 2–3 hours. The difference between the two behaviours may be between the active hunting of a hungry bird, versus the sit-and-wait-based opportunistic foraging of a less hungry bird. Once they see movement, the gaze becomes more intense, with the head cocked. The body will tend to tip forward (sometimes to the horizontal), the wings poised slightly away from the body ready for action, suggestive of a runner on a starting block. If high enough, a Buzzard will make a near-vertical drop, and at other times will launch into a slower direct glide, during which its approaching silhouette does not move against its background but just increases very slowly in size until the last second or two before striking the prey.

When hunting for full-grown birds the technique is not very different from hunting small mammals, but it is more important that the attack is stealthier. Rodents can run fast, but their progress is restricted to the ground and hampered by vegetation, making their progress much slower than a flying raptor. Birds, on the other hand, are acutely aware of other birds in the air, and many have near all-round vision, with eyes on the sides of the head to detect approaching predators. Nevertheless, some Buzzards make a habit of taking adult birds, by ambushing them on the ground. Where we had a Buzzard nest full of Woodpigeon carcasses, a Buzzard was observed gliding over long grass (Figure 3.6), so

Figure 3.6. *The approach of a Buzzard gliding close to the ground is not easily detected by prey (© Piotr Krzeslak, www.shutterstock.com).*

taking a pigeon almost before it had left the ground (Di Beaumont, pers. comm.). If a pigeon had managed to take off even a few metres in front of the Common Buzzard, it's unlikely the chase would be successful because of the pigeon's explosive acceleration. Another Buzzard near the Purbeck Hills in Dorset had been seen twice taking cock Ring-necked Pheasants with a similar technique (Johnny O'Brien, pers. comm.).

After leaving their nest in an open environment, precocial chicks of ground-nesting birds such as gamebirds or waders are as hard as small mammals to detect but are less mobile and hence more vulnerable; they may well be aware of flying predators, but as they are unable to fly they tend to freeze and use their camouflage for protection. If a Common Buzzard has seen them move, and knows where to dive, there will not be much of a chase. The same was true of an observation of a Common Buzzard catching a Common Cuckoo *Cuculus canorus*. Davis & Seel (1976) were watching a female cuckoo seemingly intent on laying an egg in a Meadow Pipit nest, a process that required intensive watching of the nest. They noticed a Common Buzzard soaring, and within 20 minutes it landed on a bush 300m from the cuckoo. After another 20 minutes it then 'took off in a low glide towards the cuckoo and, using its talons, picked the motionless cuckoo off the bush: the latter made no attempt to escape. The raptor carried its prey about 50m to a grassy mound'. Later, the observers collected feathers and one foot from where the Buzzard had carried the cuckoo. Taking young birds from their nests means that prey are even more defenceless. Karel Weidinger (2009) noted that Common Buzzards primarily visited nests with nestlings rather than eggs. That could be because Buzzards prefer to eat large chicks instead of small eggs, but could also be because Buzzards use the calls of the nestlings to pinpoint the nest.

Unless seized prey is killed quickly it may fight back or escape. Falcons have a notch on the upper mandible of the beak to aid with killing prey – it is thought to help break the quarry's neck. Other raptors, such as Goshawks and, especially, Golden Eagles *Aquila chrysaetos* have very strong feet that can cause internal injury. Davide Csermely and Giorgia Gaibani (1998) investigated how Common Buzzards and Common Kestrels attacked and killed their prey, observing attacks both on live mice and a dummy, which was made of a mouse skin surrounding a rubber pipe connected to a dynamometer to measure the pressure being exerted. The researchers had tried to achieve a situation where all the Buzzards in the experiment were equally hungry to minimise variation in the approach to the prey. Common Buzzards grabbed live mice immediately as they landed, but unfortunately weren't so convinced by the dummy and landed before grabbing it. Whether it was live mice or the false mouse that was presented, Buzzards tended to catch with one foot, not surprisingly because one foot is about as big as a mouse. Then the Buzzards squeezed the prey in bursts of a few seconds for an average total duration of 26 seconds. In comparison, the Kestrel could only exert about one quarter of the average $5.8 kg/cm^2$ pressure of the Buzzards but exerted a more continuous squeeze. The maximum pressure the Buzzards attained was $8.75 kg/cm^2$, which they compared with Harpy Eagles *Harpia harpyja*, which exert $11.3 kg/cm^2$. Common Buzzards appeared to suffocate the mouse before attempting to bite at the head. So suffocation, rather than penetration by beak or talons, is the way that Common Buzzards kill small mammals, their primary food source. This is likely true when they are trying to kill larger animals, too, although penetrating talons will also cause damage while they help grip prey. While hawking, a Red-tailed Hawk *Buteo jamaicensis* tended to grab the head of a Rabbit in one foot, which acted to immobilise it, and the chest area with the other foot, which squeezed the lungs. It is likely that Common Buzzards do the same. Although Common Buzzards may only bring the young of larger species (for example, Rabbits) frequently to the nest, the leverage in their short toes may be strong enough for bolder individuals to crush the chest of full-grown Rabbits with one foot.

As with other raptors, Buzzard's feet have a specialised ratchet-like locking system for the tendons that deliver power to the talons. This means they can hold the foot closed with the ratchet, rather than having to maintain the tension with muscles alone. Many species of birds have this locking system, which they use for perching, wading, hanging and even swimming (Quinn & Baumel 1990). Most birds have multiple locking mechanisms, so that if the foot wraps around a twig completely then the ratchet at the base of the toes can grip, but if on a branch that completely fills the foot, then a ratchet at the end of the toe will help dig the claws in and hold position. Raptors often try to hold on to, and sometimes carry, relatively large prey that their foot does not encompass, and therefore they need to apply pressure to their talons to grip and pierce the skin. So, raptors generally have their locking mechanism at the far end of each toe, near the talon. Furthermore, the raptor ratchet system tends to have larger 'teeth', which are spaced further apart than in other birds. This deep-notched ratchet is much more difficult to retract back over each tooth, compared with non-raptors, which tend to have finer teeth over a longer distance (Einoder & Richardson 2006).

Once the prey is subdued, the Buzzard takes up a mantling posture (Figure 3.7), holding its wings over the food and periodically lifting the head aloft and looking all around. The wings cover the food, hiding it from view, and the surveillance is a precaution against another predator trying to steal the food or attack the Buzzard while it is distracted. If feeding nestlings, the adult will carry the prey back to the nest as soon as possible. Carrying

Figure 3.7. *'Mantling' on food may deter a competitor (© David Dohnal, www.shutterstock.com).*

small mammals is no problem, because they are so light. Heavier species are more of a problem, and sometimes a Buzzard will need the wind to be in the right direction to allow it to gain height and carry the prey to the nest. Buzzards will take prey that is time consuming to eat up to a tree branch, or similar raised position, that gives a better view of approaching thieves and avoids attacks by ground predators.

So, it seems that Common Buzzards are well adapted for hunting small mammals that can be pounced on and offer little resistance. However, their evolution has not diverged so far from their more aggressive raptor relatives that they cannot as easily tackle larger prey, such as medium sized birds and mammals. Moreover, larger prey can retaliate effectively and possibly damage the Buzzard. This would explain why Perlman & Tsurim (2008) observed Steppe Buzzards being very cautious about approaching a strange object (a Bal-Chatri cage-trap containing prey) if they were not desperate for food. When easily killed prey such as voles, young Rabbits and birds are in short supply, farming provides consistent and easily killed alternative prey in the UK, by way of plentiful invertebrates.

Hunting invertebrates

When comparing Common Buzzards with many other birds, it seems reasonable to assume that they have not evolved for worming. Their side-to-side gait, likened to a 'drunken sailor', is far from the grace of their swooping when on the wing. Although Buzzards have fantastic distance eyesight, observations on a captive bird suggest that they may not be able to see just in front of their bill tip. This could be due to a lack of retina serving that area (as has been shown for Short-toed Snake Eagles) or possibly the margins of the visual field for both eyes intersect some way in front of the bill. This effect causes 'ballistic pecking' in Ostriches *Struthio camelus* (Martin 2015), although we will not know if it applies to Buzzards until they have been studied in the same way. From standing on the ground, Buzzards appear to assess where the worm is, then take a lunge at it. At times, they must pull the worm from the earth, in the clichéd manner of a European Blackbird's tug of war. If they can't see a worm from where they are standing, then they make a short flight (typically 10–20m) before landing, picking up a worm, and then retracing the flight path, waddling back as they collect the worms they've seen along the way. One unusual observation recorded Buzzards standing on the rear mole-board of a plough as it turned the earth in spring (Hampden Smith 1986), similar to the Buzzards following a combine harvester in autumn, but this time almost certainly for worms rather than rodents.

Common Buzzards have hooked bills, which are better for tearing flesh and skin than for poking into the earth to retrieve a juicy worm. Their large body size is more efficient if it can drop from a height, rather than taking off from the ground as they regularly have to when worming. The only external feature that makes sense for worming is having relatively small feet for running. Tubbs (1974) thought them surprisingly fast, almost thrush-like, whereas a falcon's large foot would be a hindrance, like us trying to run in snorkelling fins. When worming, Common Buzzards actually spend a lot of time standing tall and looking around, partly for worms, but perhaps also because they are vulnerable if attacked on the ground.

Of course, both rodents and worms may occur together in some habitats. An amazing photograph of a Buzzard with grass and the blood of a vole on its feet and the vole in its mouth, shows worms making a bid for freedom from the side of its beak (Plate 10; see colour section). A wide gape may help in swallowing worms, as well as small rodents. However, if worms are not nipped in the region of their anterior ganglia before being swallowed into the crop, it is interesting to wonder whether they become immobilised in some other way for easy passage from the crop to the gizzard.

Given that Buzzards can obtain enough nutrition from worms, the latter form an incredibly attractive diet because they can be locally abundant and can't fight back. So, apart from watching for other competing Buzzards, nothing could be safer than picking up defenceless worms. What's the point in all the effort of searching for fast-moving targets that might result in injury? Far from being 'lazy', it seems a sensible choice when it is available, and so something that natural selection would favour. It is interesting how many people do not regard the Buzzard as a 'wormer', but once it has been explained, the next time they are in an area with Buzzards and ploughed fields they will soon enough observe Buzzards worming. Sometimes, it is easy to miss a Common Buzzard in amongst gulls on a ploughed field. This is especially true of pale morph Buzzards, where the brown on the back blends in to the brown of the field, leaving a whitish looking outline reminiscent of a juvenile Herring Gull *Larus argentatus*. At other times, there will be no gulls, but sometimes numerous Buzzards. We have seen up to 30 at one time, but more usually there are between five and 20. In these large gatherings they are spaced at reasonably regular distances throughout a field, generally 20–100m from each other. They mix with other species as they grub for invertebrates, perhaps using plovers, waders and gulls to guide them to where worms are most abundant, and then slipping in beside them to enjoy the feast.

Although Buzzards are often seen on ploughed fields when worming, they also take worms from grassland. Whereas worms may be especially vulnerable on arable land, where ploughing has destroyed their burrows or there is a lack of grass to hide them, permanent grassland contains more worms and may be a more reliable source when ploughing has not made worms freshly vulnerable. Worming on grassland is most likely when the sward is short, so that worms can be seen, and in the damp of dewy dawns when worms are most likely to be exposed on the surface.

As well as running about on the ground, Buzzards will also use low perches, or even hang low on the wind and land on beetles, worms and other creatures. Dare (1998) described birds hovering less than 3m over particularly good invertebrate habitat, appearing 'suspended from an elastic thread as it fell and rose repeatedly then swung along the slope to find new sites'. There are many cases in the literature describing Buzzards launching from fence-posts, sometimes gliding up to 40m away to pick up a beetle before returning to the same vantage point. Weather can affect their access to invertebrates, as well as to vertebrates. Obviously, there is no point in foraging for earthworms if the ground is frozen, and so Buzzards need to move their focus to the less wary among their vertebrate prey, which may be weakened because they cannot get sufficient food themselves. While there may not be so many recorded observations of Common Buzzards feeding on invertebrates beyond western Europe, they will almost certainly feed on them where they can, in the same way that they inevitably find carcasses to scavenge wherever they are.

Scavenging

As discussed in the previous chapter, the closest many people get to a Common Buzzard is while driving along a country lane and encountering a large brown bird picking away at a roadkill. Of course, when feeding on animals that are already dead, there is no chase and the hunt is all in the search, without the need to seize and kill. Buzzards are adept at finding carcasses and exhibit interesting behaviours that are worth noting. It is rare that carcasses are the predominant prey at the nest. However, during Dare's study in Wales (Appendix 3), J. Driver found a nest on a spectacularly high cliff in mountains; each year, the ground under the nest became strewn with carcasses of sheep that had fallen from the cliff which, he surmised, likely helped feed the hungry chicks.

Most of the time, the location of carcasses are not so predictable, but humans can help with regular presentation of such food. At a Welsh Red Kite feeding station, where meat is regularly thrown out to attract these birds so the public can enjoy the spectacle, it is usually not long before 15 or so kites circle over the meat, lowering themselves little by little before making a final stoop to the ground trying to snatch the food. They turn quickly and suddenly, dropping vertically and rising again so swiftly that their movements are something of a blur. Despite being good scavengers, with more extreme features than a Common Buzzard (a longer gut and very low wing-loading), the Red Kites are reluctant to land on the ground. If their snatch misses the food, or the food is too heavy, they must rise and try again. In contrast, any Common Buzzards waiting patiently in nearby trees will swoop down, land on the ground and start feeding where the kites will not land (Plate 11; see colour section). The Buzzards cannot feed fast because of the large number of kite talons wheeling not far above their heads, forcing them to spend a high percentage of time watching for a suitable opportunity, but their persistence pays off because they start feeding long before most of the kites. Even deer-stalking can provide sustenance for a hungry

Buzzard. On a cold December day in Scotland, a ravenous Buzzard alighted on the innards of a recently disembowelled deer, as soon as the stalker started walking away with the deer carcass. It must have been very hungry, because the stalker returned from his vehicle with a camera, and stood within feet of the Buzzard as it continued to feed on the warm entrails quite unperturbed (Maxwell 2010).

Raymond Hewson (1981) studied the interaction of scavenging birds on carcasses he had laid out. He found that the Common Buzzard was one of the dominant species, always displacing not only Hooded Crows *Corvus cornix* but also Great Black-backed Gulls *Larus marinus* and Northern Ravens *Corvus corax* in the 39 confrontations recorded. This was sometimes the case even if the Common Buzzard was in a nearby tree overlooking the carcass, rather than actually feeding on it. There must be a lot of respect for those small but powerful feet with the ratchet grip. Interestingly, only one Common Buzzard could feed at any one time; if another arrived there would be a confrontation until only one was feeding. This is an interesting contrast with worming Buzzards, but when worming, although they may be relatively close, they are much farther apart than the length of a carcass. Hewson also noted that Common Buzzards preferred Rabbit carcasses to those of larger species such as sheep, unless the large animals were opened up in some way. So presumably, Buzzards are not good at tearing through thicker skins, although they will readily eat larger carcasses that have already been broached (Figure 2.8). Hooded Crows harried Buzzards, and on four occasions stopped a Buzzard landing on the food, but they never drove a Common Buzzard off the carcass. The crows often stood nearby waiting for scraps that the Buzzard had pulled off. Magpies can be quite bold, too (Figure 3.8), especially if working in pairs to distract larger scavengers.

Figure 3.8. *Magpies may compete at kills but are also potential prey (© Prokaev Vladimir, www.shutterstock.com).*

These days, we can all enjoy encounters between birds on animal carcasses by searching the internet for video clips. As with Hewson's observations, Buzzards are often dominant over corvids, but what about other species? Care needs to be taken with the interpretation of behaviour in such videos, because there is no information on what happened previously or how the film had been edited; nevertheless, they are great fun to watch. In other chapters we will discuss how Northern Goshawks seem to be dominant when it comes to nesting territories, pushing out Buzzard pairs or ruining their breeding performance, but when it comes to a fight over carrion there are videos showing that Northern Goshawks do not necessarily hold the upper hand. There may be particular reasons for this. Quite often it is an adult Buzzard that is seen deposing a juvenile Goshawk from a kill with great confidence. Likewise, we don't know how long the Buzzard has been watching the Goshawk feed. If long enough, the latter will have had a good feed, be less famished and prefer to avoid a risky encounter that could be dangerous. So, the Buzzard may have picked the moment when the Goshawk will step aside rather than stand its ground.

One amazing video shows a Common Buzzard repelling the advances of a Grey Heron four times its size. The Buzzard is feeding on a carcass, feathers raised to form a square at the back of its head. A few metres away, the heron looks interested and occasionally extends its neck to try and tweak some of the food away. The Buzzard rises up to its full height and stares aggressively at the heron, which flinches and pulls back quickly so the Buzzard resumes its feeding. Eventually, the heron moves in more assertively. The Common Buzzard launches itself at the Grey Heron, throwing its talons towards the heron's chest, before showing some ambivalence and retreating a few metres. However, before the heron can get to the food properly, the Buzzard comes back and keeps attacking the heron's chest until it can resume feeding.

Another form of scavenging is kleptoparasitism. This occurs when one individual robs (*klepto* is the Greek word for steal) another of its prey, in effect parasitising another to feed itself. Skuas, which are reminiscent of marine Buzzards, are kleptoparasites. They intercept other seabirds returning from fishing expeditions, forcing them to drop the food they have caught. Kleptoparasitism avoids the dangers of a pursuit, although it must be balanced against the threat of a counterattack from the host, which had invested time and effort obtaining the food and is likely to be highly motivated by the need to feed its young. There are plenty of instances of Common Buzzards forcing other raptors to drop their prey and taking it themselves (Busch 1997, Dickson 1997, Siverio *et al.* 2008), even taking Feral Pigeons *Columba livia* from Peregrines (Kinley & Thexton 1985) and stealing prey from harriers in East Anglia (Bloomfield 2013). Usually the host is either a much smaller species, or young, so the Buzzard is easily dominant and unlikely to suffer retaliation.

As we have described, the Common Buzzard is primarily a stealth hunter of small mammals in short vegetation, most often approaching in a low glide, or pouncing from a perch. Some of the Buzzard's adaptations, such as relatively small feet, are a constraint on what they can hunt. However, a wide gape and long intestine means that Buzzards can scavenge and exploit a wide diversity of invertebrates. With this generalist capacity, there are not many habitats where Buzzards can't find something to eat. However, that does not mean that they use the landscape uniformly or indiscriminately. In the next chapter we will look at the features that are important in a habitat, and how Common Buzzards use them.

Conclusions

1. The characteristic big broad wings of the Common Buzzard are a *Buteo* trait that would have allowed an evolutionary ancestor to exploit small mammals in the arctic tundra by providing efficient low flight over thick vegetation where there are few perches.
2. The relatively small feet for a medium-sized raptor are likely to be an adaptation for penetrating relatively dense vegetation to catch small prey, while also maintaining good leverage on talons for holding larger prey.
3. A big gape allows Common Buzzards to swallow small mammals quickly, which saves them the trouble of defending their prey from kleptoparasites.
4. The inconspicuous brown plumage when perched, together with Buzzards' ability to fly low and slow, makes them effective short-distance ambush predators.
5. Studies into the way Buzzards despatch their prey before eating suggests that suffocation by squeezing the lungs is the method most often used on vertebrates. This limits the size of prey they can subdue with their small feet.
6. While their morphological characteristics imply that Common Buzzards have evolved to hunt small mammals and young birds, they also often eat invertebrates. This is possible not only because of their efficient digestion, but also because their broad wings provide them with stable, low flight for searching and their small feet allow them to run on the ground.
7. Scavenging requires searching and then, if the carcass is big enough, defence against other scavengers. Despite their small feet and broad wings, which make them a poor match against other raptors in aerial contests, Common Buzzards appear to be more confident at repelling attacks and defending a carcass on the ground.
8. Engagement in scavenging means that Buzzards get close to humans by foraging on road kill, hunters' spoils and food offered to attract other birds.

CHAPTER 4
Habitat use

Knowing where you see a bird does not necessarily tell you where it spends most of its time, or indeed what it spends its time doing. It is relatively easy to see a Common Buzzard perching on a telegraph pole or displaying above its nest, but how do you know what it is doing the rest of the time? Trying to spot a Buzzard in woodland is far more challenging, despite it being a relatively large bird. Thankfully, at the start of the 1990s we were in a position to radio-track Buzzards in a habitat-rich area around where we lived. The radio-tags meant that we could find Buzzards when we wanted to, even if they were not conspicuous. Often, we did not see the Buzzard we were tracking, but we knew where it was by triangulating bearings on which we heard the strongest signal. Most of the tracking could be achieved from the road, especially with our long wheelbase Defender Land Rover fitted with a hydraulic mast that could be raised up to 5m above the vehicle's roof (Figure 4.1). This helped because radio signals are better heard without obstructions, and if the antenna is raised higher then there are fewer impediments, the signal becomes clearer and it's easier to achieve a good bearing for triangulation. We could also interpret approximately what the Buzzard was doing from both the signal strength and a posture sensor that made the tag beep faster if it was horizontal rather than upright. Thus, if we had a steady, slow signal, we knew the bird was perched and relatively still. If fast and constant, it was probably incubating eggs. When the Buzzard was moving, the signal strength would vary, sometimes dramatically, and the pulse rate would keep changing. If soaring, we could hear a fast signal because the bird was horizontal, but the signal would fade in and out slowly, as the orientation of the antenna changed while the bird gradually spiralled around. We spent a lot of

Figure 4.1. *In the 1990s, we were able to radio-tag broods of 1–3 Buzzard nestlings (left, Robert Kenward by © Mikhael Romanov) and track them with a 5m mast on a Land Rover (right, © Sean Walls).*

time on the road, in all weathers, recording exactly where Buzzards were perching, flying, nesting and feeding. Then we entered the data into software to look at the area and habitats being used by each bird. It felt as though we were peering into the secret lives of Common Buzzards, rather than just looking at their public appearance, and much was revealed about their ecology.

We have described how Common Buzzards can predate many different species, which gives them the potential to survive in almost all habitats with a temperate climate. Here, we illustrate how large an area a Buzzard uses, what we know about the habitats an individual requires in its home range to survive until breeding and which habitats surround the nest during the breeding season. For simplicity, we use the term 'habitat' in the sense of a vegetative assemblage as perceived by researchers, which may be classified from a satellite image as land cover or defined as a map class, but may not be how a Buzzard perceives it.

What is a home range?

Where an animal normally goes to feed itself, mate and care for young is commonly known as a home range (Burt 1943). This is a useful ecological concept for indicating the area that an individual or group requires in a particular time frame. The presence of a number of stable home ranges also indicates that an area is suitable for a population at that time. Home ranges can be described in terms of size and structure, i.e. whether the range is used evenly, or if certain areas are used more than others. Once we can define the home range, we can look at the habitats that represent different resources within the range, and examine whether all habitats are used to the same extent, or whether some are used more than others. The way to investigate all this is to find where individuals occur, and record the coordinates to plot on a map of land cover for analysis.

Before the widespread radio-tracking of birds, Tubbs (1974) explained that it is relatively easy to study Common Buzzard territories because most territorial defence takes place in the air, where they are easy to see. However, a territory is different from a home range in that the former is the area an animal defends, rather than all the area it uses. When defending a territory, animals often make themselves conspicuous in order to deter intruders without a fight, and we will examine this more in the next chapter. There are other times when it may benefit an individual to be elusive, for example, when hunting wary prey or during incursions into another's territory. Moreover, even when animals are not necessarily trying to hide, vegetation and topography may make it impossible for a fieldworker to find them at all times, especially because Common Buzzards are often associated with woodland and hilly areas. Their deliberate and inadvertent concealment therefore makes radio-tags indispensable for systematic study of Buzzard home ranges.

Before investigating how food resources and other factors may influence Common Buzzard home ranges, it is important to have a standard way of estimating the latter, so that they can be compared. First, we need to define a period of time when the tracking takes place. In Chapter 8, we will find out how young disperse from their parents' territory to form their own home range. If we include all the area over which a Common Buzzard wanders before settling, there will be a lot of landscape that the individual rejects settling in, so this is not a useful measure for the home range that it requires for food and security (if not yet for breeding). To be sure that a foraging home range is being recorded, we first have to decide whether the bird is settled or in a nomadic phase of its dispersal.

This also applies to migrant Common Buzzards, which will have a breeding home range and, if the climate in the breeding range does not support them during the non-breeding period, one or more areas in which they settle to forage in winter. Both types of home range will have been chosen by the Buzzard concerned, but the landscapes they travel through may not have offered suitable habitat in which to settle, and should therefore not be included in a foraging home range. Even if a Buzzard does not disperse or migrate, the breeding season may have different requirements from the non-breeding season, not least suitably secure nest-sites. The consequence of all this is that when investigating home ranges it is better to look at data from a relatively short period, not the entire year or indeed the lifetime of the bird.

We were interested in the home ranges that Buzzards established after learning to forage for themselves, as an indication of what they needed when they first settled. We therefore observed when they dispersed from their nest areas and picked a period after that to track them in their first autumn. To ensure efficient use of our research time and get a sufficient sample, we wanted to obtain home ranges for as many birds as we could in the same period. In and around our study area, we found that it usually took 2–3 hours to record a location for each of 12–15 birds. So, we recorded this number of birds three times daily for ten days within a two-week period, varying the time of day at which we looked for each bird, hoping to record all the habitats they visited in that season.

Habitat use

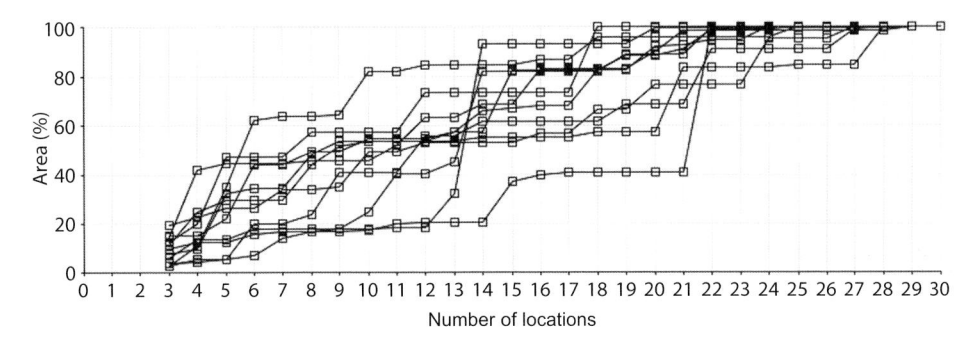

Figure 4.2. *Convex hull home ranges of 11 Common Buzzards tended to reach a maximum size, as consecutive locations were added, before 30 locations; only three of 11 increased a little in size after adding their first 25 locations.*

The resulting standard set of 30 locations was based on analysis using computer software we had started developing during the 1980s. The standard home range at that time tended to be represented by drawing the shortest line round all the locations, as if it were thread tightened round pins on a map. This was called a minimum convex polygon (MCP), now renamed a 'convex hull'. If we used a computer to plot how the home range changed after adding each location in sequence, initially the range size would increase quickly. However, after 30 locations collected during ten days there was very little increase in home range size (Figure 4.2), suggesting that the bird had visited all the areas it used. This therefore became our standard protocol for collecting home range data.

As well as helping to define how best to record standard home ranges, the availability of personal computers also provided new ways to analyse the location data. Software not only made it easy to plot and analyse convex hulls, but also to make comparisons with an earlier home range representation, namely ellipses of location density centred on the arithmetic mean coordinate of the locations (Figure 4.3b, including 95 per cent, 85 per cent and 50 per cent of the density). New ways of depicting home ranges could also be computed, including the contours plotted across a grid of location density kernels (Figure 4.3c), and cluster polygons drawn tightly around separate groups of locations that are nearest to one another (Figure 4.3d). Home ranges drawn from the same set of recorded locations look quite different for each analysis technique, and in turn have different metrics, such as area, estimated from them. Note also how the locations are not evenly spread but tend to occur as clumps with a small proportion as outliers at some distance from the main clumps.

Ellipses and contour methods are based on estimating the density of locations. The result is something like a relief map, where what looks like the summit of a hill is the most intensively used area, also where one was most likely to encounter the bird. The smoothing effect is better for producing estimates of range size from fewer locations than when convex hulls are used. However, as with convex hulls, distant excursions can greatly affect the area estimated. Moreover, the smoothed outlines can overlay areas where tagged animals were never recorded. Alternatively, the nearest-neighbour cluster method produces local convex hulls that are plotted tightly round the locations where the birds were found, tending to exclude the area between the clusters. These methods can be plotted to include 100 per cent of the locations or, by statistical exclusion of outlying locations, just a percentage of the most dense or closest locations that define a range core. Both contouring and cluster

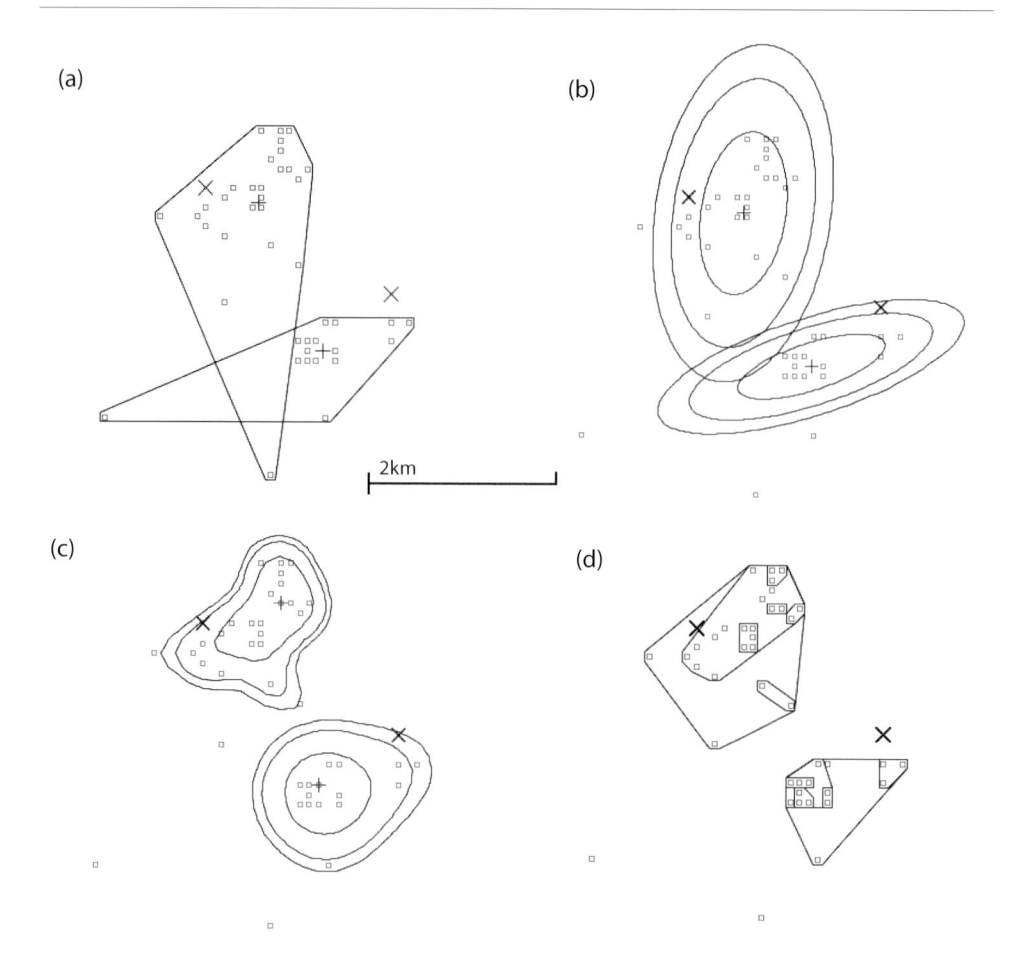

Figure 4.3. *The first two Buzzards tracked in 1990 remained close to their nests (X) in autumn, and their home ranges overlapped extensively, due to the influence of single locations (shown as small squares) that were outliers, if edge-line-based convex hulls (a) or Jennrich-Turner ellipses (b) were plotted about range centres (+). Ranges plotted as kernel density contours around 95 per cent, 85 per cent and 50 per cent of location density (c) and hulls around 95 per cent, 85 per cent and 50 per cent of locations with least nearest-neighbour distances (d) excluded outliers and fitted core locations better.*

methods can have several separate core areas and we may assume that when Buzzards travel between these separate core areas they probably fly quickly. Having described how home ranges are assessed, let's look at what knowledge we can gain by assessing Common Buzzard home ranges.

Home range size

Several researchers have plotted and measured territories of Common Buzzards based on their observations of them defending an area around their nest (Chapter 5), but very few have recorded the home ranges of Common Buzzards. Undoubtedly this is to do with the resources required to tag and track birds. In the 1990s, we were fortunate to be able to

Habitat use

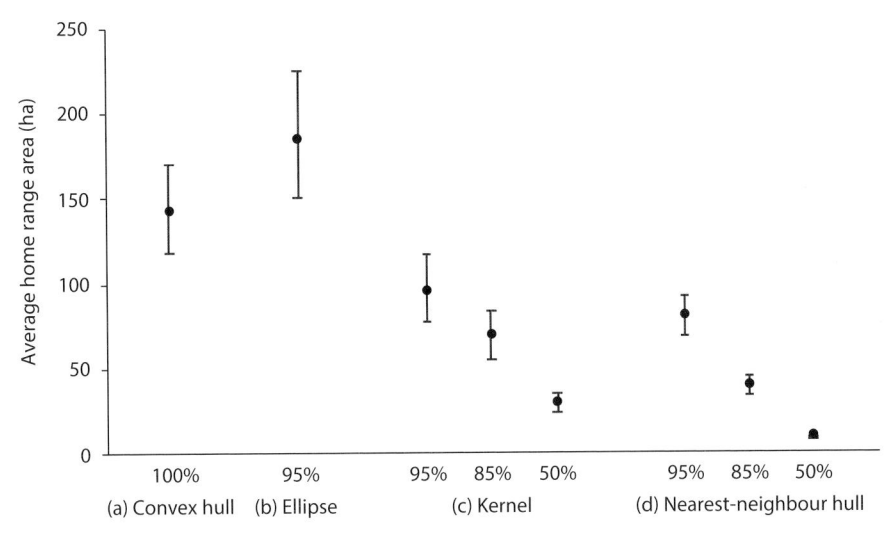

Figure 4.4. Geometric means (+/-95 per cent confidence limits) for home ranges estimated for 114 Buzzards in autumn by convex hulls around all the locations (a), by 95 per cent ellipses (b) and contours (c) based on 95 per cent, 85 per cent and 50 per cent of location density, and by local convex hulls round 95 per cent, 85 per cent and 50 per cent of the locations with minimal nearest-neighbour distances (d).

devote radio-tracking time to record more than 100 Common Buzzard autumn ranges in our Dorset study area (Appendix 3), plus nearly 30 midwinter ranges for young, non-breeding individuals. Most of the ranges were of Buzzards in their first year, although we managed to collect some of Buzzards in their second and third years. The range sizes were very variable, not least due to distant outlying locations (Figure 4.4). We used logarithms to estimate geometric mean values for our samples of ranges, because this reduced the effect of outliers.

Taking the autumn ranges and putting a minimum convex polygon around the outermost locations, these birds had an average convex hull home range of 142ha, and the expansive nature of ellipses (Figure 4.3b) resulted in an average 184ha (Figure 4.4b). Excluding about 5–15 per cent of the outermost locations, which tended to be of birds in prolonged flight and therefore on excursions (Kenward *et al.* 2001a), reduced the main part of their range to nearer 100ha (1km^2), even when using a more expansive contour, e.g. 98ha for the 95 per cent core of a probabilistic kernel contour (Figure 4.3c). However, the focal areas, e.g. covered by convex hulls around 50 per cent of locations with smallest nearest-neighbour distances (Figure 4.3d), averaged only 9ha (Figure 4.4d). The small focal area is likely because Common Buzzards appear to spend much of the day perched, and even if worming they spend many hours in a small section of a field. The ranges tended only to have one core, a single area where they spent most time, while the rest of the range was used less intensively. Perching during the day seemed to be very close to the foraging area, otherwise two cores would have been apparent. After noting that locations at night tended to be fairly central to the foraging ranges, we stopped recording locations after dark, as we were most interested in how Buzzards moved around during the day. Before considering findings from other studies, it's important to emphasise again that estimates of home range sizes of our Buzzards were hugely variable due to excursions. For example, the outermost cores including all locations could be several hundred hectares, and one range exceeded 2,000ha (20km^2) with convex hull, ellipse and kernel estimators, due to a single long-distance excursion. However, most spent the majority of their time within 1km^2.

It is interesting to compare our estimates based on radio-tags with those using different techniques and in regions where different distributions of resources may have influenced the areas over which individuals ranged. In northern Germany, a study using wing-tag re-sighting considered breeding birds to use approximately 100ha, but single non-territorial birds ranged over as much as 1,000ha (Hohmann 1994, 1995). Thérèse Nore et al. (1992) found that a juvenile would use 100–400ha through its first winter and considered this to be much bigger than the range of a breeding adult. Study birds included those that were tracked right from the nest through to the following spring, and so included some wandering between settled ranges. These findings were therefore very similar to what we were seeing with most of our birds.

In the highlands of Scotland, Weir & Picozzi (1983) mapped home ranges of single birds from observations of six visually wing-tagged Buzzards during winter, with an estimated average area of only 35ha (24–45ha). Of course, it's unlikely they had the Buzzards in view the whole time. In contrast to that very tight home range estimate, Tubbs (1974) noted that at certain times of year some breeding pairs foraged well outside their breeding territories (in one case approximately 6km from the nest). So, this would result in a bigger total range than their territory, whereas the core range could be considerably smaller than the territory in a rich worming area with a nearby roosting site. Some of the differences will be due to the region, because the size of area over which birds range to get the food they require is likely to depend on the quality of the resources as well as how they are distributed.

For her PhD thesis, Kathy Hodder analysed a sample of our ranges over a few years to investigate any seasonal or age-related differences in home range size. She found that while the range sizes of second- and third-year Buzzards were relatively similar from season to season, juveniles made many more excursions in the winter compared to the previous autumn, dramatically increasing their outermost range cores in winter (Figure 4.5). Moreover, older birds had larger autumn ranges than those of first-year Buzzards, which was particularly well demonstrated by the ranges of 29 birds tracked in both their first and second autumns, and that was true whichever home range estimate was used (Kenward et al. 2001a). Considering that we know these birds were primarily feeding on worms in autumn, it seems likely that the tendency for all Buzzards to be observed flying when furthest from the home range centre (Kenward et al. 2001a) indicated exploration rather than foraging.

There are quite a few possible explanations for the small autumn range sizes in their first year. The parents could have still been feeding the juveniles at that time, or simply allowing their young to feed with them in good areas. Therefore, the increase in the winter range estimates for the young birds could represent their need to acquire more of their own food. However, as some of the autumn ranges were of individuals that had already dispersed from their parents' care, weather rather than parental care might be more influential, by forcing less experienced birds to search further to find sufficient food in colder conditions when food is scarcer. Alternatively, as we will see below, Buzzard ranges contain more arable land in the autumn but favour pasture later in winter, which may be a response to the seasonal change in vegetation or how weather affects the availability of food within those habitats.

A similar possible resource-based reason for autumn ranges being larger in the second year is that older Buzzards of breeding age are starting to incorporate and possibly defend

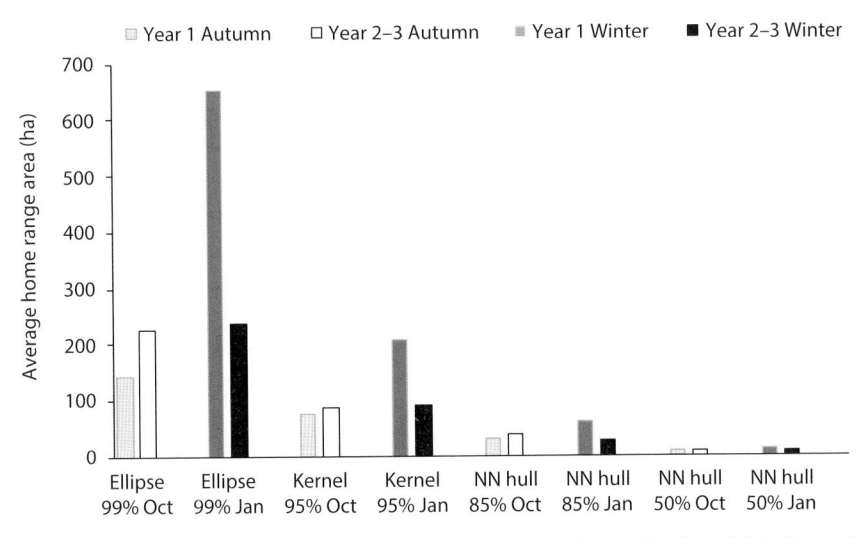

Figure 4.5. *Home range estimates for 29 Buzzards tracked in Dorset during October of their first and second years, and which were tracked again in January. First-year birds had relatively larger estimates for outermost ranges in winter, especially when assessed by ellipse and kernel methods, which are more influenced than nearest-neighbour (NN) hulls by peripheral locations. Data replotted with kind permission from Hodder (2001).*

woodland for nesting, which in our area did not seem to be particularly good for hunting compared with worm-rich fields. Where the same individuals were tracked in the autumn and the following winter, there was generally a fair amount of overlap in the ranges, so the birds did not disperse or change range completely, but rather were just going further to get the food or nesting habitat they needed. Looking at how stable the range sizes between seasons were in the older birds, the increase in range size during the first winter and its maintenance thereafter is entirely consistent with birds extending initial range cores to incorporate what they needed year-round, as foraging became harder in the winter and perhaps also to include breeding habitat in spring, and then maintaining these as territories thereafter.

These days, sophisticated tags can use a satellite-based GPS to record locations more accurately and in larger numbers than was possible for us. These can then be retransmitted via other satellites or mobile phone systems to a researcher in an office. In principle, this could enable the sampling of locations more accurately and frequently than was possible during our hurried triangulation of bird after bird. However, obtaining and retransmitting the locations requires more power than available in our four-year VHF (very high frequency) tags, and is most practical where abundant sunlight permits frequent retransmission from tags recharged by solar panels. Unfortunately, few such studies have been published so far. One exception is research in a mainly moorland area of Scotland, where the average 95 per cent cluster core was nearly 310ha (Francksen *et al.* 2016). This contrasts with the 80ha of comparable autumn ranges in Dorset (Figure 4.4) and is likely to be a function of poorer quantity or quality of resources for Common Buzzards on moorland (see below).

Although there is great variability in home ranges size, some of it due to the age of the Buzzard, another very substantial influence on range size is the habitat within the range. Common Buzzards are efficient birds, and do not appear to do more than they need to. So, if they have all they need nearby, they are happy to perch in one spot for hours on end. However, if they can't get what they need from the immediate vicinity they have to range further. So, what have we learnt about the habitats that Buzzards use and need?

Habitat use during the non-breeding season

As we have described, home ranges are smaller than we might imagine for a raptor of this size. For example, Northern Goshawks of similar body size have been found to have ranges ten times as large as those of Buzzards in similar UK farmland containing small woods. In Sweden, as the proportion of woodland increased, Goshawk ranges were found to become 3–10 times as large as in the UK, up to 100km^2 in areas of extensive boreal forest (Kenward 1982). In contrast, Common Buzzards use a relatively small area to gather what they need, so they might be expected to choose carefully where to settle. In this section we focus on what is found within an individual Common Buzzard's home range outside the breeding season, when they are not tied to any nest.

Common Buzzards are able to adapt well to the landscape, using many different habitats due to their incredibly varied diet. Small mammals live in all terrestrial habitats, Rabbits and earthworms are associated with pasture, and many other prey species inhabit woodland edge and semi-natural habitats. There are areas where Common Buzzards are not often seen, particularly heathland and young conifer thickets where the abundance of mammals is lower (Tubbs 1974, Weir & Picozzi 1983, Halley 1993, Dare 1998), although they do use moorland especially when small mammals are abundant there (Thirgood *et al.* 2003, Francksen *et al.* 2016). Traditionally, Buzzards have been considered woodland-edge and grassland birds, because while trees are good for roosting and nesting, Common Buzzards are better adapted for hunting in open habitats.

Those who have collected visual records of individual movements and habitats have mainly observed individuals that they could identify by their plumage, or those that had been visually marked. Although we know that the data are likely to be biased towards habitats in which observers can more easily see Buzzards, nevertheless they add useful information from different regions to the smaller number of more systematic studies that have used telemetry techniques. From these visual studies, Common Buzzard ranges typically contained some woodland and, in hilly or mountainous terrain, the lower valley, too. Observations by Weir & Picozzi in Speyside (Scotland, Appendix 3) showed Buzzards used the valley floodplain more often in winter, when conditions were harsher at altitude, but then moved up the sides of the wooded valley to where they could nest in summer. Birds passing through in the autumn (those that set up a foraging area for at least 20 days) tended to be found more in agricultural land (Weir & Picozzi 1983, and from the same study area, Halley 1993), but were also probably easier to see there than those in woodland. In the New Forest, which is mainly woodland and heathland with smaller areas of grazed grassland, Tubbs (1974) thought Common Buzzards were predominantly associated with woodland and that they could get sufficient food from within the woodland during the breeding period. However, they also used the most open areas, and individual Buzzards (or pairs) developed regular feeding places well outside the defended territory, sometimes several kilometres away in small patches of agricultural land in and around the forest.

In a French study, which radio-tracked some Common Buzzards during the autumn and a few through the winter until the next spring, Nore *et al.* (1992) found the movements were quite different between the territorial adults they observed and seven Buzzards tagged as nestlings. The juveniles spent a fair amount of time exploring different places as they dispersed away from their natal nest, before they settled into a home range in more arable areas with high invertebrate abundance, which they shared with other Buzzards that lacked breeding territories. During a poor vole year, the study observed that Buzzards preyed mainly on field crickets (subfamily Gryllinae), rather than the worming observed in the UK. Otherwise, the Buzzards were found where voles were most abundant and on open areas where it was considered easier for the Buzzards to see and catch voles.

In our study in southern England, we not only described ranges, but we were also able to assess which land-cover classes were used more than expected by chance, and vice versa. From the birds' perspective, the former represented habitats likely to be most useful for them while the latter were avoided land covers. We had three different maps for the analysis (Plate 12; see colour section). The Farm Survey was a detailed record based on Kathy Hodder visiting all areas and visually checking what was in a field or within a woodland boundary. She also spent a lot of time counting Rabbit burrows throughout the study area. This map was more precise than the Land Cover Map of Great Britain (LCMGB), prepared in the Centre for Ecology and Hydrology using satellite imagery, for which images of light reflectance classes (visual colours and infra-red) in summer and winter seasons were associated with particular types of land cover (Fuller *et al.* 1994). The LCMGB could not be as accurate as the field survey (Plate 13; see colour section), partly due to image interpretation (Fuller *et al.* 1994) and also because the map was prepared from 1989 data, whereas we tracked the Buzzards mainly during 1990–94, by which time some land cover may have changed, e.g. different crops may have been grown (compare arable areas on the two maps in Plate 13; see colour section) or woodlands felled, etc. Nevertheless, the satellite images had immense advantages in that no time was needed for mapping and, moreover, the map was available on a 25m resolution for the whole of the UK. Thus, the modelling of Buzzard requirements that we conducted locally could be extrapolated onto a wider area beyond the study area, or potentially even beyond the Buzzard's national distribution. As expected, results from using the LCMGB did not differ substantially from the detailed Farm Survey maps but were generally less clear-cut. A third map was of soils (Plate 14; see colour section), produced originally by the British Geological Survey from geological data. This map had some interesting predictive ability for where Buzzards nested, probably because soils affected not only the surface vegetation, but also the activities of burrowing Rabbits.

We were able to compare the map classes that occurred in the cores of the 114 observed autumn home ranges with what would have occurred if the same core outlines were thrown to a random point on the same satellite-based land-cover map 999 times. We conducted this analysis for the smallest cores, because the larger ranges mostly included almost all map classes, whereas hulls round the 50 per cent of locations with closest nearest-neighbour distances averaged only 9ha and were therefore likely to omit the land-cover classes seldom used by Buzzards. The clearest results were for the observed core ranges to include pasture, meadow and deciduous woodland significantly more than randomly placed ranges, and for range cores to exclude human settlements (Figure 4.6).

Unlike the satellite-based LCMGB, which covered the whole of Great Britain, the field survey map could only be used for those Buzzards (about half) that had home ranges entirely within the study area. This made analysis more challenging, but it seemed that this

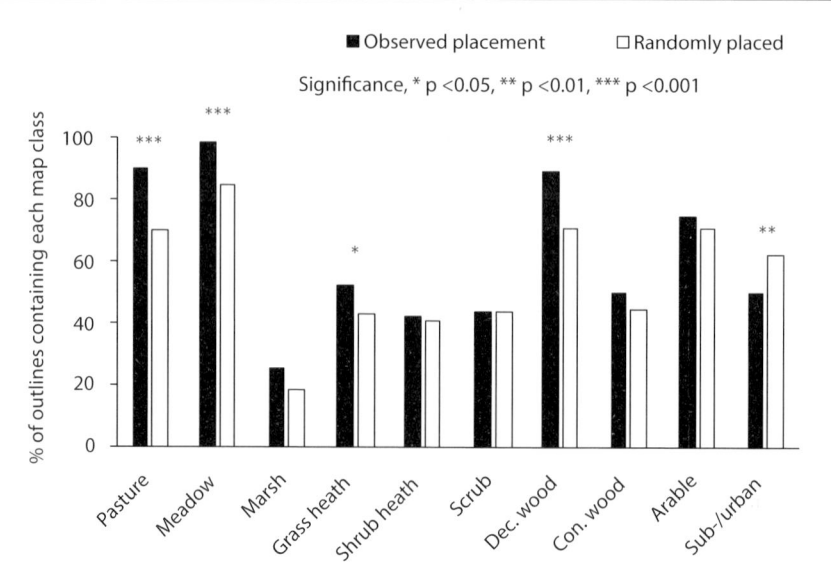

Figure 4.6. *Comparison of land-cover classes within 114 observed home ranges, estimated as hulls around the 50 per cent of locations with closest nearest-neighbour distances in autumn, compared with randomly placed similar outlines on the Land Cover Map of Great Britain (LCMGB).*

subset of birds made less use of heathland and conifers, while also showing especially strong use of arable land in the autumn but not in winter. On the map of soils, the stand-out result was a strong association of ranges with fine loam soils, with coarse loam soils also used disproportionately more than non-loam soils by older Buzzards (Hodder 2001).

Meadows and pastures (such as long-term grass leys) usually have more, and also bigger, earthworms than arable fields, but in autumn the freshly harvested crop fields and ploughed areas provide bare earth on which it is easier for Buzzards to see earthworms, especially when these are freshly displaced from their burrows. Later in the winter, autumn-sown crops will have grown up a little and frosty weather can keep the worms below the surface, which may explain why Buzzards used arable fields less at that time.

Mixed woodland was used disproportionately more than coniferous woodland, so it seemed that Buzzards favoured areas with deciduous trees, although we had few fully deciduous woods. There were usually conifers scattered in among the broadleaved trees, so mixed woodland emerged from the analysis as the most favoured habitat. Outside the breeding season, such habitat is needed for safe roosting at night and shelter from inclement weather during the day. Indeed, there was some evidence that Buzzards spent more time inside woodland during colder weather. Younger birds may also sometimes prefer not to be conspicuous to territorial adults (Hohmann 1994). While we know that deciduous woodland is likely to provide a greater diversity and more abundant potential prey, there was not much evidence that our birds were hunting in the woodland. There are no small birds' nests to raid during autumn/winter, and small mammals are likely to be quite well hidden under the undergrowth and fallen leaves. There will occasionally be food in woodland, such as a sick or starving young squirrel, but hunting tended to be out in the open, or at least along the edge of woodland. So it is conceivable that the use of woodland had more to do with mixed/deciduous woods having better perches along the edges of small copses, and a less restricted view across open fields, at least once the leaves have fallen, compared with the large conifer plantations.

Anthropogenic features such as building and quarries were on the whole avoided. Common Buzzards are starting to appear in some urban areas, but in rural regions they can avoid areas with much human presence. As we saw in Chapter 2, Buzzards can be remarkably tolerant of some human practices, following ploughs and harvesters and attending Red Kite feeding stations. However, they still seem to be wary of human activity that does not present feeding opportunities. Even roads, which can provide roadkill for scavenging, were generally avoided within home ranges (Hodder 2001); this may surprise people who enjoy seeing Common Buzzards standing on roadside telegraph poles, but for the few who tolerated or even favoured such sites, a larger majority of Buzzards were found elsewhere.

More recently, we have developed computer-intensive analysis for estimating a minimum quantity of various land-cover classes required within the range (Kenward 2001). Resource-Area-Dependence Analysis (RADA) works on the assumption that if a resource found within a land-cover class is important and the land cover is spread out in small parcels, then the animal will need to increase its foraging area in order to have enough of it in the home range, whereas if there is sufficient within a more contiguous area, then the home range can be smaller (Figure 4.7). Therefore, by plotting the percentage of land cover within a home range against the home range size (both log transformed), we can detect when the ranges with a high proportion of a particular land cover are small, whereas those with a small proportion are larger. The steeper the decline in range area with increasing abundance of the land-cover class, the stronger is the dependence on that class. Moreover, by also determining which range outline gives the strongest dependence, and hence the area of the land-cover class within the outline, it is possible to estimate the average area of the resource-yielding land class on which the individuals in the population concerned show greatest dependence. When applied to the Buzzard ranges we had collected, there were two important habitats: RADA predicted that Buzzards in Dorset were dependent on 14ha of long grass meadow

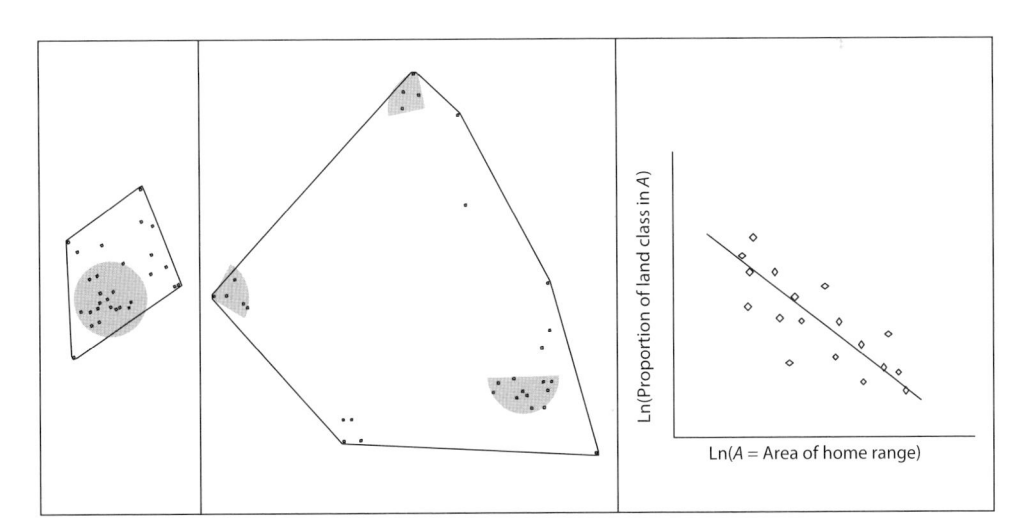

Figure 4.7. *Two home ranges (left and centre), which include a similar area of a land-cover class (grey). In the centre, the land class is in three fragments, so the land-cover class is a much smaller proportion of the home range area. A plot of the area × proportion relationship for many such home ranges is shown on the right.*

and 0.54ha of rough ground. Interestingly, home range size did not show any dependence on arable land, which may be because it was an abundant map class, and therefore never at a limit. In fact, arable land seemed to be a 'poisoned chalice' from which young Buzzards were more likely to disperse later (Kenward *et al.* 2001d), so maybe the lack of dependence shown by the RADA was a better indicator than comparing use with availability of land cover. Unfortunately, there are no equivalent quantitative studies of non-breeders from other areas with different habitat compositions. Nevertheless, the analysis agrees well with what we would expect from their diet in our study area.

Eduardo Moreas Arraut *et al.* (2015) used RADA to create a model of how a population of Common Buzzards would settle on a landscape of particular land-cover classes. The model was built on the basis that a Buzzard needed a patch of wood, or at least a tree in which to roost, plus 14ha of meadow and 0.54ha of rough ground, as just described. Using the LCMGB of our study area, a simulated Buzzard would first search the matrix of 25m × 25m pixels for one classed as woodland. Then the 'virtual Buzzard' would search

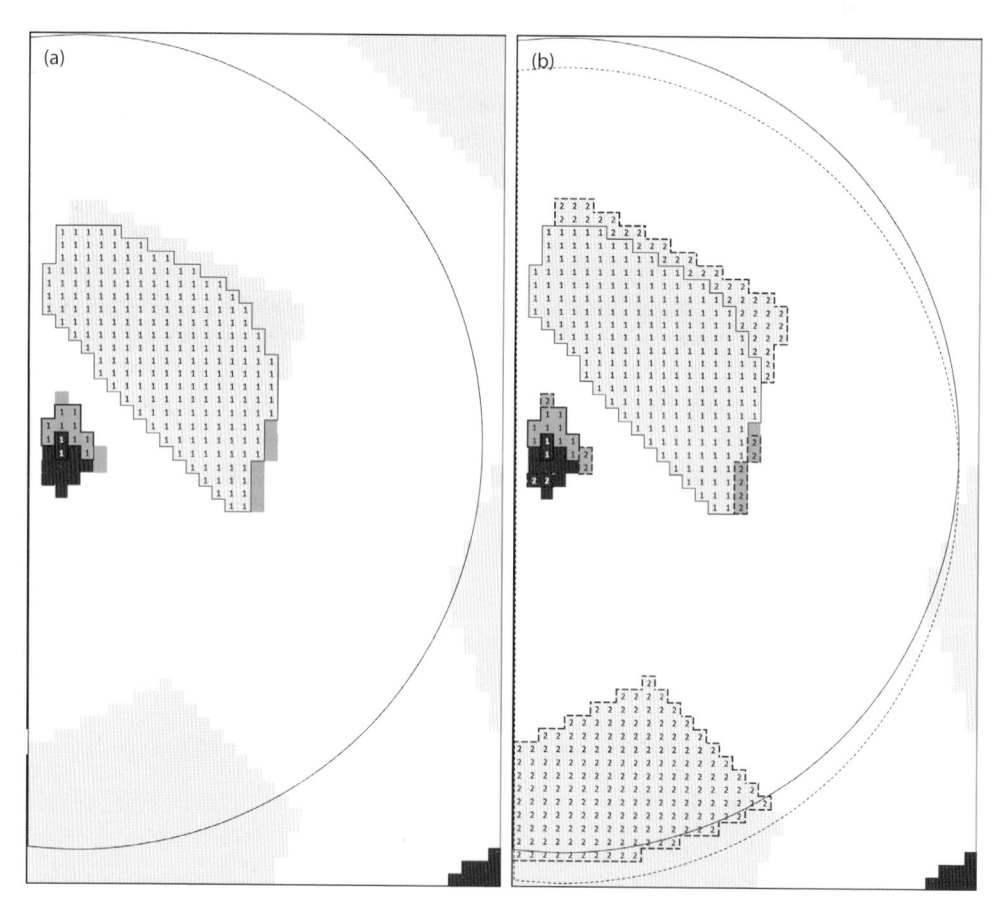

Figure 4.8. *Simulated home range establishment by a virtual Buzzard (1) requiring two pixels of woodland (dark grey), nine pixels of rough ground (mid-grey) and 224 pixels of meadow (pale grey) within a home range not exceeding 800m radius (a), after which one other virtual Buzzard (2) with the same requirements can settle in the same wood (b), with its land-cover requirements and comparable required habitat and home range circumference outlined with dotted lines.*

within a maximum range core radius for a pixel of rough ground. If it could not find one, then that site was abandoned, and another woodland pixel was used as a start-point to search for a rough ground pixel. If an appropriate pixel was found, then all eight of the surrounding pixels were examined, and if they too were rough ground, and not already occupied by another virtual Buzzard, then they were included in the simulated range of the virtual Buzzard. In fact, 1ha contains 16 pixels of 25m × 25m, so only nine rough-ground pixels were needed altogether, but if there were not enough the simulated Buzzard moved to one of the neighbouring rough-ground pixels and repeated the process. On each iteration, the computer routine would check whether the virtual Buzzard had the minimum rough ground. If not, it would continue trying to incorporate rough-ground pixels. If none of the neighbouring pixels were rough ground, then it would search for another within the maximum distance from the wood pixel and start incorporating neighbouring pixels in the same way (Figure 4.8a).

Once sufficient rough ground was incorporated, the process was repeated, looking for and incorporating meadow pixels. Having found enough rough ground and meadow near the woodland, the simulated Buzzard could settle, and a home range could be drawn around the pixels it had occupied, making them unavailable for future simulated Buzzards (Figure 4.8b). Then another Buzzard could start searching, with fewer core woodland and rough-ground pixels to choose from, because of those already taken by previous simulated Buzzards. The process was repeated until no more simulated Buzzards could fit in the area. The whole process was then repeated 100 times, and average values used to compare home ranges sizes with those recorded from real Common Buzzards and to estimate a carrying capacity (where no more Buzzards could squeeze in) for the area.

It was impressive just how much the pattern of simulated Buzzard ranges resembled the pattern of home ranges actually recorded from the radio-tagged Common Buzzards in the same landscape. You can see an example in Figure 4.9. Of course, the values used for the minimum rough ground and meadow of simulated Buzzards had been obtained from real birds in the same area. A better test would be to model new areas and then record the pattern and density of Buzzards there. However, the fact that their ranges could mimic so closely based on so few habitats made it likely that these habitats were indeed the most important, and that RADA was really identifying minimum amounts. Moreover, the simulated ranges not only followed the pattern of habitats, but they were also very similar in size and shape to real ranges. The core areas, up to 85 per cent of the modelled home ranges, were a very good match for the core areas of real home ranges, which were approximately 56ha. If values such as the search radius or the amount of a habitat that was required were altered in a sensitivity analysis, then the factors that made a bigger difference to range size were associated with the meadow, rather than rough ground. This is not surprising, because 14ha of meadow corresponded to 25 per cent of the area in a 56ha range core, whereas the rough ground only amounted to 1 per cent. The power of the model is that it can be applied to other areas to predict densities, both where Buzzards do occur and where they don't, which would help forecast the eventual population of a colonising or restored species that requires a variety of habitats in a complex landscape.

Another interesting aspect of the RADA analysis was that it indicated some divergence in the foraging strategies of Buzzards. For the population of tagged birds as a whole, the placement of ranges showed significant avoidance of suburban areas, which were mainly large gardens in settlements of 50–500 people and one town of about 3,000 people.

The Common Buzzard

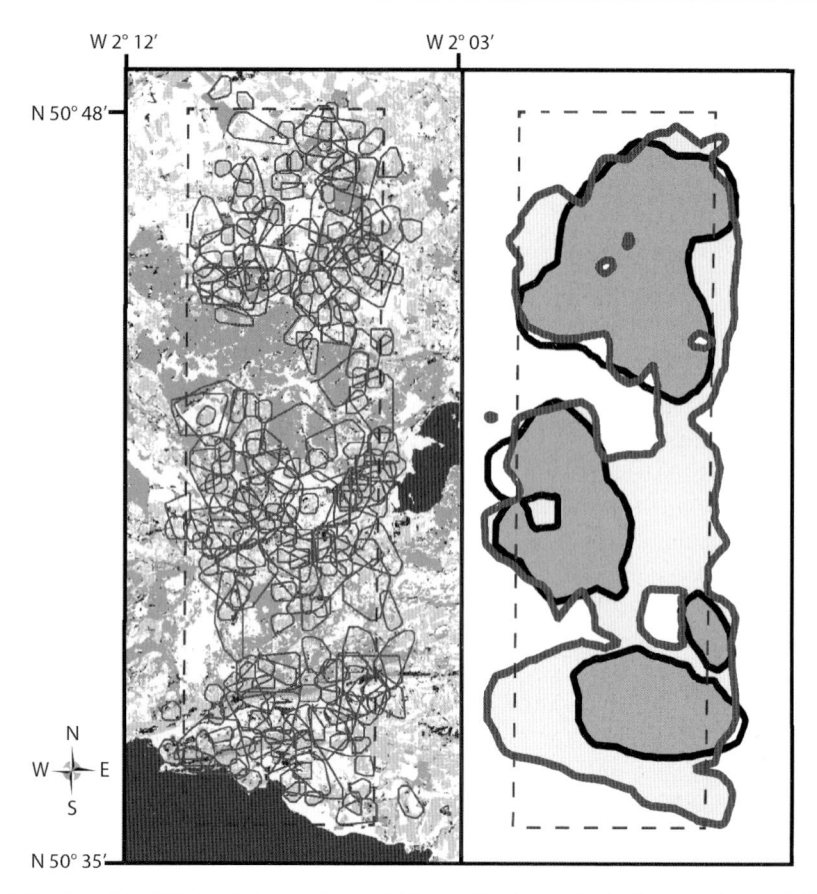

Figure 4.9. *One of the 100 simulations used to model Buzzard settlement in the Dorset study area (left) and the expected 95 per cent settlement boundary from the 100 simulations (right, pale grey), showing also a similar settlement boundary for the observed Buzzards (dark grey). The maps are 10km wide and the dotted study area 6km wide. Kindly drawn by Eduardo Arraut.*

However, about one-fifth of the 114 autumn ranges included the suburban habitat within the cores, and in those cases the smallest ranges were those with the highest proportion of suburban habitat. Therefore, a small proportion of the tagged birds were showing a dependence on this habitat, which would have provided worms on lawns and small paddocks, as well as small rodents associated with houses (Kenward *et al.* in review).

On the Continent, it can be especially difficult to investigate which habitats are favoured by resident non-breeders in autumn and winter, due to influxes of Common Buzzards from areas with more severe weather. The presence of wintering birds is not consistent from year to year, and appears to be dependent on the harshness of the weather at the breeding grounds, or where food is most available to migrants, and possibly on the way the winds are blowing in the intervening regions. With this type of migratory irruption, birds do not necessarily settle for any length of time in one place. Thus, the concept of a 'home range' is not easy to apply with seasonally nomadic birds. Rui Lourenço (2009) found that there were many Buzzards in Portuguese rice fields in November, but their numbers steadily declined through to January. Even if it were possible to trap and radio-tag a non-resident

Common Buzzard in a wintering area, there would be a high chance that it would not stay in the area to be tracked, and the costs of keeping track of it further afield by manual radio-tracking are prohibitive. Again, GPS tags will give a more complete picture when they are applied to Common Buzzards in such areas. Without tracking technology, other techniques have been used to assess how Buzzards use different habitats. For example, where transects have been conducted during the winter, Common Buzzards have been very elastic in their habitat use (Kasprzykowski & Rzępała 2002) and can occur in good numbers in agricultural land (Wuczyński 2005, Nikolov *et al.* 2006, Lourenço 2009). When there is a diversity of habitats, they usually seem attracted to grassland (Schindler *et al.* 2012).

Breeding season habitat

Having seen what habitats Common Buzzards favour during the non-breeding periods, what do they use most when trying to raise young? Habitats used in the breeding season not only have to provide the food required to feed the breeding pair and their growing brood, but also to deliver sufficient protection for the eggs and young in the nest from cold and wet weather, or from other animals that are keen to eat or destroy them.

We can assess where Common Buzzards breed at several scales. The macroscale of the whole species will be covered in Chapter 10 on demography, and the microscale of which tree and where the nest is built is covered in Chapter 6, concerned with nesting. Here, we look at the mesoscale or landscape scale, presenting the general habitat needed around the nest to sustain the breeding pair and their nestlings.

As with other aspects of Buzzard behaviour, it is worth considering how the information was gathered. Crucially, it is important to understand that it is difficult to find Common Buzzard nests. Therefore, many projects have used territorial display flights to establish that there are pairs attempting to breed in particular woodlands. Additionally, Common Buzzards have a habit of maintaining more than one nest, with an obviously redecorated but inactive spare nest, distracting field-workers from another better-concealed site in which eggs are laid. Even without undetected early failures, to confirm breeding it's necessary to see chicks in the nest or at least the characteristic 'whitewash' of faeces that are squirted out from the nest (Plate 15; see colour section). If the nest is in woodland with little lower canopy or ground vegetation, such as a Scots Pine *Pinus sylvestris* plantation, the white faeces are obvious on the forest floor. Unfortunately, for many other woods, under-canopy branches or climbing plants can hide the nest from view, and also catch the whitewash so that it's not obvious from the ground.

This means that it is sometimes impossible to find all active nests. The proportion of nests found will depend on the topography, habitat, tree structure, other vegetation, prey abundance and behaviour of individual Buzzards and the skill and experience of the searcher in the area. Moreover, as well as not using individual nests consistently, pairs don't necessarily breed every year. There is a particularly relevant article exploring this by Asko Lõhmus (Lõhmus 2003), which recommends that one should not take one year in isolation to study sites used for nesting. The association of nests with particular good foraging habitats can be more apparent in low prey years, whereas in years of abundant food the placement of nests may be less constrained. So, single-year studies may not be conducted in a year when site-use was typical. On the other hand, if studies across years consider only each site, pooling data from different years and not also taking frequency of use into account, then the favouring of one kind of nest habitat in particular years could be masked by years when habitat use was more varied than usual due to the abundance of food everywhere.

For these reasons, all comparisons between areas and habitats will be imperfect. In our case, radio-tags on breeders helped to find nests, especially of first-time breeders and on occasion in sites where we would not otherwise have looked (such as a lone oak *Quercus* sp. in a hedgerow). Furthermore, our searches of woodland were fairly thorough and backed up by observations of territorial display and young birds calling during an annual mop-up process. Therefore, we feel that there is sufficient consistency between ours and other multi-year studies to draw general conclusions about breeding season habitat while yet again appreciating the diversity and adaptability of Common Buzzards.

While territories are not necessarily the same as home ranges, they are all we have as a representation of the ranging behaviour of breeding Common Buzzards in the breeding season. A review that sought to define a typical territory size for the Common Buzzard from historic literature, based on observations of displaying birds, would probably tend to propose around 100ha, or 1km^2. However, the size of the territory that breeding Buzzards defend is very variable indeed, even within the same landscape. For example, Dare (2015) worked in two areas of the UK (Appendix 3). In lowland Devon, where it was warmer and the vegetation more lush, the average territory size was 142ha, whereas in more mountainous north Wales it was 220ha, almost twice as much. However, there was a lot of variation in size in both study areas (53–223ha in Devon and 60–340ha in Wales), and therefore plenty of overlap in range size between sites. It was far from a simple story. The largest territories were found at the highest altitudes, especially when there were few if any trees. Similarly, from a separate study, an increase in distance between nests was significantly correlated with increasing altitude for upland areas of Wales, according to Ian Newton *et al.* (1982). These independent results are consistent with the general hypothesis that there would have been less abundant prey on the upland moors, so the birds need to cover a larger area for sufficient food. However, comparisons are even more complicated than for home ranges during winter, because observations are based on when birds were conspicuous, and hence may be influenced by both territorial behaviour and the open nature of the land cover.

Nevertheless, the tendency for home ranges of breeders to be smaller in agricultural areas is corroborated by studies outside the UK. In the Netherlands, Hans Van Gasteren *et al.* (2014) found home ranges of three breeding GPS-tagged Common Buzzards near a small airport were very similar to those in our study; 46.6±24.0ha. The ranges appeared to be defended territories and only overlapped slightly. Another GPS study in Estonia shows the home ranges of three breeding Common Buzzards were much bigger, covering 2–3km over varying landscapes (Väli *et al.* 2015). They have far more locations than we collected with radio-tags, and this helps to emphasise the time these Buzzards spent along the habitat edges.

As another component of the habitat, there may be other Common Buzzards that influence where an individual Buzzard ranges. There is now very good evidence from Robin Prytherch (2013) that the size of breeding territories decreases as the number of adult Buzzards increases in a study area. When he started his study near Bristol (Appendix 3) in 1987 the average territory size was about 130ha, a decade later it was near 90ha, and 20 years later in 2007 it had reduced to 60ha. Some territories ended up being as tiny as 20ha. This has implications when considering the habitats that Buzzards require for breeding.

One implication is that those birds with the smallest territory will probably be going outside their defended territory to feed, like Red Kites that have a small defended territory around the nest, but much larger communal areas where they can feed (Carter & Grice 2000). Another implication is that, unlike the foraging ranges defined by where an individual travels to extract food, the boundaries of a territory may be influenced by features in the landscape that are more related to territorial defence

It is not just the presence of other Common Buzzards that may affect areas used during breeding; other raptors can have quite an effect, too. Achim Kostrzewa (1991) investigated interference competition between raptors nesting in the same area, specifically Common Buzzards, European Honey-buzzards and Northern Goshawks. If Goshawks started nesting in an existing Common Buzzard territory, then although some Buzzard pairs tried to continue, they often ended up moving their nest further from the Goshawk nest or abandoning completely. A contrasting example comes from Tenerife, where Benehara Rodríguez *et al.* (2010) found Common Buzzards were nesting particularly close to Barbary Falcon *Falco pelegrinoides* nests. This was probably because of the scarcity of nesting places, with both species trying to use the same crags, but it could also have been due to Buzzards kleptoparasitising the falcons. So, the social behaviour and the presence of other raptors can influence where Buzzards nest, perhaps concealing other relationships with the landscape and vegetation.

In our own study area, Kathy Hodder (2001) looked at what was around the nest at different scales. As we were concentrating on the movement of juveniles, we did not have the time to map territorial display or the home ranges in summer of the few tagged breeders in our area. Therefore, without knowledge of the territory boundary, Kathy Hodder first assessed the habitats in the field survey and LCMGB within 250m, 500m, 1km and 2km of each nest (Plate 13; see colour section), and compared them with what was available in the whole study area (Figure 4.10). Rather like the winter home ranges, mixed woodland was favoured in buffers of all radii around nests, but especially close to the nest. Rough vegetation, arable land and grassland were weakly avoided, as were grassland and roads near the nest. Conifer plantations tended to be avoided, although in mixed woodland conifers were often used for nesting (Chapter 6). Nests also tended to have less open heathland and buildings nearby, land covers that were also avoided in the winter ranges, but unlike the autumn ranges the nests were placed with arable land under-represented nearby. However, this result may have been influenced by the nature of the arable crops in our area being mainly cereals (including maize); it is known that proximity to Oilseed Rape *Brassica napus* can have benefits for Common Buzzards (Panek & Hušek 2014), and Alfalfa *Medicago sativa* for other raptors (Heroldová *et al.* 2007).

In another analysis, Kathy Hodder compared land covers in the same 250m, 500m, 1km and 2km radii around the nest with a density index for all nests within 2km. Additional habitat variables were used in the analysis, including Rabbit burrow density and the length of the line surrounding the various habitat patches. Unsurprisingly, considering what we know about their diet, Rabbit burrow density was very significantly related to nest density, most strongly within 1km of the nest. A good food source was clearly important, whatever the vegetation. Individual land covers were not so obviously correlated with nest density unless they were combined into composite habitat groups, but boundary lengths of grass ley, arable and grassland were important. In fact, the total boundary length around all habitats and roads was significantly correlated with nest density. This may well have been

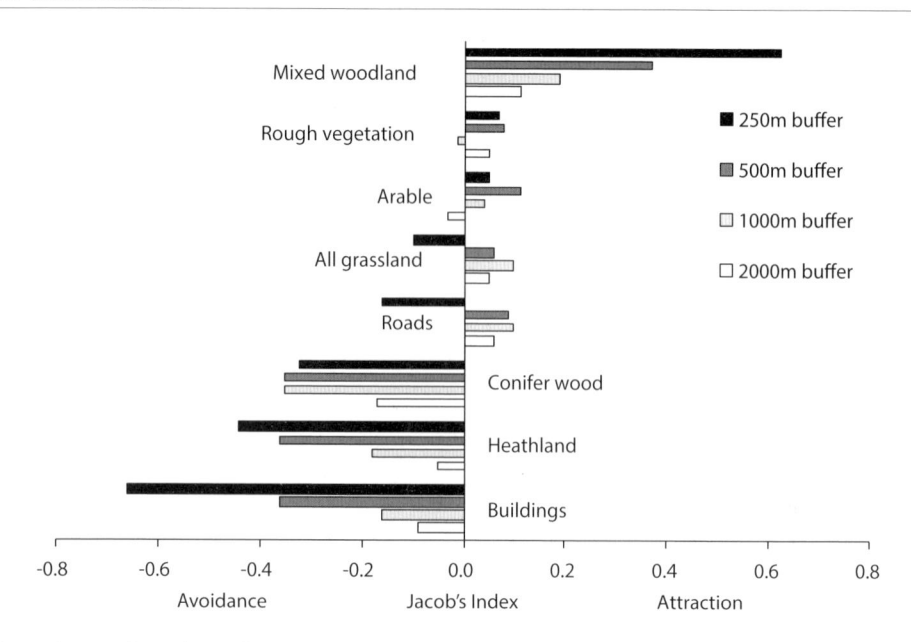

Figure 4.10. *The tendency of 35 nests to be placed in particular field survey map classes is shown by values of -1 to +1 of Jacob's Index. Positive values indicate that a higher percentage of a particular map class was found within a circular buffer of the nests, compared with the percentage of that map class within the study area. Negative values indicate under-representation of that map class around nests. Data from Hodder (2001).*

because, in cultivated areas, Rabbits tend to maintain burrows in the field boundaries, or because boundaries provided perches, but it may also have been an indication that lots of small patches of habitats were better than large areas of the same habitat. Woodland boundary length was not particularly correlated with nest density, despite the tendency of Common Buzzards to nest near woodland edge. This suggests that although being near such edges probably aided nest access through the trees, woodland edge was abundant enough not to influence nest density in our area. Another factor strongly related to nest density was the presence of loam soils (Plate 14; see colour section), especially fine loam within 2km of nests, while conifer and heathland map classes were associated with soils that were sandy and had the lowest presence of nests.

The combination of mature mixed woodlands with the pastoral grazing that was found in our region also seems optimal for nesting Buzzards in other studies (Rooney 2013). While territories may include lowland heathland or upland heather moors in some cases, such territories usually include other habitats. Nesting density tends to be a lot lower in areas dominated by heathland, where there is little biodiversity and low prey abundance (Austin & Houston 1997, Dare 2015). Although individual Common Buzzards can be reasonably tolerant of humans, past studies have shown that Buzzards tend to build nests further from anthropogenic features such as paved roads, buildings, orchards and other plantations of unfavourable trees (Penteriani & Faivre 1997, Bustamante & Seoane 2004, Löhmus 2005, Sergio *et al.* 2005, Zuberogoitia *et al.* 2006). There are, of course, exceptions. From our own study, in general there was an avoidance of buildings, but we did track down a radio-tagged Buzzard that nested right in the middle of an active paintball camp. This was out of our study site, so we did not know how other nests were distributed in that area. The nesting process may have commenced in the season before the start of the regular noisy gatherings of running, shouting paintballers. Nevertheless, the birds were successful in their first year of breeding and stayed at least one more year to breed again, before we were unable to continue tracking. In another part of our study area, Buzzards are now breeding in large rural gardens; we will discuss how they may adapt to suburbia in Chapter 11.

Some studies have investigated the 'dampness' of the woodland, which of course affects the vegetation. This is a little difficult to compare between sites, because there is no quantitative measure of dampness – it is what the researcher considers damp, and that could be lush vegetation, abundance of standing water or perhaps seasonal flooding. There doesn't seem to be a general rule, with some finding association with these wet areas (Jędrzejewski *et al.* 1988) and others avoidance of them (Lõhmus 2005). This is another indication that the tree species and the ground within the woodland is unimportant; the wood is just providing a structure in which to place a large nest while food is gathered outside.

So, some habitats are more important than others for nesting Common Buzzards but, as with the winter ranges, an important factor in the landscape is habitat heterogeneity. Buzzards like to nest where different land covers meet, rather than in monocultures. As Colin Tubbs wrote in his monograph: 'Diversity appears to be the key feature'. The importance of habitat heterogeneity is mentioned by others, too (Picozzi & Weir 1974, Dare 1998), which is why Austin & Houston (1997) and Hodder (2001) specifically measured the boundary length of areas around nests compared with areas without nests in the very different landscapes of Scotland and lowland England. Where you have smaller parcels of habitats, the boundary length between them increases, and Buzzards were more likely to nest in such areas. In Spain, a study by José Sánchez-Zapata and José Calvo (1999) and another by Bustamante & Seoane (2004) found that the length of forest edge was an important feature for determining where Common Buzzards were likely to nest. Newton *et al.* (1982) found Buzzard nests at higher density in mixed farmland in central Wales than hill sheep farming country or forestry plantations, which have much larger land compartments. In Estonia, more diverse landscapes had better productivity in poor vole years than more homogenous landscapes, perhaps providing a clue to the driving factor in this choice (Lõhmus 2003). Lastly, Rooney (2013) has found heterogeneity to be the single biggest explanatory factor for the rate at which Buzzards are colonising the different areas of Ireland.

As well as the diversity of habitats, this association with boundary lengths is also worth considering not just from the viewpoint of having lots of small patches, but also in terms of what the boundary itself might be providing. For example, a hedgerow, fence line or woodland edge provides perches above the surrounding land from which to hunt. These boundaries can also have areas of bare ground or shorter vegetation, for example, a track-side or verge, and while that doesn't give good habitat for the prey, it provides areas where small mammals, lizards and snakes are more vulnerable to a hunting Buzzard as they cross the more open areas, in the same way that arable fields make earthworms more visible. The transition habitat can also be important, for example, providing areas of weeds between crops, or rougher grassland that increases the abundance of invertebrates and small mammals. Or it may not be the actual habitat in the transition area that increases the prey, but the fact that the prey species also become more abundant in areas with more than one habitat to sustain them.

Some studies report a maximum altitude for Common Buzzards nesting in their area, usually either due to a lack of nesting sites, or of food, as the vegetation turns into moorland or bare rock. Arianna Aradis *et al.* (2012) suggest that there is a big reduction above 1,200m in Italy, although we know that sometimes Common Buzzards can nest at 1,550m above sea level (Penteriani & Faivre 1997). Generally, nest density reduces at higher altitudes (Newton *et al.* 1982, Weir & Picozzi 1983, Austin *et al.* 1996, Penteriani & Faivre 1997, Selås 2001). Studies in areas where there is a diversity in relief tend to find that Buzzard nests are most likely in those areas with more relief (Penteriani & Faivre 1997, Selås 1997, Sergio *et al.* 2002), maybe because they can find places sheltered from the full force of the elements (e.g. in ravines and other nooks and crannies). On the other hand, the most rugged areas may also provide updraughts that make flying more efficient, and it could be this that the Buzzards find useful. Of course, we should not overlook the alternative explanation for an apparent increase in abundance, namely that conditions are making the Buzzards easier to see and count.

When there are steep slopes and cliffs the aspect can be important for temperature control. Where this has been recorded in warm southern Europe, Common Buzzards tend to nest on slopes or crags facing north-east (Cerasoli & Penteriani 1996, Penteriani & Faivre 1997, Zuberogoitia *et al.* 2006). This aspect catches the early morning sunshine after the cooler night, but avoids the sun when it gets too hot later in the day. Vincenzo Penteriani & Bruno Faivre (1997) found that the nests on these slopes were also in the densest and coolest canopy of large Beech *Fagus sylvatica* trees. Alternatively, in more northerly latitudes, Selås (1997) found that nests in colder Norway were oriented more south-east, so would have warmed up most immediately and for longer after sunrise. There, Buzzards can lay eggs while the weather is still freezing, so it is an advantage to catch the sun's warmth as soon after the cold night as possible.

The strongest theme to emerge from all the studies is that whether foraging to survive during the winter, or settling in a breeding territory, Common Buzzards are found in areas with small plots of more varied habitat, probably because these provide the best hunting opportunities. In the next chapter we will look more at how those areas are defended.

Conclusions

1. Home range sizes are very variable but are typically 100ha (1km^2) or less in farmed areas with woodland, smaller than one might expect from such a large bird.
2. Use of the home range is patchy, with maybe as little as 10 per cent of the total range being used intensively for foraging during the non-breeding period.
3. During autumn and winter the home range is tightly focused on the hunting and roosting areas. In southern England that can be on the grassland and arable fields where earthworms are the primary prey, and deciduous or mixed woodland for roosting.
4. Juvenile Common Buzzards tend to be found in arable areas for their first autumn. However, the size of home ranges in autumn, which is smaller for juveniles than older Buzzards, depends on the content of grassland and rough ground.
5. A model used to predict the size, shape and placement of Buzzard ranges, based on grassland and rough ground content, matched home ranges recorded from wild Buzzards and the overall density of Buzzards.

6. While breeding, Common Buzzards defend territories. These contain what is needed to feed the territorial pair and the fledglings they hope to raise. Instead of just being focused on feeding, these territories also need suitable nesting sites.
7. Mature mixed woodland, together with nearby pastoral grazing, appears to be optimal nesting habitat in Dorset, where Rabbits are a major food for nestlings.
8. Whether in the breeding season or not, the areas that Common Buzzards use tend to have small patches of different habitats. They often use the edges between habitats, for instance by perching in a tree at the edge of a woodland overlooking fields.
9. Heathland and large areas of coniferous monoculture are avoided for foraging and nesting.
10. While Common Buzzards can use anthropogenic landscapes very effectively, they have tended to avoid direct association with buildings; however, a few Buzzards in Dorset have been found foraging and breeding in large gardens.

CHAPTER 5
Territoriality and nest defence

Although it is hard to spot the most exciting hunting behaviours in Common Buzzards, their spectacular courtship and territorial displays are there for all to see. They positively revel in showing off, and will shout about it to attract more attention. It's difficult for birds to defend a borderless sky, so they need to let everyone know that they own a particular airspace and all that lies under it. Admittedly, not all humans notice, even those that love wildlife. The display may be high and cover great distances quickly, so binoculars are helpful for a full appreciation and enjoyment. However, once you catch sight of the aerial roller-coaster or the vertical plummeting to the nest, you are instantly hooked and will search to see it again and again.

Having investigated what Common Buzzards eat, how they catch their prey and the habitats they use in the winter and for nesting, we can now describe how they defend these resources.

Territoriality

Ever since people began surveying raptor populations, by systematic recording of nests for falconry nearly 1,000 years ago in the Domesday Book (Yalden 1987), it has been apparent that many species of raptor space their nests regularly. Exceptionally, there are some colonial nesting species and a few rare instances of species that sometimes nest in colonies and in other areas have single nests. For instance, Sergio (Sergio *et al.* 2003) showed that whether a population of Black Kites nests in colonies or not depends on the predation risk, in his case the presence of Eurasian Eagle Owls *Bubo bubo*. Of course, even with the regular nest-spacing of most raptors, there are some habitats in which particular species would not be expected to breed. In this case, the nests appear clumped into areas of good habitat, but more locally within that good habitat the nests are spaced more regularly.

Raptor biologists study nest-spacing by recording Nearest Neighbour Distances (NND). As you would expect from the big variation in size of home ranges and territories (Chapter 4), there can be a considerable difference between the minimum and maximum NND for Buzzards. For example Picozzi and Weir (1974) recorded 0.5–2.7km, varying with year, and we had values of 0.5km to 4.2km for the 203 clutches laid during 1990–97 in Dorset. Given a set of nest locations, software can calculate a 'G' statistic, which indicates how regularly nests are spaced in a landscape. G is calculated as the ratio between the geometric and the arithmetic mean of the squared NND between used nests. Territorial birds, with regularly spaced nests, have G values higher than 0.65, whereas communal species that nest in colonies, or birds with restricted nesting sites and are clumped, have low G values nearer to zero (Brown 1975). As would be expected from their overt territorial behaviour, the minimum G value reported for Buzzards is 0.65, in Tenerife (Rodríguez et al. 2010), and they have been recorded with a G as high as 0.96 in Italy (Penteriani & Faivre 1997). From our data, using the 203 nests from 1990–1997, our Common Buzzards had a G value of 0.74, which is comparable with other records from the UK (0.67–0.75, Dare & Barry 1990) and Germany (0.76–0.77, Kostrzewa 1991), and indicates moderately regular spacing. However, was this more regular than expected by chance?

To test that possibility, we used the R statistic of Clark and Evans (1954), which divides the observed average of NNDs with the value expected if the observed number of nests is expressed as a density in the study area. R is less than 1.0 if a predominance of short NNDs indicates clumping, and greater than 1.0 if distances tend to be longer than from random spacing. In our case, the distances tended to be longer than 1.18km expected by chance (Figure 5.1), but we also noticed that R (and G) varied between years according to how many nests (between 21–30) were active in the study area. Indeed, in years with fewer nests the regular spacing lost its statistical significance (Figure 5.2). It was only with the occupancy of all sites that the regularity of spacing became very strong. It is likely that the unsuccessful nests still had a territorial pair keeping the regular distribution, although they were not

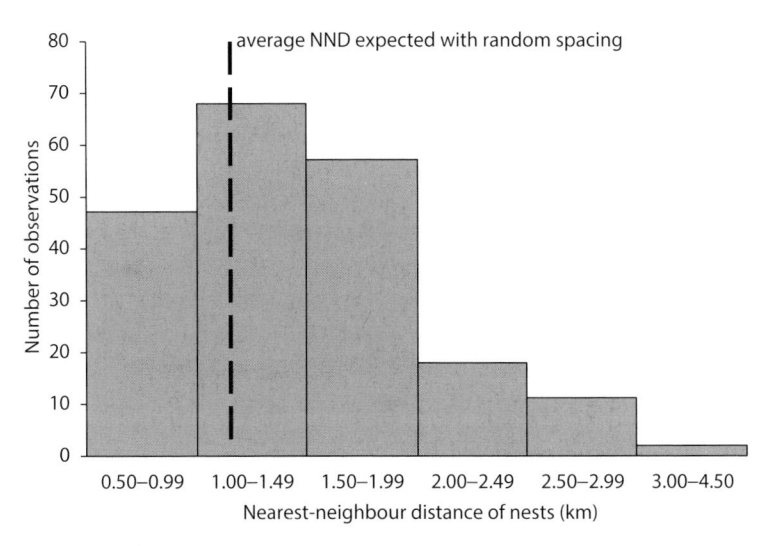

Figure 5.1. *The nearest-neighbour distances of occupied Buzzard nests in the Dorset study area were longer than expected if spacing had been random (shown by the vertical dotted line), indicating that spacing tended to be regular.*

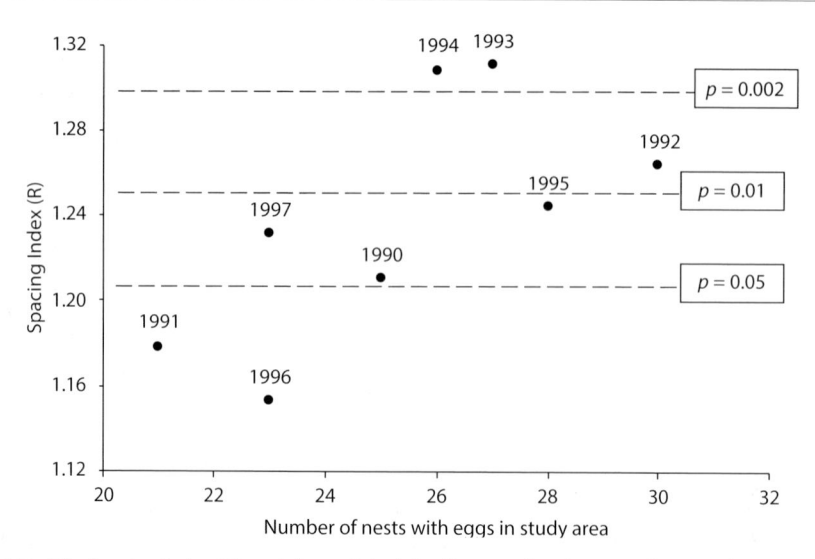

Figure 5.2. *The Spacing Index (R) and the statistical significance of its departure from random spacing tended to be greater in years when nest occupancy was high in Dorset.*

included in the analysis because our criterion for nesting was the laying of eggs. It would be interesting to see whether R values (or G where area is not defined) become higher as Buzzard density increases beyond the 4–5km^2 per pair as seen in our Dorset study area during the 1990s.

Regular spacing could be due to optimal division of resources in a placid way, like waders spacing themselves on a mudflat recently exposed as the tide recedes. In the case of Common Buzzards, which not only space their nests but also call at and chase away other Buzzards, it is very apparent that nesting pairs hold defended territories. Of course, habitat may sometimes disrupt this pattern for individual pairs. For example, there may be an isolated wood on a ridge top, with surrounding land able to support two pairs of Buzzards. In this case, they may nest on either side of the wood and use different valleys, but their nests are clumped together due to a lack of other good potential nest-sites. As a population, though, we can expect them to be as regularly spaced as the landscape will allow them.

Where Common Buzzards have been studied visually, it has sometimes been possible to overlook several nest-sites and the associated territories from a high vantage point. From there, fieldworkers have observed the local Buzzards to identify prominent perches where the territory holders stand to mark their presence and look for intruders. At times, territory owners need to chase an intruder, or meet the neighbour seemingly 'to confirm' where their adjoining territorial boundaries abut. By joining the dots of where neighbours meet and display to each other and where recognisable individuals restrict their activities, researchers have published approximate territorial boundaries for studies in southern England (e.g. Figure 5.3) (Tubbs 1974, Dare 1998, Prytherch 2013), Scotland (Weir & Picozzi 1983, Halley 1993), and central France (Nore *et al.* 1992). It is a surprise that there are not more studies with mapped territories, although this is likely due to how time consuming this is. Tubbs (1974) noted how he often had to spend over six hours at a stretch watching from a suitable vantage point, and it was not uncommon to spend that time without hearing or seeing a Buzzard. He explained how the conditions have to be right for soaring and was

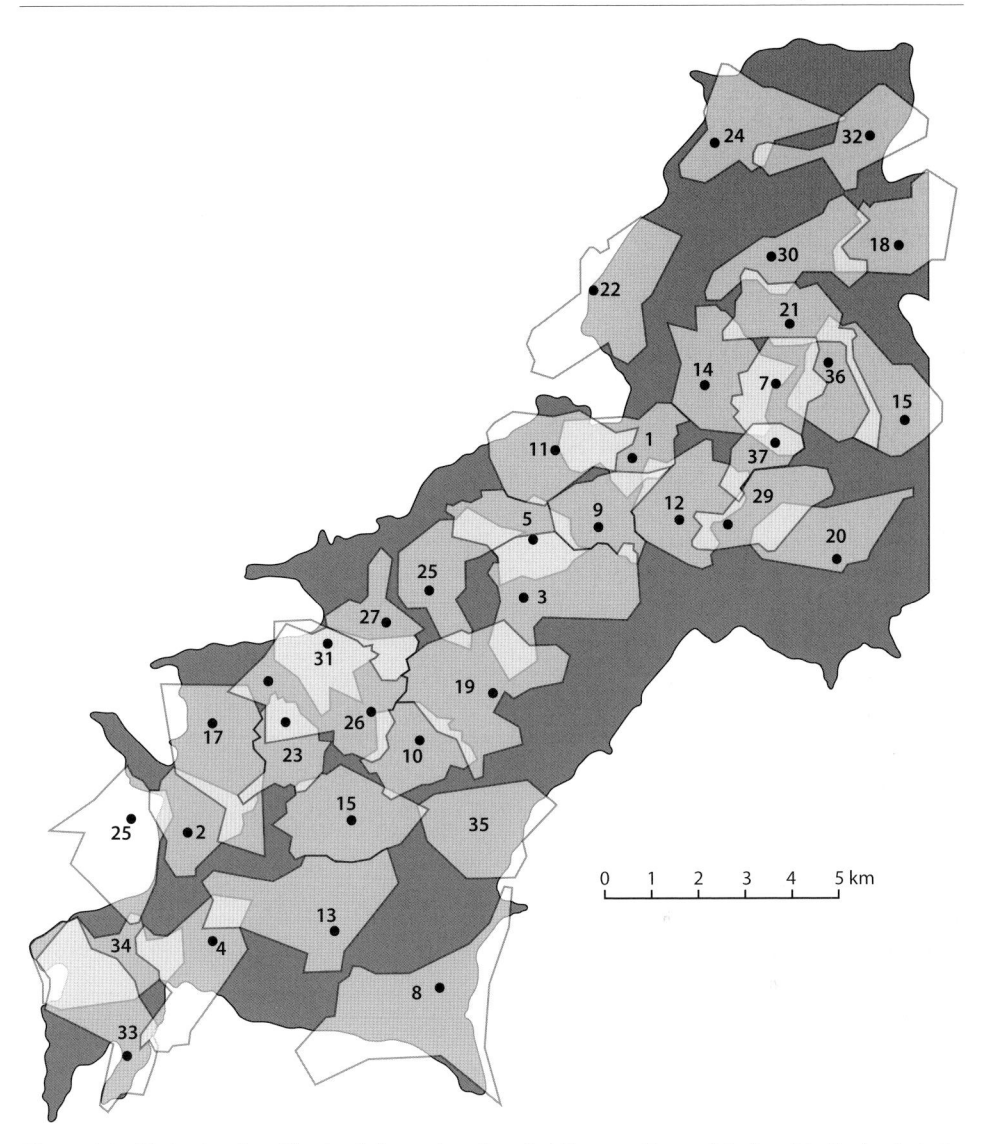

Figure 5.3. *Territories plotted by visual observation of marked Common Buzzards in Speyside, Scotland, during 1969–72. Dots show nests in numbered territories, for which a change in shading depicts areas of overlap. From Weir & Picozzi (1983), reproduced with kind permission from* British Birds.

honest in noting 'An element of subjective judgement in defining territory boundaries was thus inevitable'.

Territorial boundaries tend to follow hilltops, woodland edges and other features (Dare 1998). The highest ground tends to be towards the edge of the territory, almost never in the centre. So, typically, a territory would cover a valley rather than hang both sides of a ridge. This may be because Common Buzzards spend much of their time perched (therefore near ground level). Perches high on the landscape at the edge of a ridge give them the best possible view of the territory, in a good position to spot any intruders that need confronting (Prytherch 2013). It is also important to remember that our perception depends on where the human observers interpret the lines between conflict sites, rather than necessarily what the Buzzard pair considers their territory. If observers interpret territories from known aerial confrontation areas, this may also explain why the high ground is on the edge of the territory. When watching aerial displays and combats, it is apparent that there is a distinct advantage to being the higher bird in a contest. The bird above can dive on the bird below and has its talons pointing down to the adversary, whereas the bird below can only defend by flipping over to bare its own talons momentarily, before having to right itself and fly on. We also know from previous chapters that Common Buzzards have evolved to use updraughts efficiently, rather than to be power-fliers. Therefore, if a Buzzard is to gain height quickly it will place itself in whatever updraught is available. In flat areas, these may be thermals, which are columns of rising warm air, but Buzzard territories are often in hilly areas where thermals are less stable. In these landscapes, most of the updraught is orographic, that is, it comes from the wind being redirected upwards when it reaches a hill or mountainside. Moreover, early in the breeding season, when territories are being vigorously defended, the sun's radiation is often not strong enough to create thermals in Northern Europe. Buzzards may therefore start a display with a strong powerful flight from a lookout perch, but they will quickly speed to where they can gain height fastest, and in many cases that will be the ridge of a cliff or hill. Thus, Common Buzzards fight their boundaries over updraughts, which occur over ridges. Because of the advantage of using all the free lift energy available, the combat arena may be away from what the Buzzards are defending. So a Buzzard may perch along the edge of what it is defending but then move out to meet an intruder and have the dispute on a better flying ground. Marina Cerasoli & Vincenzo Penteriani (1996) found common 'rendezvous points' were associated with defended nests; however, they were also in areas with the best updraughts, and the average distance was nearly 800m from the nearest nest.

With the boundaries that are drawn (and drawing is as good as any other estimate we are going to get), typically these territories are around 100ha, but can be as large as 200ha or as small as 60ha (Figure 5.4); they are usually stable over time if the population remains the same. Dare (1998) found territories were consistent between years and observed that boundaries hadn't changed much between his original study (1956–1958) and a decade later. The territorial patterns were even recognisable when he revisited in the 1990s. Likewise, in the late 1980s, Duncan Halley (1993) re-examined a district that Weir & Picozzi had studied in 1968–1972, using the same techniques so that there was the best chance of a robust comparison. There had been a 57 per cent increase in the number of territories, but no significant change in territory size. The new territories were mainly due to colonisation of new ground. Tubbs (1974) noted that not only were the same boundaries defended each year, but the territory owners even used the same boundary marker trees.

More recently, long-term studies of territories have suggested that vacant areas have been filled and, with no new ground to colonise, more Common Buzzards have fitted into the same area. Buzzard densities seem to have hit new highs and territory size has declined, showing dependence on density. For example, Julian Driver & Peter Dare (2009) working in Snowdonia, North Wales, saw the density of breeding pairs double from when they started, along with shrinking territory sizes. Prytherch (2013), who has studied Common Buzzards colonising an area near Bristol from a very low density, has seen many large territories contract as the number of breeding pairs increased eight-fold in the period 1982–2012. The plasticity of a territory appears to depend on the original starting density of

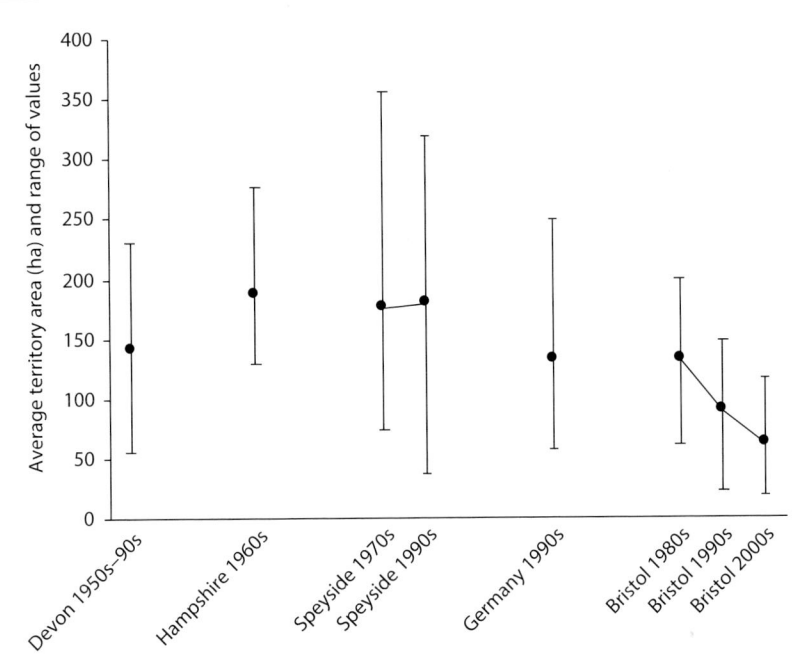

Figure 5.4. *Breeding territory sizes estimated from visual records of displaying Buzzards in five study areas, showing values (mean and range) in different time periods for Speyside and Bristol. Data sources in Appendix 2.*

breeders, how many more Common Buzzards and other territory-holding raptors arrive, and what the landscape can hold. At some density, which will probably depend on the food and nesting opportunities, the territories can't get any smaller and no more birds can breed. Before this density is reached, there may be effects on breeding performance. Prytherch, who has one of the longest running datasets on Common Buzzard territory use, has seen the productivity reduce from 1.89 to only 0.56 chicks per pair (Prytherch 2013). This could be partly due to the division of a limited food resource and partly to poorer territories being adopted because the best areas are already taken. It could also be a result of interference competition, through the extra hassle created by closer neighbours and non-breeding birds that distracts parents from the job of feeding young. It is likely to be a combination of all these factors.

Watching the recent recolonisation of Common Buzzards in Suffolk, where breeding began in 2005, Dare (2009) thought that overt territorial behaviour was less common in the early years when the density was low and the birds spread thinly. Having watched the dramatic change in territory size over time in his area near Bristol, Prytherch (2013) has raised an interesting question. He wondered why Buzzards were motivated to defend such large territories in the past, when they could make do with smaller territories more recently, and in fact their foraging range was only a small part of the whole territory. His most plausible theory was that Common Buzzards needed to meet their neighbours in order to establish the territory, and if their nests were further apart they had to go further to do that. This is quite the converse of considering territories merely as what's worth defending, and hence smaller than the home range throughout which an animal roams. The idea is, though, in keeping with moving to rendezvous points that suit neighbours contesting their territorial boundary, such as ridges with updraughts. There will reasonably be a limit to how far it's worth travelling to display ownership. On Dartmoor, Dare (2105) found that where large areas of moorland separated some nests, the neighbours would not meet to establish a territorial boundary, presumably because the threat was much reduced and not worth the energetic expense of flying too far away from the area supplying their needs, or running the risk of potential trespassers on the other side of their territory.

Non-breeding birds may move around an area looking for weak territory owners to oust. Sometimes, it seems as though non-breeders cruise around territories in small groups 'looking for trouble', because if they don't get a robust response there may be an opening to obtain a territory (Prytherch 1997). Territory owners are liable to expend much energy defending their territory and guarding their mate from extra-pair copulations, as well as feeding the incubating females and then the nestlings, and protecting them all from predators. Therefore, it is better to avoid unnecessary combat and energetically expensive displays. This is probably why neighbouring territory holders do not spend the whole time fighting each other over a boundary. Instead, they seem to respect the boundary and rarely overtly cross into a neighbouring territory except to chase an intruder away (Sylvén 1978, Hohmann 1994, Dare 1998, Prytherch 2009).

For resident Common Buzzards, breeding territories are defended year round (Cramp & Simmons 1980, Newton *et al.* 1982, Prytherch 2013). Certainly, from our own experience of living among nesting Common Buzzards and taking tame hawks (Common Buzzard and Northern Goshawk) out to fly, there is no time of year when the local Common Buzzards do not get vexed by intruders. Nevertheless, Weir & Picozzi (1983) thought the winter territory to be smaller (105ha) than the breeding territory (175ha), and Sylvén (1978) recorded the same for some of his ranges. Dare (2015), too, described how those with bigger ranges that included open moorland tended only to use the lower core area during winter. However, he thought that there was minimal aggression between neighbouring territories during winter, with more of a passive respect for the territory boundaries and of other birds virtually hiding in the landscape. Tubbs (1974) also considered there to be little territorial activity from November to late February. However, he also noticed that some territorial birds commuted daily to agricultural feeding grounds outside the territory, sometimes being absent for eight hours a day, so this must have reduced the defence of the territory considerably. If the territory holders had to go so far to feed, perhaps there wasn't anything within the breeding territory worth defending at that time of year. The lack of aggression during winter in these earlier studies may also have been a result of Buzzards at the lower densities not wasting energy on vigorous defence all the time.

During the nestling period in two different years, Boerner & Krüger (2009) investigated the responses of Common Buzzards to a stuffed male Buzzard of the same or different morph to the breeding pairs. As soon as the stuffed Buzzard had been presented, alarm calls were played in order to attract attention, and indicate that the intrusion was hostile. The owner's reaction, assumed to be predominantly the male's, was unlike the response to a predator (see interspecific nest defence below). Buzzards reacted more strongly when the intruder was of the same morph. Intriguingly, the relationship was even stronger if the same morph as that of the mother of the territory owner was used. They suggested that territory owners had obtained the strongest definition of their species by imprinting on their mothers.

Some studies have also described territories of single unpaired individuals, often in brief winter territories, which tend to be smaller than those of breeding pairs (Tubbs 1974, Weir

& Picozzi 1983, Hohmann 1994, Schindler *et al.* 2012). In the New Forest, single birds did not advertise their territory as much as a pair (Tubbs 1974). In Scotland, Weir & Picozzi observed only single juvenile birds being territorial in the absence of adult territory holders. They found them in a place where there was evidence of poisoned Buzzards, and therefore never a stable territorial pair to prevent the singletons from defending a patch. Ulf Hohmann (1994) studied both sedentary and migrant birds in the same area of Germany. The wintering sedentary single birds that set up a territory did not enter neighbouring territories, but arriving migrants didn't 'know' the boundaries and so were frequently chased from territories by resident breeders. Some single birds held winter territories over a very small area of good feeding, which had no roosting trees. So, at dusk they flew low and inconspicuously into distant woodlands within a breeding territory, and typically left before dawn the next morning. If their departure was delayed, for example by a hard frost or particularly foul weather, then as they left they were more often spotted by the resident territory owner and fervently chased away. In a recent study in Scotland, all 25 separate roosts of non-breeders contained only a single bird, so there didn't appear to be any communal roosting (Francksen *et al.* 2016c).

Tolerance

Although the Common Buzzard has been regarded as a classic example of a territorial raptor, which keenly keeps out all intruders, quite a few researchers have noted that not all Common Buzzards are expelled from a territory. Indeed, some Common Buzzards appear to share parts of their territory with particular Common Buzzards, but not with others. This was recorded even in older studies, when Common Buzzards were less common in the UK. In recent decades, there have been regular sightings of many Common Buzzards worming on a field. Near our study area at the end of the 1990s, it was possible to see more than 20 in the same field, and when disturbed more emerged from surrounding trees, something we would never have imagined even a decade before. This raises a question: when there are more Common Buzzards than available territories, what happens to the non-breeders? Are they tolerated or exiled into less desirable sites?

Tubbs (1974) observed areas where two or more different pairs frequently overlapped their hunting grounds, which were outside the defended nest territory. He also suggested that juvenile Common Buzzards had to go to a 'no-man's land' containing poor-quality habitat unsuitable for breeding territories. On the other hand, Picozzi & Weir (1983) found that some non-breeding individuals shared areas with territorial pairs, whether they themselves were territorial or not, and Dare (1998) thought that the juveniles may be tolerated through the autumn and winter. Hohmann (1994) saw single residents tolerating transient migrants, whereas breeding pairs would quickly evict them from their territories. We decided that these contrasting ideas about juvenile relationships with nesting pairs were worth further investigation, using the autumn home range data that we had collected from our tagged Buzzards (Chapter 4).

Unfortunately, we had no pairs of breeding birds radio-tagged to help outline their combined territories on the study area. In the early years, we attempted to catch some of them but, despite all their territorial bravado, adult Common Buzzards turned out to be very cautious about venturing into the traps that we were permitted to use. Moreover, many

The Common Buzzard

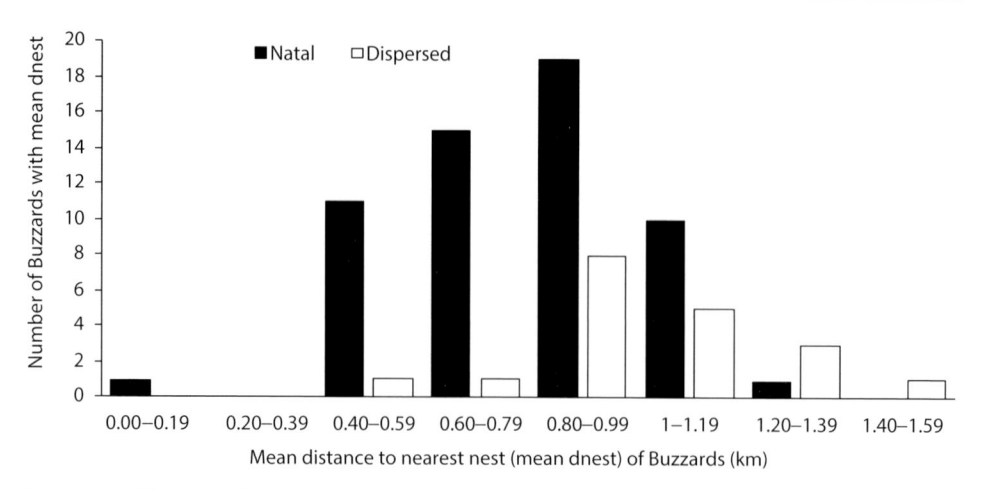

Figure 5.5. *The mean distance to their natal nest in their first autumn was less for Buzzards that had not dispersed than the distance to the nearest nest (in this case non-natal) for birds that had dispersed. Data from Walls & Kenward (2001) with kind permission from Elsevier Ltd.*

of the birds we did tag as juveniles tended to disperse away from the study area, or not to breed until after we had finished collecting home range data. Nevertheless, we had data on the ranges of many first- and second-year birds (Chapter 4), and we knew which nests they came from. Therefore, we could look for how the ranges of known individuals overlapped, armed with knowledge of their age and whether they had dispersed from their parent's territory. This turned out to reveal much about Common Buzzard social organisation (Walls & Kenward 2001). Our first step was to estimate an activity centre for each juvenile Buzzard that remained in the study area during autumn and winter, around which the routine tracking locations were most dense. We then looked at the distance between that activity centre and the nearest nest, which we treated as an index of adult Buzzard presence. We considered those juveniles nearest their own nest not to have dispersed, whereas those nearer to another nest were considered to have dispersed. Juvenile Buzzards nearest their own nest had geometric mean distances appreciably less than those nearer other nests during their first autumn and winter (Figure 5.5).

This suggested that breeders tolerated their own young nearer to their nest more than they did young from other pairs. While the difference of a few hundred metres between the groups does not seem far for a bird that can range over several square kilometres (Chapter 4), it was understandable. If nests were spaced as tightly as possible, the resulting pattern would be roughly hexagonal, with each nest surrounded by six others. At the density of nests within our study area, the average gap between nests would be 1.8km, so that non-breeders settling between nests would only be able to move around 900m from one nest before getting closer to another nest (Figure 5.6). Thus, the dispersed juveniles were actually putting themselves about as far away as they could from the 'strange' nests. In the following autumn, even those who stayed near their own original nest had moved out to around 800m from the nest, a similar distance to the dispersed birds. When it came to their second winter, however, the few that remained nearer to their own nest were again back to 460–700m from it, significantly closer than dispersed birds. The parents continued to be tolerant!

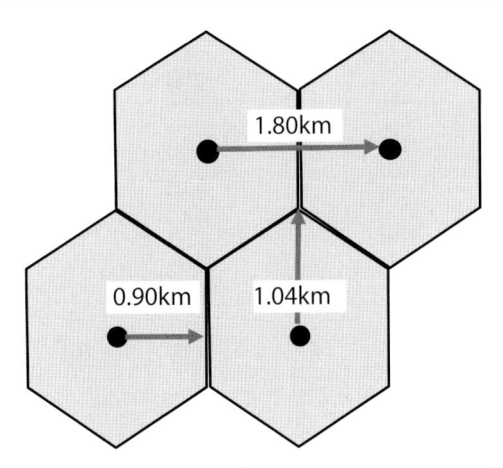

Figure 5.6. *With an average distance of 1.80km between nests in Dorset, birds with home ranges spaced evenly between nests should average 0.90–1.04km from nests; average distances of range centres from nests were 0.86–1.10km.*

Plate 14 (see colour section) shows that Buzzard nests were not spread evenly throughout the study area. Therefore, we also checked that the juvenile birds that stayed within the study area were not being banished to 'no-man's lands' of poor habitat, such as open areas with few trees or lowland heath, which was also not used for nesting. Comparison with random locations across the study area told us that juveniles were not restricted to areas of poor habitat. Instead, the non-dispersed birds remained within the nesting areas and had a highly significant positive attraction to the nests (Walls & Kenward 2001). Moreover, although the dispersed birds had a weaker attraction to nests, it nevertheless tended to be positive and indicated that they too were remaining in the sort of habitats suitable for nesting. A recent winter study by Schindler *et al.* (2012) recorded seeing Buzzards more frequently outside breeding territories from the previous summer, and concluded that territorial sedentary breeding pairs displaced non-breeders and wintering birds into suboptimal habitats. However, their smaller study area (20km^2, compared with our 120km^2) and lack of marked birds makes it hard to conclude anything about the significance of normal territorial defence for use of habitats. Also, as we have shown in our study area, what is best for breeding is not necessarily optimal for foraging during the winter. Indeed, there may be some cost to staying mainly within the territory all winter in order to defend it, if the feeding is better outside.

As well as looking at the spacing behaviour of wintering Common Buzzards with regard to nests, we also investigated relationships between neighbouring home ranges of these non-breeders. We presumed that any territoriality would be demonstrated by lack of overlap between core home range areas, but that results might differ depending on which home range model is used, and that the differences might tell us more about their home range use and territory defence. For example, Figure 4.3b (in Chapter 4) shows how the 99 per cent contours overlapped each other, but the core 85 per cent outlines do not overlap. So we used ellipse home range estimates as a representation of how territories might be perceived by a bird that considered everything within a certain distance of its home range focus as being within its territory. At the other extreme, we used nearest neighbour cluster hulls to investigate whether these non-breeding Buzzards were merely defending particular patches of important resources, for example a Rabbit warren, plus a roosting copse that may be some distance away. We also picked an intermediate and commonly used home range model based on plotting contours round kernel density estimates (see Figures 4.3–4.5 in Chapter 4 for an explanation of the home range estimators).

Overlap could only really be analysed between birds that were sufficiently close to one another, so we used each neighbouring duo of birds that overlapped home ranges, whether the range estimates were based on hulls using all locations, or where there was a 99 per cent chance of finding the Buzzard within a contour or ellipse based on location density. None of these duos were breeding together, though there is a small chance some may have been breeding with other Buzzards. We then compared the amount of overlap between neighbouring duos for each 10 per cent reduction of outermost locations (or location density for contours and ellipses). If there was a territory, we would expect the overlap to suddenly drop to near zero. We looked for differences in the amount of overlap between neighbouring duos of ranges from Buzzards of the same or different sexes (we expected less sharing with 'same-sex neighbours'), age and relatedness (whether birds were siblings from the same nest in the same year).

Relatedness and age of the adjacent birds had the biggest effect on how much the home ranges overlapped. When Common Buzzards were juvenile, there was no dramatic drop in home range overlap for siblings and even their central areas often overlapped. On the other hand, unrelated neighbours had virtually no shared area once we removed the outer 20 per cent of the ranges, whichever home range model was used. So, siblings seemed to tolerate each other much more than other neighbours. When they reached their second year, even if they were siblings, there was virtually no overlap for 70 per cent cores of kernel ranges or for 90 per cent cluster hull ranges; this was equivalent to an average core territory of approximately 56ha, whichever of these home range models was used (Figure 5.7). Ellipses did not show this pattern, probably because they are poor estimates for territory boundaries that are often delineated along landscape features (Dare 1998, Prytherch 2013). We concluded that siblings can tolerate each other before breeding age, but when older there is a core of unshared territory for all unmated Buzzards, although there might be overlap in the periphery of the range.

Combining the information on distances to nests and home range overlaps, we can now say that in general, although exceptions are probably always likely, Common Buzzards tolerate each other if they are related, either as offspring in the breeding territory or siblings in their first year. Dispersed young are not banished in groups to very poor habitats, but may individually be in slightly suboptimal habitats for nesting because they avoid or are kept at the periphery of the main territories of breeding birds to which they are unrelated.

The sex of these pre-breeding juveniles had no detectable added effect on either home range overlaps or distance from nests. That would clearly not have been the case for breeding adults, but we never had both members of a pair tagged on our study area. However, as the home ranges of older birds tended to be slightly larger and may well not have overlapped completely in their defended cores, one might expect territories defended by breeding pairs on average to be larger in terms of food resources, even before possible extension to convenient topographical features for display flights.

Although all these findings help us to understand the territorial structure and spacing of Buzzards, they do not explain the frequent winter observations of large numbers of Common Buzzards worming in the same field. Such birds are not all related, and it is

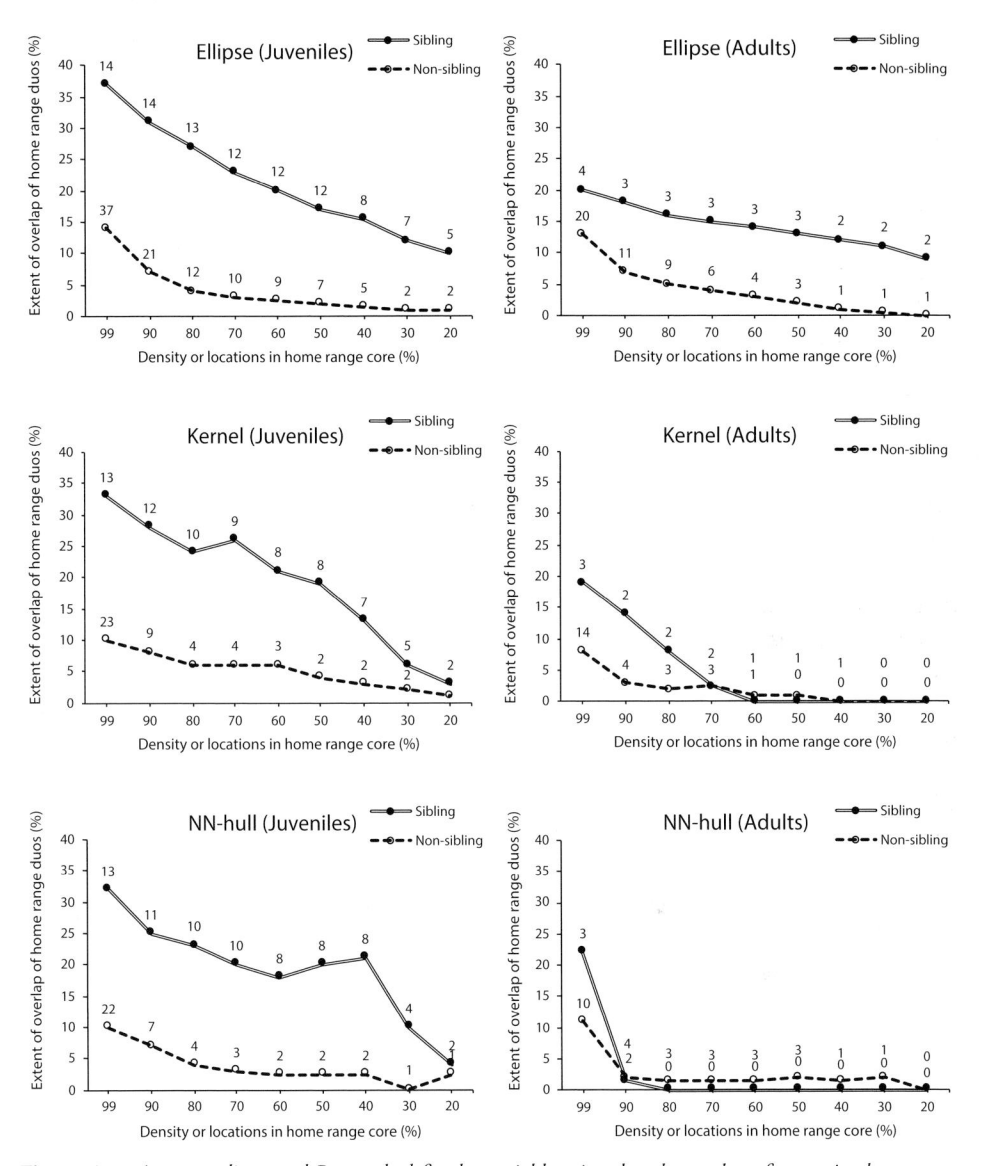

Figure 5.7. *Among radio-tagged Buzzards defined as neighbouring duos by overlap of expansive home range ellipses, siblings had extensive range overlap as juveniles but core ranges with more intricate outlines were separated as adults for both siblings and non-siblings. Numbers of duos are shown above data points, and plots cannot exceed 99 per cent of density distributions. Data from Walls & Kenward (2001) with kind permission from Elsevier Ltd.*

doubtful they are all juveniles, although a project to specifically look at ages in these groups, for example from photographs, could clarify this. Looking at their regular spacing on the field, as though they have a 20m invisible buffer around them, and their very independent arrival, they have not arrived at the field as a flock, but as individuals. Far from being poor habitat, such a field has rich pickings and could be one worth fighting for. The same can happen at other particularly good feeding sites, such as Red Kite feeding stations, where

Buzzards may congregate with the kites. Weir & Picozzi (1983) noted an area with a large number of wing-tagged birds together over an outcrop that had a particular abundance of Rabbits; it could therefore support many Buzzards that nested further away. Buzzard worming happens at a time of year when there are no breeding commitments (after the young have fledged and before nest-building begins) so it is not worth defending the fields as breeding territories. The fields may therefore even be outside the breeding territories.

Another personal observation, although anecdotal, offers some insight into how there can be so many Common Buzzards feeding in the same field. While tracking a juvenile Buzzard that had not dispersed in mid-winter, we noticed that the young bird screamed frequently at its parents, making a begging call for more food. Despite the youngster drawing attention to itself with its discordant begging call, the visibly present adults did not attempt to drive it away. Given our evidence of adults tolerating their own young, that was no surprise. Any other Buzzard that came into view was called at by the parents, and if it came too close then it was seen off briskly, as we would also expect. In the early afternoon, another pair of Common Buzzards landed on the arable field and started worming. One parent quickly swooped down and attacked one of the intruders, preventing it from worming. It then chased the other bird away, during which time the first intruder returned to the same patch and started to worm again until the pursuing adult returned and again threatened it, landing on the ground beside it to show full intent. At that point, other Buzzards started to gather on the field and began to feed. It was obvious that the local breeding pair was not going to be able to keep all the wormers off that field, so they appeared to give up defending the area for the rest of the day. They had simply been overwhelmed and the territorial system had broken down, albeit temporarily. In spring, crops start to grow and worming is no longer an option on arable land. With no abundant resource to hold all those Common Buzzards in one spot, the temporary gatherings disperse and those territory holders that have been inundated during the winter can re-establish their territory.

Similar changes in territorial behaviour in the winter have also been described for other birds, after an early study of Pied Wagtails *Motacilla alba yarrellii* on Port Meadow in Oxford (Davies 1976). In the study, individual wagtails defended a territory along a river bank, while flocks of other wagtails used local abundances of food that fluctuated quickly at the edge of temporary pools on a water meadow. However, there were times when a territory holder would go off to join the flock to take advantage of the temporary food glut, especially if food was short on its own territory. Moreover, when food became abundant within a territory, the flocking birds invaded and the territory became indefensible. Interestingly, the Pied Wagtail retains a fully insectivorous diet during the winter, and the size of the animals it eats means that it must do all it can to feed as constantly as possible. Likewise, the Common Buzzard in the UK is unusual as a territorial large raptor that switches to relatively small invertebrates during the winter, and requires a large number of worms.

Another contravention of territory can occur when related individuals help with to rear young. Prytherch (2013) made a rare observation of nest-helping in Common Buzzards. In 1993, he thought he had found a male courting with two females. In fact, the second female turned out to be the offspring of the resident pair from the previous year, which helped to provision nestlings over two breeding seasons. This is a known phenomenon in other birds, including raptor species (Newton 1979a, James & Oliphant 1986). Helpers are related, and so are investing in the persistence of their genes to future generations. Nest-helping is unlikely to be a frequent behaviour in Common Buzzards, given the amount of time people have spent watching their nests without reporting it, but perhaps nest cameras will reveal more examples. For example, Jason Fathers (2006) filmed two female Eurasian Sparrowhawks incubating and raising young in the same nest.

An interesting corollary of Buzzards being able to recognise and tolerate their own young, but having different relationships with 'strange' young, is a contrast this provides with Northern Goshawks. Kenward *et al.* (1993) found that 23 per cent of 77 Goshawk nests monitoring through radio-tagging were involved in nest-switching, in which dispersing young hawks visited 'strange' nests for long enough to be detected there in checks at 2–3 day intervals, and therefore presumably long enough to be fed. Despite Common Buzzard nests being much closer than those of Goshawks, such nest-switching behaviour was never detected during similar monitoring of Buzzards in Dorset. One wonders whether the diversity of plumage colour and patterning in the 'Buse Variable' is a convenient enabler for individual recognition, or perhaps an adaptation for recognising kin and strangers.

Interspecific nest defence

Other Common Buzzards are not the only threats and competitors in the landscape. In particular, other raptors want to use the same space for nesting and feeding. Intraguild predation (e.g., raptors killing raptors) is very widespread. A review by Fabrizio Sergio & Fernando Hiraldo (2008) checked 39 empirical studies on 63 raptor populations and found that Northern Goshawks consistently affected the territory occupation rate of Common Buzzards. Such a strong influence shows how the presence of other predators can influence the demography of a species. It is worth noting that Common Buzzards are more often attacked by other raptors than they are the aggressors (Chapter 9).

Common Buzzards share their distribution with many species of raptor, with which they compete for food and nest-sites; some are a significant predatory threat. The two most dangerous raptors are Eurasian Eagle Owls and Northern Goshawks (Lourenço *et al.* 2011b), although eagles, too, could be important where they coexist with Buzzards. Goshawks will take over Common Buzzard nests (Kostrzewa 1991, Krüger 2002a, Hakkarainen *et al.* 2004) and if there are Goshawk nests nearby more young Common Buzzards are killed (Figure 5.8). In Finland, Common Buzzard nests within 1km of a Goshawk nest were less likely to produce young (32 per cent failure vs. 8 per cent failure) and overall caused a 20 per cent reduction in fledgling numbers (Hakkarainen *et al.* 2004). It wouldn't be surprising if the Goshawks also killed juvenile Buzzards after they had left the nest, but in the absence of radio-tagging the disappearance of a young Buzzard is difficult to differentiate from dispersal. So, a 20 per cent reduction in success pre-fledging is only the start of the effect on breeding success; the end result is likely to be even lower productivity. In Finland, it was also interesting that apart from predation, and possible interference competition from being harried while hunting, there also seemed to be exploitation competition over nest-sites. More than half of new Goshawk nests were on old Common Buzzard nests. In comparison, about 25 per cent of new Common Buzzard nests (and also 25 per cent of new European Honey-buzzard nests) were built on nests of smaller raptor nests. Nevertheless, a pair of Common Buzzards can sometimes drive off a single Goshawk (Kostrzewa 1991).

Krüger (2002) has specifically studied the relationship between Common Buzzards and Northern Goshawks in Germany, including some very interesting experiments. In one, he

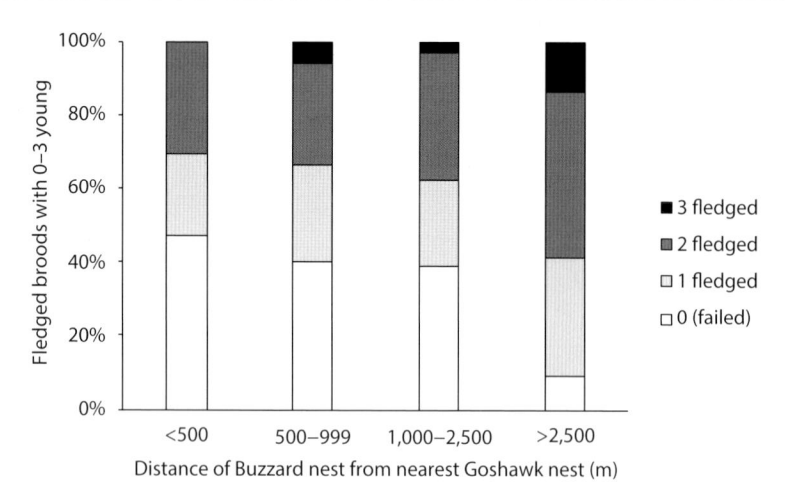

Figure 5.8. *Kostrzewa (1991) noted that the further 276 Common Buzzard nests were from Northern Goshawk nests, the greater their productivity in terms of young fledged per brood.*

placed a stuffed Goshawk within 20–50m of some active Common Buzzard nests and played Goshawk calls for 45 minutes every third day. Fledging success at the control sites (where no stuffed Goshawk was presented) was 75 per cent, but for those that experienced the stuffed predator the success was substantially lower, at only 33 per cent. So, the stress of simulated Goshawk presence nearby had a dramatic effect on the breeding success of Common Buzzards, and the next season they were significantly more likely to move their nest elsewhere. Krüger considered whether it might simply be due to lost hunting time if the breeders were standing guard while the stuffed Goshawk was presented. He tried to compensate for this by feeding some nests to see whether the effect could be reduced, but the results were inconclusive. When exposed to these conditions, the breeders most likely to fledge young had a greater presence of both mother and father at the nest plus increased calling at the stuffed bird. The least successful breeders seemed to disappear as soon as the first calls were played. Most Buzzard attacks on the stuffed Goshawk were not followed through, with only one of 14 Buzzards trying to make contact. In comparison, breeding Goshawks had to be prevented from attacking a Goshawk dummy before they destroyed it.

The study also showed that Common Buzzard nests were less often attacked by corvids if a Goshawk was nesting nearby, which could have been an advantage if the effect of the Goshawk was not so negative in other respects. Intriguingly, productivity of Buzzards in our study area was relatively low if corvid remains were found in the nest, perhaps due to the harassment observed (Hodder 2001). The breeding habits of Red-breasted Geese *Branta ruficollis* in the Taimyr region of Russia illustrate the potential benefits of having a predator nearby. In the population studied, geese could either nest on islands or on riverbank cliffs. Nests on cliffs were more susceptible to mammalian predators. The geese only nested on cliffs if there were sufficient raptors nesting nearby to chase off ground predators such as foxes. During poor lemming (*Lemmus lemmus*, *Myopus schisticolor*) years, when there were fewer Rough-legged Buzzards and Snowy Owls *Bubo scandiacus*, geese moved to the offshore islands to breed away from mammalian predators (Prop & Quinn 2003). Goszczyński (2001) noticed that the productivity of Common Buzzards and Northern Goshawks nesting in the same wood were correlated, despite the fact that the latter did not rely on voles as did the Buzzards. He suggested that in poor vole years, Buzzards turned to avian prey, so overlapping more and therefore interfering more with the Goshawks' diet.

The Eurasian Eagle Owl is another species that regularly kills other raptors. Common Buzzards will attack eagle owls all year, to the extent that in one study around 40 per cent of targeted Buzzards could be caught in nets around a live owl (Zuberogoitia *et al.* 2008). Eagle owls are known to provoke a strong response (often panic and flight) in any bird that is under its watchful gaze. They also have the advantage of better sight during twilight, owing to their enormous (and beautiful) eyes, making them particularly lethal for diurnal raptors that can't see well as soon as the light starts to fade. When breeding Buzzards were presented with a stuffed Eagle Owl within 50m of their nest, 16 light-morphed males were significantly more aggressive than the 38 intermediate or dark males. The opposite trend was true for females, with the four dark females being much more aggressive than the 50 other females (Boerner & Krüger 2009). There is variation in how Buzzards will defend. Some call and circle above, while others, like the dark females in this case, attack and make physical contact with the intruder.

Less dangerous raptors of a similar size can battle hard for a nest-site, especially if migratory and arriving after resident Common Buzzards have started on their nest. This has been observed with both European Honey-buzzards (Kostrzewa 1991) and Black Kites (Sergio *et al.* 2002). We have witnessed Common Kestrels and Peregrines (and once a Marsh Harrier *Circus aeruginosus*) attacking Common Buzzards. However, their nest-site requirements are so different that this is probably more to do with competing over hunting areas, and trying to ensure that their young are not taken by Common Buzzards. It is inevitable that there are conflicts with Rough-legged Buzzards where the two species occur together. In areas where Golden Eagles coexist with Common Buzzards there appears to be a differentiation in spatial use, with Golden Eagles dominating the higher areas and Common Buzzards restricted to the lower ground, with little overlap (Dave Anderson, pers. comm.). We know that Buzzards may use high moorland less than lower ground, even in areas without Golden Eagles, so more work needs to be done to establish how much of this is due to antagonism rather than separation based on habitat preferences.

Looking from the opposite perspective, Buzzards can also have a very real effect on the reproductive success of competing raptors by attacking less able nestlings, as was shown in a video clip of a webcam on an Osprey nest in Scotland (http://www.bbc.co.uk/news/uk-scotland-18611742). An adult Osprey is seen incubating two chicks that are quite large, with feathered wings and downy bodies. The adult flies off the nest and within six seconds a Common Buzzard comes in and snatches one of the chicks that had not even raised its head. The Buzzard is off the nest in less than a second with its bounty. The camera was in a fixed position and therefore could not reveal whether the Buzzard was spotted by the parent Osprey and perhaps lost the chick, but the chick did not return to the Osprey nest.

Territorial behaviour

Now we understand what Common Buzzards need to defend and what they are defending it from, we can describe the behaviours employed when these birds are at their most conspicuous. They are especially spectacular because they are big enough to observe easily from a distance, even without binoculars at times, and they are vocal, too.

Although we describe Common Buzzards 'attacking' each other throughout this section, it is rare for physical contact to take place; Buzzards mainly bluff, and even if they dive at a bird they will pull out a few metres from their target. Nevertheless, they do sometimes make that contact and occasionally fight to the death. In two cases, we have found dead birds where the likely cause was another Common Buzzard. In the first case, we found a radio-tagged Buzzard on its back under some Willow *Salix* spp. bushes. Some of the chest feathers had been plucked and were lying nearby, but no meat was eaten. This is common in raptors that kill birds of the same species. The plucking appears ritual, as if to confirm that the bird was dead but with no wish to eat the meat, perhaps because of disease risks that are associated with cannibalism. On the other hand, recent reviews suggest that is not such a risk (Rudolf & Antonovics 2007), and it doesn't seem to stop female Goshawks from eating males during the courtship period (Kenward 2006). Nevertheless, we did not know which raptor had killed the bird, and there were certainly aggressive interactions with Peregrines in the area. On the second occasion, it was more convincing because the dead Buzzard had feathers from another Buzzard still in its tightly clenched talons. The bird also had talon wounds, including a punctured eye.

From his familiarity with individual Common Buzzards, based on observations of up to 1,200 hours a year over 20 years, Prytherch (2009) is convinced that males spend all their reproductive life in one territory and will defend it to the death. As far as we know, there are no records of individually identifiable breeding Common Buzzards changing from one breeding territory to another (Weir & Picozzi 1983, Prytherch 2009). Territory switching may therefore be a rare occurrence in a stable population of resident Buzzards. In comparison, 35 per cent of marked young male Eurasian Sparrowhawks were shown to switch territories, and as many as 60 per cent of young females could switch (Newton 2001). Prytherch has actually witnessed attacks resulting in injured Buzzards, as well as birds very probably killed by other Common Buzzards. He considers that a weak territory holder may endure many weeks of attacks from intruders before it is eventually deposed.

Territories are defended year-round, but the main display period is through March to mid-April, after which defence becomes less conspicuous during incubation. Displays pick up again from September, while juveniles disperse, and then diminish in late autumn. During the winter, intruders are quickly expelled, or ignored if they travel through rapidly enough.

Calls

Many walkers will be alerted to a territorial Common Buzzard by a piercing, elongated call, as described in Chapter 1. This sound typifies the 'call of the wild' for UK films and TV programmes, although the higher-pitched trilling cry of the Red Kite is now used quite often, including anachronistic application to Scotland for a popular series set in the mid-20th century! Until the reintroduction of Red Kites in the 1990s, Common Buzzards were the only diurnal raptors most people in the England would have heard; Common Kestrels and Eurasian Sparrowhawks are far less vocal. Most of the time the call is aimed at other Common Buzzards, but sometimes it is the human walking near a nest wood that will provoke the call. If the person goes too close to the nest then the pitch will change, and the call will sound more slurred as if the bird is more worried, in the same way that human voices wobble when anxious, and the calls will be more frequent. We were particularly aware of this, because certain Buzzard pairs got used to our annual visits to their territory, perhaps remembering that we would find their nest and climb to it in order to measure and tag the nestlings. Pairs that recognised us tended to quickly change to the 'concerned pitch' as soon as we were within sight, rather than escalating as we got closer to the nest. There is a vulnerable time, when the female is sitting on eggs, when calls are fewer. If people get too close, the female is likely to slide off the nest and move quietly to another part of the wood, but as the chicks age, then the parents get more vocal again.

Flights

When walkers look up to find the source of the calls, they are often disappointed. This may be because the bird is calling from a perch within woodland, the vegetation making it hard to pick out a brown bird, despite its large size. However, just as frequently the Buzzard can be so high in the sky that it is difficult to pick out with the naked eye, especially when directly above the searcher. Some males spend many hours soaring above their territory, with or without the female, allowing them to observe any activity that might endanger the nest or give away the presence of prey, while at the same time probably advertising ownership. Common Buzzards are most dramatic when they perform their multiple stoops. They close their wings and drop almost vertically in a long stoop, before pulling out their wings to rise with the momentum. Without a single flap, they often rise half the distance they have dropped, before closing their wings again for another drop and rise, continuing with more undulations of decreasing amplitude. There can be as many as 20 undulations (Tubbs 1974), although 3–5 are more common (Prytherch 2009). These flights have been aptly described as 'roller-coasters' (Weir & Picozzi 1975), after the fairground attractions that make the same movement to scare passengers (Figure 5.9), or a 'victory' display because it is often performed after seeing off an intruder (Prytherch 2009). Such undulating displays are common in many other raptors (Brown & Amadon 1968). The displays are most impressive and common in spring, when males are re-establishing their territory for the breeding season, but they commonly occur through the summer. During the nest-building phase, the male may carry foliage or prey in its talons while performing the roller-coaster. The rituals are performed more aggressively (faster, with talons lowered) if there is an intruder in the territory. Roller-coasters are sometimes performed by two males parallel to each other along a common boundary (Weir & Picozzi 1975), but more often over the nest wood, and may be finished by a particularly vertical plunge to the nest. The best time to watch for these displays is in the middle of the day when thermals become strong enough to aid the display (Tubbs 1974). Weir & Picozzi also found soaring more likely when the temperature was above the seasonal norm, when wind was Beaufort force 1–4 and when there was partial cumulus cloud cover (an indicator of thermal air currents). Certainly, weather has quite an impact on how vigorous the display is.

Common Buzzards usually gain height by circling on thermals or other updraughts, on windward slopes or woodland edges. This is an efficient method for them to watch over the territory, but there isn't always a strong thermal conveniently placed. When trying to repel an intruder, the territory holder seems to seek to rise as quickly as possible, before the trespasser begins to feel comfortable. So there is nearly always active wing-flapping at the

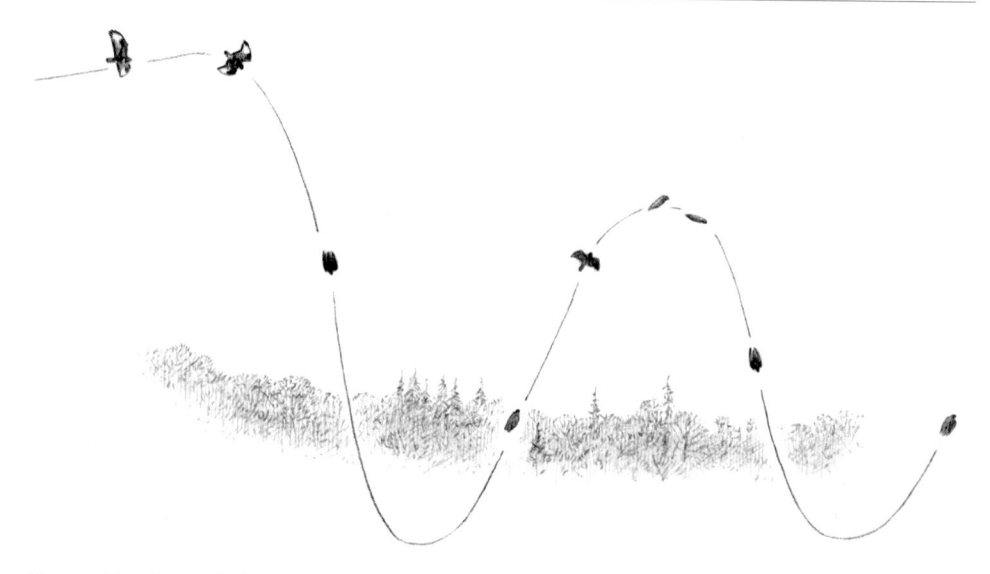

Figure 5.9. *A typical Common Buzzard roller-coaster display usually ends with a dive to the nest. Redrawn from Robin Prytherch (2009), with kind permission from* British Birds.

beginning, even if just to get to some fast-rising air more quickly and gain advantage over the other bird. This active flight, with purposeful wingbeats, often looks more powerful than the usual rather leisurely flapping. There will also be 'display flapping' mixed in with this, a series of 4–7 beats with the wings raised much higher than usual in a 'deep wingbeat'. The latter is also used when patrolling the territory boundary (Mebs 1964, Tubbs 1974). Calling is common during this 'display flap' (Prytherch 2009), perhaps proving to the intruder that the territory holder still has the energy to call while demonstrating the power of its flight – as well as warning the intruder to flee before it is too late. Sometimes, the legs hang down, probably indicating a readiness to use the talons against the intruder, but having the added effect of making the bird look bigger and more powerful. If there is any lift, then the powerful display will continue until the environmental elevator (orographic or thermal updraught) is reached, when outstretched wings are better at catching the lift. The wingbeats in a display flap are probably pronounced in order to reveal the white underwing, which stands out clearly against the typical green and brown landscape into which Buzzards often fade; the white flashes give intruders a lot of warning that the territory holder is on its way. When flying a tame Buzzard, the local breeding Buzzards will call but they are hard to spot with the naked eye until they raise their wings high and the flash of white underwing immediately gives them away. Weir & Picozzi (1975) thought that the deep wingbeat behaviour was more to do with courtship than aggression, as suggested by Demandt (1934). However, they also noted that if intruders showed their underwing when being chased, it appeared to provoke a more aggressive attack from the resident, which suggests it is recognised as aggressive behaviour. Generally, trespassers will try to go unnoticed, keeping low, making short flights and using the vegetation as cover.

While on en route to meet an intruder, the power of a displaying territory holder's wingbeats may depend on the status of the bird they are challenging, more effort being made if the intruder is an adult (Prytherch 2009). The resident will hold their wings out as far as possible, spreading the slotted primaries to their full extent, and thereby making the bird look as big as it can while spiralling up. In contrast, the intruder may hold its wings in, closing the tips and holding them slightly back, to demonstrate submission. Often, the intruder will leave as the adult rises, and the adult will follow the bird, as if escorting it from the territory from below. Once the bird has crossed the boundary, then the owner may display, or simply return to an obvious sentinel perch where it makes a rather heavy landing, thumping the perch as it grabs it, in a pumped-up display of strength.

If the intruder shows no signs of leaving, then the resident will rise to be level with or above it and engage in combat. First, it will adopt an assertive flight posture: the head will be pushed down and turned up to give a humped neck appearance, with the wings outstretched and the tail fanned to make it look bigger; the legs are sometimes lowered to indicate how strongly the resident feels (Prytherch 2009). The deep wingbeats of the display flap will again be used to initiate a strike, but as the bird gains on the intruder the wingbeats become shallower, and in the final approach the wings will be lower than usual before going into a glide for the last few metres. When virtually on top of the intruder, the resident will swing its talons out, often resulting in the rival turning and flashing its yellow legs and feet in self-defence (Figure 5.10a). A chase across the sky tends to involve long swoops punctuated by bursts of zig-zag movements and tumbling when the resident catches up with the intruder.

There are times (though very rare) when the resident and intruder lock talons and cartwheel in the sky, plummeting earthwards (Hume *et al.* 1975, Heywood 1986, Marsh 1989, Towill 1999). Prytherch (2009) and Tubbs (1974), both of whom spent a great deal of time observing and recognising individuals, believed that such talon grappling was almost always a resident attacking an intruder, but that it could be misinterpreted as courtship by those who see the phenomenon but can't recognise individuals. However, it may be possible to observe a cartwheel between a male and female in a pair, as evidenced by an observation of a food pass that momentarily turned into a cartwheel before the female glided down to a perch and started feeding (Dickson 1998). If it had been an attack against an intruder, the bird receiving the food would not have been able to casually glide down and start eating. In

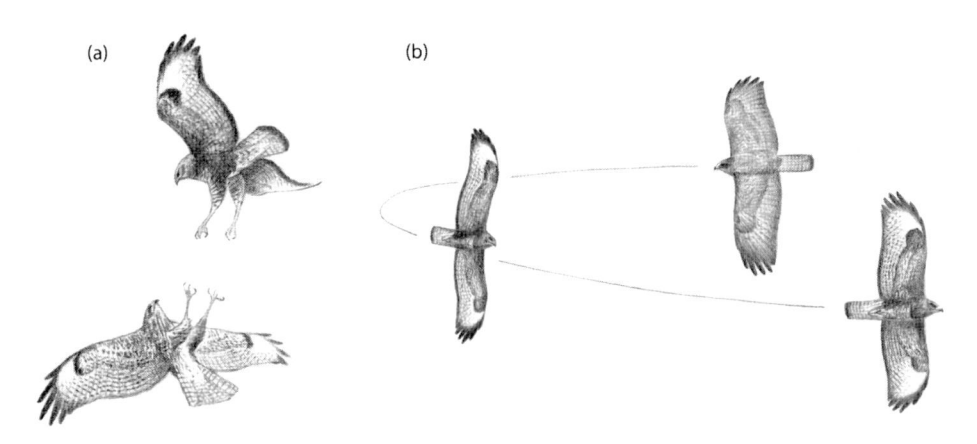

Figure 5.10. *On the left (a), a Buzzard turns over to defend against a territorial attack from above. On the right (b), a 'display' bank is a less severe threat. Redrawn from Robin Prytherch (2009), with kind permission from* British Birds.

fact, compared with many other raptors, especially harriers, Buzzards do very little food passing that might result in talon grappling, although Weir & Picozzi (1975) did observe Common Buzzards passing nesting material via a mid-air drop-and-catch during the pair-forming period.

The chase often extends out over other territories, whose owners may be drawn up to see the intruder off. The presentation of talons to force the intruder to turn over can be confused with the 'dive-on, turn-over' courtship behaviour described in Chapter 6, but an aggressive attack is identifiable if the other bird flees. If the intruder is chased for several kilometres over other territories then, as the defender returns to its own territory, it will be escorted but not attacked by the neighbouring territorial birds which probably recognise the individual. Mostly, it is the male that gives chase, with the female remaining below, 'wing waving' (see perched behaviour). However, sometimes the female will come up with the male to make a two-pronged attack on the intruder.

If a patrolling territory owner happens to come across an intruder perched, or feeding on the ground below, then of course it does not need to gain height because it is already in the perfect position to attack. Rather than endanger itself, as it would do with physical contact no matter how good the starting position, the resident will often call a warning several times before attacking. As well as lowering the risk of damage, this call may also solicit help from the neighbouring territory. It is very difficult to say whether the bird is purposefully asking for neighbourly help, or whether the calling incidentally alerts the neighbouring pair to the presence of an intruder that they, too, need to drive away. Whichever is true, neighbours do seem to 'help' escort interlopers away from surrounding territories.

Prytherch (2009) describes the typical close approach to an intruder as a 'display bank' (Figure 5.10b): the owner flies at the intruder, but then banks at the last moment, turning itself so that the outstretched wings are vertical (i.e., the bird is at 90° to normal flight). By spreading the wings like this, the resident looks as big as possible and makes a good display of its white underwings, as well as presenting the talons to the intruder. Just as with hunting, when the intruder is on the ground a resident may position itself in the line of the sun, gaining a good view of the intruder but making it hard for the intruder to see its opponent clearly. If the resident mounts a full attack, an intruder on the ground may turn over on its back and present its talons in a 'fallen angel' posture (Plate 16, see colour section, and see ground conflicts), waiting for the stoop to pass before trying to escape, rather than be caught struggling to take off when the stoop is at its fastest. There are times, however, when there is no warning before an attack, as observed when regularly flying a tame Buzzard within established territories. This is an artificial situation and this response could be to do with the intruder being a repeat offender, or that the tame Buzzard looks much weaker. A tame Buzzard cannot spend as much time flying as a wild bird and so is never going to be a match in flight. Without the presence of a human, the tame Buzzard could have been killed.

Territorial ceilings

Territories appear to have a ceiling boundary above which territory holders will not attempt to evict intruders. The behaviour was first described by John and Frank Craighead (1956) for a closely related American *Buteo*, Red-tailed Hawk. The ceiling probably exists because it would waste too much energy trying to gain so much height when the bird may simply be

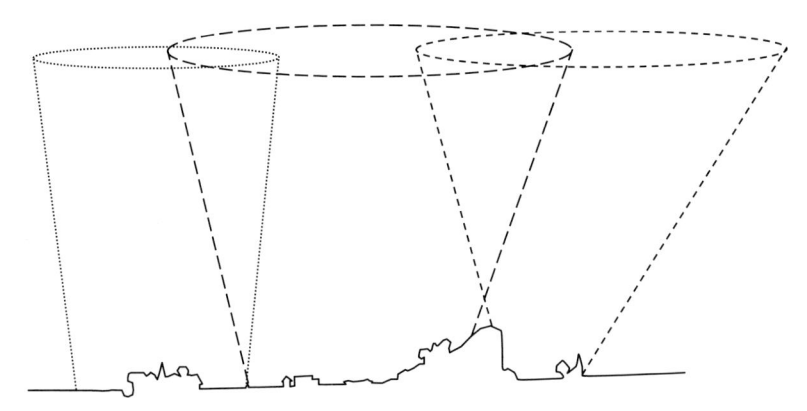

Figure 5.11. *Territories of three different individuals can be visualised as inverted cones with a seasonally variable ceiling above which intruders are not challenged. Redrawn from Newton (1979), after Gargett (1975).*

passing over and posing no threat at all to the territory. In fact, Common Buzzards seem to rise to a great height before leaving their territory in order to exceed the ceiling over other territories, for example, when going to winter feeding grounds (Tubbs 1974, Weir & Picozzi 1975). For Common Buzzards, the height can be over 1,400m above ground level (Prytherch 2009), depending on the area, the weather and possibly the stage in the breeding cycle, although of course the weather will change with the season.

However, rather than imagining a home range as a vertical cylinder with a lid on, raptor biologists tend to think more of inverted cones, where the defended area gets larger with height, and therefore inevitably there is also more overlap at height, until the ceiling is reached (Figure 5.11). This idea is based on behavioural observations of raptors going further into other territories at height, and also driving off intruders getting near their territorial ground area, as it is perceived by us. Likewise, when chasing an intruder away, the pursuit will extend a long way into other territories. In fact, the territory could rise vertically, but a territory holder may be behaving proactively by intercepting a potential intruder before it gets to its territory. Obviously, it is very difficult to know how the individual perceives it all, and there is likely to be a lot of variation between individuals of the same species. What it is important to understand is that although we have concepts of spacing and territoriality, there is a great deal of plasticity depending on the circumstances – habitat, orographic and thermal lift, season, density of intruders and so on.

Perches

Not all territorial activity takes place in the sky. Some Common Buzzards have very conspicuous perches that in themselves advertise territory ownership. We often spot them in dead trees, where the leaves don't obscure the bird, on pylons or on telegraph poles. Several such prominent perches can form boundary markers, which a patrolling Common Buzzard will fly between (Dare 1961, Weir & Picozzi 1975). The same markers can be used from year to year, and sometimes neighbouring territories will have marker perches very close to each other from which male Common Buzzards will repeatedly call at each other (Tubbs 1974, Weir & Picozzi 1975), although Prytherch (2009) suggests that rather than calling at each other, they are communally calling at an intruder, and we have witnessed similar. It should be noted that the most conspicuous perches will likely be just outside or on the edge of a wood, and high, which would be consistent with boundaries following linear ground features, like the hilltops that provide orographic lift. The alignment of marker perches and perceived territory boundaries is not coincidental, and not all the prominent perches are on the edge of the territory (Tubbs 1974); some will be lookouts nearer to the nest tree. There is a sense that males use the perches near the boundary more, and females those nearer the nest, although both birds of a pair will perch in the same marker tree at times. These conspicuous sentinel perches are often used immediately after display flights.

High and open perches are the most efficient way for a territory holder to both survey the surroundings and advertise ownership, and so are self-evidently important. Hohmann (1994) found that breeding territories contained high perches, but territories of single birds or transients didn't necessarily have them. Indeed, only the more dominant birds use the most prominent perches. Juveniles or low-ranking Common Buzzards may not use high perches, presumably in order to avoid aggression (Weir & Picozzi 1975, Probst 2002). Sylvén (1978) sometimes found individuals 'were dominant at a perching site towards some individuals, mostly distant neighbours, but not against others'. Prytherch (2009) observed that single birds' displays and defence were relatively low key.

It is possible to interpret the mood of a Common Buzzard from its perching posture, by observing features common to many other raptors (Brown & Amadon 1968). A relaxed Buzzard will stand on one leg, with the feathers roused out and covering the other leg (Plate 1b; see colour section). The head seems to sink into the shoulders, giving the bird an overall shorter and rounder appearance. Alternatively, when hunting from a perch, the feathers will be drawn in, making the Buzzard look sleeker, and its focus will be on the ground in front, possibly making the bird lean forward a little, or into a horizontal stance if about to launch an attack. The point is that when hunting the focus is usually on the ground.

If another Common Buzzard (or potential predator) flies over, then the first indication we see will be a tipping of the head so that it can focus on the advancing bird above. If the bird above warrants more attention, then the neck will be extended and the feathers start to flatten around the bird, giving it an elongated appearance, though still upright. When the perched bird is dominant, it is likely to show a 'wedge-head', where the feathers on the nape, crown and cheeks are raised to form a slight crest at the back of the head, increasing its size to look aggressive (Plates 16, 17; see colour section). This is the equivalent of 'raising the hackles' in dogs, or a human wearing a war mask.

From the upright postures, the bird will move first to a slanted posture and then to a horizontal pose. Taking up the horizontal position is often combined with squirting a white plume of uric acid from behind, presumably making itself comfortable (and reducing mass) before taking flight into what may develop into aerial combat. If it is anxious about the advancing bird then its head will be held high, the wings held half out (ready for flight) and the bird might (if it has time) straighten and bend its legs several times as if ambivalent about leaving. If it is being attacked, it will then swiftly fly off. On the other hand, if the other bird is intruding, the perched bird will hold its head in and level with the body, possibly calling as a warning. By adopting a horizontal stance but trying to keep its eyes on the intruding Buzzard in the sky, it will raise its head, causing the nape feathers to bunch up and form a hump. Again, this serves to make the bird look bigger (like a bouncer on a door puffing out his chest when he doesn't want someone to come in), and the bird will enhance this effect by holding its wings out and dropping them slightly. If at that point the intruder starts to retreat then the bird will straighten up again, having completed an 'assertive bow' (Prytherch 2009). This may happen several times. On the other hand, if the intruder does not start retreating, then the territory holder will likely launch itself and start the display flap while calling, before rising to meet the intruder.

Another response to intruders, more often exhibited by the female, who may remain perched or land heavily on a perch while the male heads for aerial combat with the intruder, is 'wing waving', described most clearly by Prytherch. In this, the bird takes the horizontal, hump-necked posture and flaps half-folded wings in fast and shallow beats. This may be a display in itself or showing ambivalence about flying to join in combat.

Mates can perch very close to one another, and often roost near each other even in mid-winter (Dare 2015). Prytherch (2009) has observed them regularly perching within 30cm of each other, and describes how the second bird to arrive often makes an assertive bow before settling down. He also records how an older female, settling by her adult daughter, tipped her head on its side by 90° while looking at the other Buzzard. He had the impression that this behaviour 'switched off' any aggression her daughter might have felt. The same behaviour can be observed when approaching a tame Common Buzzard with food, and it is easily interpreted as searching for food, like someone peering around to see if it can see something it wants. However, in retrospect it could have been a submissive gesture to its effective (human) parent providing food. Trained Northern Goshawks can turn their heads completely upside-down in this way, making them appear quizzical, and the same behaviour is also seen in falcons. Observations of this nature in the wild are rare, so it is difficult to make a more confident interpretation. It would be useful to collect more systematic records of factors that may contribute to this endearing behaviour, perhaps with the help of other falconers who have more opportunity to observe these traits.

Ground conflicts

Aerial chases can lead to two Buzzards continuing the conflict on the ground, and sometimes birds land next to each other at a carcass or in a good worming field. Weir & Picozzi (1975) produced great descriptions of different Buzzard stances, which are classic pictures of raptor behaviour. A dominant bird will make itself look as big as possible by adopting the 'full angel' posture: holding its wings up, out and slightly behind, pulling its head back and creating a wedge-head, and puffing out the rest of the feathers (Plate 16; see colour section). A bird not quite so sure of itself will make a 'half angel' posture with the wings only half open. Two equal combatants may go into a stand-off, facing each other in a 'griffon posture' (looking like Griffon Vultures *Gyps fulvus*), where the wings are held out further sideways than up. At other times, a Buzzard trying to protect food under its feet will mantle, bringing the wings forward and over the food, lowering the head for protection and spreading the tail out to hide the food (Plate 17; see colour section).

A sub-dominant Buzzard that doesn't want to give up on what it has, or can't see an escape, will take the 'fallen angel' posture: it falls on its back, tail folded underneath its back, with open wings, and presents its feet, prepared to strike. It puffs its feathers out, partly to increase its size, but probably also because the attacker may harmlessly strike feathers rather than its body. Prytherch (2009) believes that the fallen angel position is only adopted by juvenile or very sub-dominant individuals, and that it appears to turn off the aggressive response in the attacker. If the ground fight leads to locked talons, then it is likely they will utter the 'chittering' call that Buzzards make when distressed (Weir & Picozzi 1975). At other times, Common Buzzards may land a short distance away and approach by walking in an aggressive 'neck-hump walk' (as often performed by vultures): the wings are held out and slightly down, the neck is extended and humped, with the head down. Alternatively, the wings can be held right down, virtually dragging along the ground, but with the head more upright. In both cases, the approach is aggressive and usually causes the other Buzzard to move away from the food.

Gender differences

Descriptions of how a pair of Common Buzzards deals with an intruder almost inevitably consider that the male takes the initiative, gets closer to the intruder and is more likely to make a strike (Tubbs 1974, Boerner & Krüger 2009, Prytherch 2009, Dare 2015). However, Prytherch warns that it is not always easy to decide which bird of the pair is the male. Like many other birds, the male needs to impress the female, so he must defend a good territory that the female will choose to be in. However, there are times when females get involved, especially if the intruder is a female (Tubbs 1974). Staffan Andersson and Christer Wiklund (1987) tested the nest defence of Rough-legged Buzzards, a species where male and female can be identified more easily. When either a human or a stuffed Snowy Owl was near a nest, only the male made attack flights. The amount of calling was similar between males and females, but the males met the intruder more often than females, and females did not pursue a fleeing intruder in the same way as males. They proposed that this is a potential selection pressure for reverse sexual dimorphism; being smaller, the male would be more manoeuvrable in an attack, but inevitably it will be a combination of both manoeuvrability and size, and the power that brings, which will provide the best defence. Observations at the nest suggest the larger female intimidates the male partner rather than the other way around.

Again, the practice of flying a tame Common Buzzard among wild breeders in our Dorset study site has brought opportunities for close observation. Unfortunately, the wild breeders were not marked, so it was not possible to know for sure which bird was the male or female, but the observations did demonstrate the differences in how each Buzzard of the pair behaved. Often, only one bird approached, and it is likely from what others have written that this was the male. The other would hang back in the wood. On one occasion in winter, a pair approached the tame Buzzard that had perched in an open Poplar *Populus* spp. tree with no leaves. One bird dropped suddenly to within 5m of the tame Buzzard, pulling up vertically and then flying on. The other bird was much more tentative, landing in the top of the tree about 10m above, and calling, but not approaching more closely. Likewise, the neighbouring pair on different occasions had approached together, but only one bird swooped close to the tame Buzzard.

Pairs spend lengthy periods soaring together, seemingly to advertise their presence as territory holders, whether they are nesting successfully or not. Undoubtedly, some of this also represents courtship behaviour. It is not difficult to understand that some displays, including cart-wheeling as mentioned previously, may have both defensive and courtship functions, because such flying can demonstrate both power and agility for either aggressive purposes or to impress mates.

Conclusions

1. Territorial defence includes exciting aerial displays, with which people can connect because the flights are frequent and highly visible during the breeding season.
2. Aerial contests often use rising air along ridges; a territory drawn as a map joining these conflict sites may differ from a home range based mainly on foraging areas.
3. Territories, like home ranges, are very variable in size and may also be influenced by where neighbouring Buzzards are nesting. As nest density increases, territories get smaller and can become no larger than the core home ranges of foraging Buzzards recorded by radio-tracking, despite having larger territories recorded historically, when densities were lower.
4. The most dramatic and predictable aerial defence is during the breeding period, but resident Common Buzzards defend their territory year-round and therefore influence where juvenile and wintering Common Buzzards can forage and roost.
5. Adult Buzzards are tolerant of their own young in their territories and foraging areas during the winter, and siblings may share such areas in the first winter. Very rarely, offspring from a previous year help adults feed subsequent young.
6. Territorial behaviour may temporarily break down at rich feeding sites in winter. Up to 30 Buzzards have been seen on and around one field rich in earthworms.
7. As well as competition with their own species, Common Buzzards defend their territory from other large raptors, such as Northern Goshawks and Eurasian Eagle Owls, which prey on them.
8. Potential rivals entering the territory owner's airspace are first threatened vocally, then shown deep wingbeats, flashing the white underwing, before the territory holder rises to escort the invader from its territory or to attack it. During such attacks, Common Buzzards sometimes kill each other.
9. If the intruder is very high, and appears to just be passing through the territory, it may not be challenged. This avoids energy expenditure by the territory holder for an intruder that will soon be out of the territory.
10. Territory holders use conspicuous high perches to demonstrate their ownership, watch for intruders, call at ingressions and launch attacks if an intruder is persistent. While on these perches, the intent of the individual can be interpreted from the posture and head shape.
11. It is thought that the male territory holder does most of the territorial defence, with the female providing support where needed, although females can take the lead if the intruder is a female Common Buzzard. However, the experimental evidence is minimal, so more research is needed to quantify sex differences.

CHAPTER 6
Courtship and nesting

Despite all the showiness of the courtship displays, finding Common Buzzard nests takes patience, perseverance and skill. Unlike cliff-nesting birds that are on display all the way as they fly to a ledge, woodland-nesting Buzzards can enter the canopy some distance from the nest. Therefore, watching an obvious breeding pair does not guarantee that a field worker will rapidly find a nest. Buzzards don't always choose the same nest, so just popping in to check a regular nest quickly during incubation can develop into a season-long search. After several searches post-hatching there is a feeling of elation when you can eventually spy a cryptic nest, or relief tinged with embarrassment when young can be heard screaming in July from the wood where there really didn't seem to be an active nest at all. After many years, you become attuned to watching the pair, then slowly walking through woodland closely scrutinising suitable trees, knowing just how far most Buzzards like to nest from the edge and at what sort of height. You get to know how to interpret the behaviour of the birds near the nest; for example, an incubating female that will drop down the trunk and fly off horizontally, low to the ground, trying to keep herself and the nest inconspicuous. Later in the season, you can find the spotted white flecking of the nestlings' excrement projected from the nest to the undergrowth below, increasing in diameter as the chicks grow. When there are chicks, both parents will be far more vocal as you near the nest, almost like they are shouting 'warmer, Warmer, REALLY HOT', or if they are too busy at the nest to notice your approach, there are a few occasions when you can hear their quiet contact calls, or the 'peeps' of young chicks.

Courtship

The aerial displays Common Buzzards perform as their ritual courtship are very similar to the territorial defence we described in Chapter 5. While the initial pairing may be rather short-lived and perfunctory, pairs will go on advertising their union continuously, although always edged with a certain friction between them, especially when the female is tending her young in the nest.

While a few studies have described the arrival of pairs to establish a new territory (e.g., Weir & Picozzi 1983), only Robin Prytherch has spent sufficient time in the field observing known individuals to have recorded the behaviour of territory holding single males being assessed by prospecting females. Despite his extensive fieldwork, there are still very few observations. Although the territorial defence of single birds is relatively low-key compared with defending pairs, the behaviour of an incumbent solitary male changes when a female arrives, appearing agitated and more aggressive. After flying with exaggerated wingbeats, the male settles to observe the female. The female is bigger, and so is usually dominant, and if prospecting for a male will be quite untroubled by the male's display, or even fly at him and bump him off his perch. The male will then make mock attacks, flying fast at the female and then banking away from the potential mate, who again does not respond to his advances. Indeed, she seems to be testing him, to make sure that his territorial ability is good enough for her, so he will try harder, sometimes knocking her off her perch. Prytherch observed a male that was seemingly too timid in his attacks and didn't manage to keep a female to whom he displayed. When the next female turned up, he was much more aggressive, and he seemed to 'win her over' because she stayed. Having gained a partner, the male will feed the female from about a month before egg-laying, throughout incubation and until the chicks are old enough and the female can start hunting again without risking the eggs or chicks cooling while she is off the nest. Copulation will start when he initiates courtship feeding, early in March, and continue even after the last egg has hatched (Cramp & Simmons 1980, Rooney 2013), as found in many raptors (Negro & Grande 2001). The peak mating period is in early April, when Buzzards copulate more than once an hour, conspicuously on exposed branches, poles, crags or even pylons, and often during or straight after territorial disputes. The female solicits copulation by head lowering (although not in every case). The male hovers above her, then lowers onto her back, making a quiet *eez-ka* noise (Prytherch 2009).

A pair of Common Buzzards calls to each other during displays and when they perch close to each other, typically with *Ca-au*. When near the nest and wanting the male to bring food, the female makes the *nya-nya-nya-nya* food begging call of the juvenile. What is less often heard is a nasal sounding *quack-quack*, which is only audible for humans from within about 30m. Once accustomed to the noise from a tame Buzzard it is possible to hear wild Buzzards making the same call, but extremely rarely, when very close; observers not familiar with the sound probably would not attribute it to a Buzzard.

Courtship displays are similar to territorial displays and can be performed together. The pair will rise on a thermal by soaring, most often circling in the same direction and less frequently counter to each other, before sailing about the territory together. Where the gender of the individuals are known it seems that the male leads, and the female follows (Weir & Picozzi 1975, Prytherch 2009). The soaring will be punctuated with display flights and chases. These can involve a 'dive-on-turn-over', when the male shallow dives on the female, lowers his feet, and the female turns over flashing up her own feet, but typically without either bird extending the talons. The feet rarely touch and the female carries on soaring confidently in the territory, rather than fleeing like an intruder. Weir & Picozzi describe 'wing touching', in which the male will fly so close to the female's back that their flapping wings touch, although Prytherch has never seen this activity in all his years of observation, and nor have we, so this may be a rare individual trait.

These displays may be a way for the male to advertise his fitness. There is no published evidence yet for this in buzzards, but there is for related raptor species. Beatriz Arroyo *et al.* (2013) have characterised display flights of Montagu's Harriers, which have a similar display involving plunging and undulations, although they perform an even more manoeuvrable display than Buzzards in terms of twisting and the steepness of the stoop. The study found that these 'skydances' could be a good indicator for both the quality of the male and the site he was defending. Montagu's Harriers breed in colonies, and the pre-laying frequency of display flights was found to be higher in larger colonies. The display was also more vigorous at sites and in years when vole numbers were high, although it did depend somewhat on the weather. Presumably, females could assess the relative quality of the site for nesting from the displaying males, depending on the conditions in which they were displaying, and in the same way it seems likely that the best individual male harriers would have more time to display and more energy to be more vigorous in their displays. Although such a site or mate selection process hasn't been studied in Common Buzzards, and of course it is not as easy to separate the sexes as with the dimorphic plumage of harriers, it seems reasonable to presume that these displays have evolved to be a true indicator of quality, on which basis a female can choose her mate. However, the ultimate test may be provisioning, which would explain the start of copulation coinciding with courtship feeding.

As well as the scope for aerial acrobatics to be a test of a male, female Buzzards do make choices on appearance. Krüger *et al.* (2001) found that Common Buzzards were more likely to choose mates of the same morph as their mother. More generally, where morphs are not so obvious, mate choice could be based on other physical indicators of quality. For example, large birds such as Buzzards cannot replace all their feathers during one moulting season, and if there are insufficient nutrients when growing feathers then the moult will not advance as far during the summer. Just after each moult, it is possible to see the proportion of feathers that have been replaced in the last season, with new feathers looking fresh and rich in colour compared with the faded feathers from previous years. Particularly noticeable are the brown contour feathers of the back, as well as the primaries and secondary flight feathers (Plate 3; see colour section). Less easy for us to see, but feasible for a raptor's sharp eyes to notice, are the stress bars that form if there is a severe sudden shortage of food. Another visible trait is cere colour, which we know is used as an indicator of current health and used for mate choice in other species (Plate 2; see colour section). Humans may also not be as able as Buzzards to see some differences clearly, because birds could be using UV light for some of the assessment. This has been demonstrated far more effectively in Montagu's Harriers (Mougeot & Arroyo 2006), but the same is likely for Common Buzzards.

Common Buzzards are renowned for pairing with one partner and defending their territory, so we tend to think of them as monogamous. There have not been many documented records of Common Buzzards having more than one mate, although Picozzi & Weir (1974) did record seven cases of males supplying food to more than one female (in different nests). Prytherch (2013) also found a male copulating with both females that provisioned nestlings at one nest, although one of them was an offspring of the resident pair. There have also been cases of polyandry. Rafael Barrientos (2006), studying the island subspecies *Buteo buteo insularum* on Fuerteventura, recorded a female mating with two distinguishable males. The trio appeared harmonious, with one male going to incubate the eggs while the female copulated with the other within obvious sight. At first, they incubated four eggs (a relatively large clutch size), but they abandoned the nest before they had any chicks, so perhaps that state of affairs was not sustainable and therefore unlikely to be inherited as a trait. However, better success has been observed elsewhere.

In 2009, a nest in Ireland containing five nestlings was quite a surprise for Robert Straughan and Eimar Rooney (2010). The following year there was an even more remarkable six young, but having seen the exceptional brood size the previous year, this time they were ready to take blood samples for DNA tests of paternity (Rooney 2013). As well as the nest of six, they sampled 53 nestlings from 26 other nests. Only the nest with the six young provided conclusive proof that there were at least three adults (more likely four) contributing to the nest – proving for the first time that it isn't only monogamous pairs that produce viable young. Since then, nest cameras deployed by George Swan (pers. comm.) in Cornwall, UK, have also observed three adults simultaneously provisioning fledglings at the same nest. In this instance, the relationship of the third bird to the breeding pair was unclear. It could, as Rooney (2013) had observed, be an example of polygamy or, in keeping with the observations of Prytherch, it might have been a non-breeding juvenile from the previous year's brood. Really, we should not be so surprised these days, because the introduction of genetic analysis has found that more than three-quarters of supposedly monogamous bird species have extra-pair fertilisations (Griffith *et al.* 2002).

Nest-sites

Although nest-sites of Common Buzzards are as variable as their plumage and diet, the majority are reasonably similar. The essential requirements are protection from predators and weather, combined with access to the nest and food nearby. Nests tend to be high in trees (Plate 18; see colour section) or on cliffs, which gives protection from many mammalian predators, but if these threats are very low Buzzards can nest on the ground, like Rough-legged Buzzards do on the steppes. Most people consider Common Buzzards to be tree-nesting birds because more people live in lowland areas, where the presence of woodland is critical for breeding Buzzards (Rodríguez *et al.* 2010), than live near cliffs. Therefore, we will start by looking at the more common tree-nesting sites, before looking at where Buzzards nest on cliffs.

The most common characteristic of Buzzard nests-sites reported in the literature is that they are found in relatively small woodlands, or at least that they are close to woodland edge (Tubbs 1967, Joenson 1968, Picozzi & Weir 1974, Nore 1979, Kostrzewa 1987, Jędrzejewski *et al.* 1988, Hubert 1993, Goszczyński 1997, Penteriani & Faivre 1997, Dare 1998, Krüger 2002b, Sergio *et al.* 2005, Turzański & Czuchnowski 2008). If nests are in larger woodlands, then the preferred woods have a relatively long edge (i.e., they are irregular in shape, which increases the amount of edge to internal area). For example, Javier Bustamante and Javier Seoane (2004) found that Buzzards liked 'forest inter-digitated with open areas'. Goszczyński (1997) showed that nest density decreased as the size of a wood increased, even though bigger woods have more nests, presumably because the nests were around the edges and the interior was not being used. One piece of research found Buzzards further from the edge than usual (mean 660m, Zuberogoitia 2006). This was in a large area of managed plantation forest with no pastoral edges. The nests were restricted to small patches of more mature trees and may have been influenced by competing woodland raptor species, such as Booted Eagles and Northern Goshawks. The density was also much lower than in a comparison site in the same study.

Given that Common Buzzards hunt mainly in more open areas, nesting near the edge of woodland gives them quick access to hunting grounds, and shorter distances to carry prey back to the nest, reducing the energetic expenditure during the demanding brood-raising period. On the other hand, they need a secure place to nest, and it might seem more secure to put the nest further into the wood where it is more hidden. However, that may be a very human perspective. We find it harder to locate nests within woodland, but for flying nest-raiders like corvids or raptors, or mammalian predators that can climb trees and work more on scent and sound, it may not make much difference where the nest is within woodland. In fact, in terms of protection it may be better that the adults hunt in fields as close to the nest as possible so that they can monitor the nest and return quickly should it need defending.

It also seems that woodland is not really necessary, because Common Buzzards sometimes nest in the open. For example, we had one nest in an isolated hedgerow tree (Plate 19; see colour section), as have Tubbs (1972), Roche (1977), Hubert (1993) and Dare (1998). In an area of Northern Ireland, where there are relatively few woods, hedgerow trees were the most common location: 30 per cent in hedgerows, compared with 21 per cent on forest edge and 13 per cent in small clumps of trees (Rooney 2013). Likewise, in North Wales, Dare (2015) found nests in isolated trees or groups of trees on otherwise treeless moors, and the same is common in Scotland (Plate 20; see colour section). That said, we had only a few isolated hedgerow tree nests out of hundreds of woodland nests in Dorset. Hedgerow nests are mentioned rarely in the literature, so it is likely that if there was an unoccupied wood sufficiently close, then a pair of Common Buzzards would use that, because a group of trees will provide better protection against wind-driven rain. Indeed, Włodzimierz Jędrzejewski *et al.* (1988) noted that Common Buzzard nests were in denser stands of trees within woodlands, possibly affording greater protection against Northern Goshawks, just as Eurasian Sparrowhawk nests are more confined to thicker woodland when Goshawks are present (Newton 1986). On the other hand, in coastal woodlands in Italy, where rain is less frequent and there are more large raptors, Salvati *et al.* (2001) found that Buzzards did not occupy woodlands less than approximately $0.5km^2$.

Another aspect of nest placement that seemed common in our study was that if the woodland was on a hillside, then the nests were most often towards the lower edge of the wood, not the upper. Only Penteriani & Faivre (1997) have recorded this in the literature, but it is probably a common characteristic elsewhere. At first, we thought having the nests so near the bottom of the wood surprising, because more often Buzzards would be seen perched high up in the wood, and we assumed that they would be perching close to their nest. However, these high perches were probably better lookouts for territorial birds scanning for intruders, whereas it is likely that the lower part of the wood is not only better protected from the weather, but also requires less effort to carry in a heavy prey item captured on low ground. In contrast to our area, Dare (2015) found that in the wooded Snowdonia valleys, nests were positioned one-third of the way down from the top edge of the wood. This is not altogether surprising, if we suppose that nests are placed as close as possible to the hunting grounds. Then, if the slopes are wooded right down to the stream or river that etched the valley, as is often the case in Devon, the Buzzards are far more likely to be hunting on pasture or moorland above the wood, so they would be closer to top of the wood, and nearer the higher territorial perches as well. On the other hand, these nests near the top of the wood do not support a hypothesis of Buzzards wanting to nest lower for added shelter. So, perhaps the energetics of food delivery is more important than shelter from the weather. Indeed, Brüll (1964) noted that Northern Goshawks tended to have relatively easy flight lines along tracks or through more open woodland to their nests, probably because flying through dense woodland is dangerous for a bird with limited manoeuvrability when carrying heavy prey. This could also help explain why Buzzards tend to nest near woodland edges.

As woodland does not supply the majority of Common Buzzards' food, they do not appear to select woodland types based on tree species (Jędrzejewski *et al.* 1988, Selás 2001, Sergio *et al.* 2002, Bielański 2006). However, trees have to be of a size to support a large nest and its inhabitants. More mature trees not only have larger branches to support nests but also branches and trunks more widely spaced, which is easier for a Buzzard to fly through. Therefore, some studies report that only stands containing trees over a particular age are used (Hubert 1993, Selås 1997, Bielański 2004), and this can be more important than the size of the woodland (Zuberogoitia *et al.* 2006), provided those old trees are sufficiently near to an edge. It would make an interesting study to look specifically at these theories to see if they might affect where Common Buzzards, or other raptors, build woodland nests. However, it would need comparable measurements and rigorous analyses from many diverse sites to test the theories properly.

Cliffs are more commonly used at higher altitudes, partly because there may be fewer and shorter trees, and also because mountainous regions often have more cliffs. There has been some debate in the literature about whether Common Buzzards 'prefer' to nest on trees or cliffs when both are available. Brown (1978) and Tubbs (1967, 1974) suggested that cliffs would only be used as a last resort if suitable trees were not available. However, Geoffrey Fryer (1986) found Buzzards nesting on cliffs when he considered there were abundant suitable trees nearby, corroborated by the fact that in a few territories both cliff and tree nests were used in consecutive years. In one case, there was a particularly recognisable female associated with both the cliff and the tree nest in different years. This suggested that the same pair could even swap between the two, although we don't know that it wasn't a different male in each case. Nevertheless, he did find that Buzzards never used cliff nests for more than one consecutive year. This could suggest that the cliff nests were never quite satisfactory, since there were plenty of tree nests in the same area that Buzzards used from year to year, in one case at least nine years in an oak in the same study area. The opposite trend was true for Sergio *et al.* (2005), where cliff nests were more likely than tree nests to be used in consecutive years, but notably this was in an area where 76 per cent of the nests were on cliffs, so the latter may have been the better sites in that area. Alternatively, Buzzards may have had an imprinted preference from the type of nest from which they fledged, as is the case for Peregrines. Michael Holdsworth (1971) working in North Yorkshire (UK, Appendix

3) thought most pairs preferred cliffs and used trees occasionally, although some used mainly trees while having cliff alternatives. Each territory consistently contained more than one nest-site. Some sites were definitely used more frequently than others, but it was rare for any to be used consecutively for more than two years.

As with nest-building in trees, cliff-nesting Common Buzzards are looking for a structure that is relatively inaccessible to predators, and other aspects of the cliff are not so important. While the open nature of cliff sites will tend to facilitate the detection of, and defence against, flying predators, access by mammalian predators is a particular problem on cliffs. Graham Austin (1992) found that cliff nests invariably had an overhang or vertical face immediately below the nest, presumably making it safer from mammalian predators trying to climb up. Just as Peregrines and Lesser Kestrels *Falco naumanni* use cathedrals and other buildings as artificial cliffs, so Common Buzzards use pylons (Melde 1983), or ruined houses, for example, in Doñana, Spain (Castillo-Gómez & Moreno-Rueda 2011). However, unlike some other city-dwelling species, Common Buzzards don't yet appear to use occupied buildings. In our analyses of land-cover types (Chapter 4), human settlements tended to be avoided. However, it is not impossible that over time Buzzards may start to inhabit built-up areas (Chapter 10). Indeed, we have seen Common Buzzards in our nearby suburban parks and perched on lamp posts in built-up areas.

If predators are sufficiently scarce, then Common Buzzards can nest successfully on the ground in open areas. In Finland, one nest was found on a 4m-high boulder (Solonen 1982). Nests have even been seen among seaweed on a beach, on sand dunes and even under a gorse bush (Hardey *et al.* 2009).

Nest-building

Tree nests

Most of the quantitative literature suggests that Common Buzzards are not in the slightest bit fussy about the species of tree they nest in (Tubbs 1967, Joenson 1968, Picozzi & Weir 1974, Fryer 1986, Jędrzejewski *et al.* 1988, Hubert 1993, Jędrzejewski *et al.* 1994, Cerasoli & Penteriani 1996, Selås 1997, Dare 1998, Krüger 2004a, Bielański 2006, Turzański & Czuchnowski 2008, Prytherch 2016). Some studies have given percentages of nests in particular species, but not shown selection, because there were no tests against what sort of trees were available. For example, oaks supported 50 per cent of nests in central France (Nore 1979), 78 per cent of nests on the French Saone plain (Roche 1977) and 44 per cent of nests in Somerset, UK (Prytherch 2016). In Finland, 66 per cent of nests were in spruce (Solonen 1982), and according to a report in Germany, 70 per cent were in Beech trees (Hubert 1993). In our Dorset study area, the majority of nests were either in tall pines, of which the highest were in plantations of Corsican Pines *Pinus nigra*, or in oaks, which were mostly somewhat lower (Figure 6.1).

Any selection seems likely to be a local phenomenon. This was illustrated by Iñigo Zuberogoitia *et al.* (2006), who found Common Buzzards selected pines over eucalyptus trees in a northern Spanish site, but in southern Spain there was no significant difference between the use of pine or eucalyptus. If there is selection, it is most likely to be because

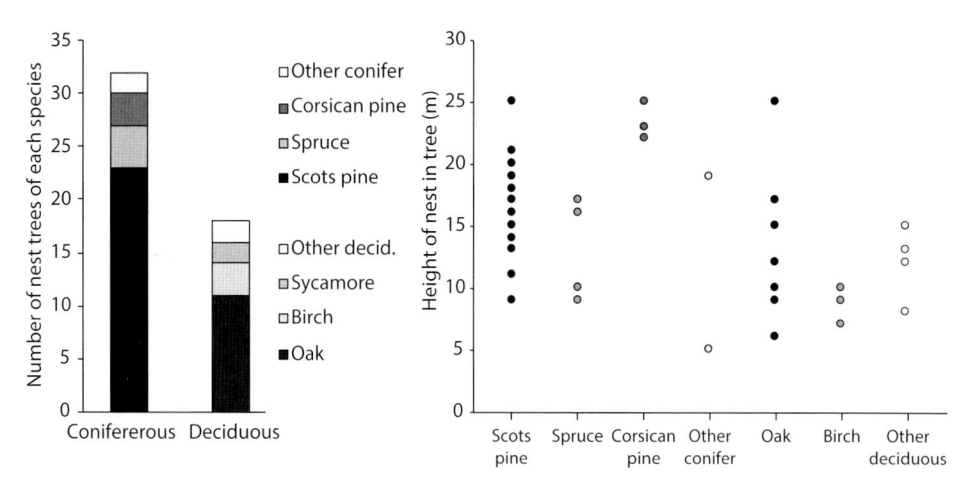

Figure 6.1. *The nest trees (left) used by Buzzards in our Dorset study and the heights of nests (right) in them.*

particular tree species are providing an important physical characteristic not offered by the other trees available. In the last example, the pines could have provided more cover, in terms of denser foliage, than the eucalyptus because the northern area was exposed to greater rainfall. The requirement for foliage to provide cover for the nest has been proposed by several researchers. Christine Joubert (1989) thought that Common Buzzards selected evergreen trees because deciduous trees had no leaves to hide the nest when it was being built in late winter. Sergio *et al.* (2005) reported that nest trees were more likely to have Ivy *Hedera helix*, affording visual protection when building the nests in spring before the leaves had grown on the deciduous trees. More recently, Prytherch (2016) reported 52 per cent of 108 nests in northern Somerset having Ivy around them, and that those with Ivy were slightly lower than corresponding nests without. When searching for nests, Ivy certainly hides nests very well from researchers on the ground! However, contrary to expectation, survival in Ivy-covered nests was significantly lower. Buzzards can nest in the most surprising of tree species. For example, Dare (1998) found one in a Rowan *Sorbus aucuparia* and another in a hawthorn *Crataegus* spp., and we've found nests in rather flimsy Silver Birch *Betula pendula* trees. However, while the species of tree is not of overriding importance, most Common Buzzards probably choose individual trees quite carefully.

For instance, Buzzards do tend to choose the more mature trees (Selås 1997, Bielański 2004, Zuberogoitia *et al.* 2006), presumably because these are both higher, making the nest harder to scent or further to climb for mammalian predators, and the larger branches provide sturdier foundations for the structure. There can be clear selection for trees damaged by wind or disease, because lost branches often provide a more spacious hole in an otherwise quite dense canopy, and sometimes part-broken branches or strongly leaning trunks provide a better platform on which to start making a nest. Use of damaged trees was common in our study area, typically in areas with Corsican Pines. When healthy, these pines not only have very dense branches, but also many branches angled down slightly, making it difficult to build a stable nest. However, if damage to the trunk had created a fork or other protuberance, then that could make enough of a horizontal platform to gain purchase for nest-building (Plate 18; see colour section). Unfortunately, deformed trees in modern plantations can be a scarce resource because they are vulnerable to felling by forest managers who want uniform,

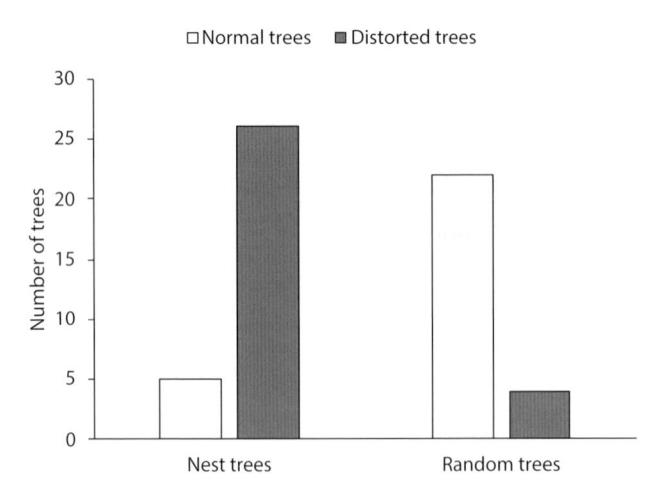

Figure 6.2. *Bielański (2004) noted that pine trees used by Common Buzzards (n = 17) and Northern Goshawks (n = 10) at nest locations in a Polish forest tended to be those most distorted through leaning, bent or forking trunks, when compared with trees picked randomly from the same woods.*

straight trees (Figure 6.2). In the UK, where a lot of government forestry land is now primarily used for recreation, there are plenty of deformed trees.

Although there is a tendency for Common Buzzards to build their nests high in mature trees, as with their selection of tree species there are always exceptions. From those articles that give nest heights, the averages are typically between 11 and 15m, with low nests at about 4m and the maximum height only restricted by the height of the trees (Tubbs 1967, Picozzi & Weir 1974, Hubert 1993, Cerasoli & Penteriani 1996, Dare 1998, Sergio *et al.* 2002, Zuberogoitia *et al.* 2006). Over half are in the top third of the tree (e.g., Hubert 1993) but not in the top canopy (Bielański 2004) because the nests are most often where the main trunk of the tree forks, leans, bends or has strong side branches (Cerasoli & Penteriani 1996, Zuberogoitia *et al.* 2006, Bielański 2006, Turzański & Czuchnowski 2008).

Being close to the trunk is very helpful for researchers climbing to the nest because nests out on small branches are a good deal more risky and difficult to reach. Of course, these two aspects are not unrelated; Buzzards build large nests, and there is much better structural support next to the main trunk, plus shelter from the wind blowing from some directions. Nevertheless, some nests can be further out on branches, provided the lateral branch is substantial enough to hold it. Fieldworkers visiting nests then have to shuffle along the branch in an ungainly fashion, often after climbing higher to attach a safety rope. A study in Poland found nearly 90 per cent were in a fork with the main trunk of a deciduous tree, but in conifer trees 22 per cent of nests were more than 1m from the trunk (Bielański 2006). As an extreme, nests can sometimes be found 3m, or even 5m from the main trunk (Zuberogoitia *et al.* 2006). One of our nests far out on the side branch of a Beech was simply too dangerous for us to access to measure and mark the young.

As well as the lack of structural support, another reason given for not nesting in the top canopy could be concealment from predatory corvids or large raptors flying above, combined with poor visibility from below provided by the lower branches and understorey of the surrounding trees. There will still be access horizontally through the canopy of the large overstorey tree selected for nesting (Hubert 1993, Bielański 2004). Marina Cerasoli

(1996) reported that nest trees had 16 times the crown volume compared with other trees in the area. On the other hand, some nests can seem very exposed. Particularly memorable for us was climbing a dead Scots Pine on a hillside because it felt very unsafe, with potentially rotten branches and a long way to fall. Of course, this was a tree with no foliage for cover from below, but it was a real exception.

Choice of nest tree may also depend on existing nest structures. Common Buzzards are known to use the nests of other birds. Harri Hakkarainen *et al.* (2004) found 26 per cent of Buzzard nests had been occupied by a different (smaller) raptor in the year before, and Dare (1998) found Buzzards adopting a small Carrion Crow *Corvus corone* nest. From our observations it also seemed possible that some nests were built on old Grey Squirrel dreys. On the other hand, Buzzards have to put up with their nests being taken over, especially by Northern Goshawks and Black Kites, as we'll discuss below when considering competition and nesting success.

Nests on cliffs, buildings and the ground

Whereas Common Buzzards will nest in trees at the base of a hillside, if ledges permit they tend to nest nearer the top of a cliff (Sergio *et al.* 2005, Rodríguez *et al.* 2010). Just as Buzzards build nests high in trees, they build high on cliffs probably to make the nest as inaccessible as possible for ground-based predators. Sergio *et al.* (2005) recorded more information that supports this: the cliffs chosen by Buzzards were higher, wider, more elevated and steeper than was commonly available in the same area. In contrast, Davis & Saunders (1965) give descriptions of individual cliff nests on Skomer Island, UK, where there were no mammalian predators. The nests showed considerable diversity, from simple protrusions to roomy, grassy ledges, old Northern Ravens' nests and ledges 'devoid of any nest material', under a slight overhang and on cliffs with different orientations. One site received particular attention: 'Two west-facing ledges near the cliff-top with tall dense cocksfoot grass screening them and one with a lush growth of Thrift *Armeria maritima*, Red Campion *Silene dioica* and Yorkshire Fog *Holcus lanatus* about two thirds down the gully, facing north have been occupied'. In Tenerife, 42 per cent of Buzzards used cavities on cliffs as opposed to ledges (Rodríguez *et al.* 2010), but the use of cavities for nesting is not widely reported. Cliff cavities may have been safer than ledges on the friable rock of that volcanic island.

One case has also been recorded of a Common Buzzard nesting on an abandoned building. Carlos Castillo-Gómez & Gregorio Moreno-Rueda (2011) found a nest on a wall of a semi-collapsed, solitary house in south-east Spain. The nest was only about 3m above ground, lower than we expect from tree nests. The nest seemed relatively loose, maybe a first breeding attempt, but it raised three chicks to fledging. Unfortunately, the wall was dismantled before the following breeding season, so it was not possible to see whether the pair would try breeding there again. The surroundings were flat, intensively managed arable land, so this was the only raised structure for the birds to nest on other than Geolit, a large science and technology park 500m away, and there was an abundance of Rabbits, hence the success of the nest. So, yet again the flexibility of the Common Buzzards allowed success where other species might not have managed.

Recently in the UK, there have been online reports from the Scottish Raptor Study Group that Buzzards in Easter Ross have nested on the roof of a disused building as well as on the ground at the edge of a drainage ditch. Nesting on the ground seems to be more frequent in northern areas where there is a low density of mammalian predators, especially in areas where these may forage less frequently, for example on moorland as described in 1920 by Arthur Brook, or on a beach (Hardey *et al.* 2009).

How many nests?

When Common Buzzards use artificial nest-sites, it's usually because there aren't alternatives, and therefore there may be only one nest in a territory. More commonly, Buzzards build several nests per territory, as is apparent from almost any study (Tubbs 1967, 1974, Holdsworth 1971, Picozzi & Weir 1974, Fryer 1986, Dare 1998, Reif *et al.* 2004a, Prytherch 2013), so it is a widespread phenomenon. Tubbs (1974) found that alternative nests tended to be in the same species of tree as the nest tree. Not all nests in a territory are necessarily maintained to the same degree, and so some may just be fringe nests from previous years to which the Buzzards had made some repairs. Typically, with sufficient searching, a territory will have about three alternatives within 200m of each other. It is difficult to be more accurate, because it is not easy to know whether alternative nests are missed in the searches. Some other species with precise nesting requirements, but which are not dependent on vegetation, will use the same nest (or at least the same colony position, e.g. many seabirds or Sand Martins *Riparia riparia* and Common Swifts *Apus apus*) from year to year. Still, using a different nest each year is probably the norm for the majority of bird species; most birds are so small that they have to cope with vegetation change at the micro-habitat level, such that site characteristics have changed by the next breeding season, anyway.

Common Buzzards, which probably evolved from a species that had to be adaptable to locally fluctuating prey abundance (Chapters 1 and 2), might need to be able to change nest-site from year to year according to the feeding conditions. However, from a purely energetic point of view, if resident in the same area, building more than one nest seems a big waste of energy that could be better spent hunting for food. Thus, other explanations must be sought for building alternative nests. Some have speculated that multiple nests are to do with territorial display, providing a permanent visual display to warn any other Buzzards that are thinking about using the territory that the area is already occupied, rather like scent-marking in mammals. So, perhaps this behaviour is part of a suite of activities that defend a territory, like the great aerial displays into which Buzzards also put a lot of energy (Chapter 5). Given the effort needed to build nests, perhaps the number within a territory is an indicator of the number of years that the territory has been successfully used. A male may concentrate on the main nest, and just repair and decorate old nests with some fresh greenery, an activity that requires relatively little effort but highlights the number of nests and indicates to the female what a good territory the male has. On the other hand, Carmelo Fernández & Paz Azkona (1993) found that the chance of Golden Eagles *Aquila chrysaetos* changing nests between seasons was correlated with the number of nests they maintained (on average, four, with some having seven alternatives), and the number of nests was purely a factor of nest-site availability. It is worth noting that changing nest didn't alter productivity of these eagles.

In fact, nests can last for many years, so Common Buzzards do not need to build many nests each year to accumulate the handful that are normally spotted by human observers. For example, a population model run by Jiménez-Franco et al. (2014a), based on many years of observation of nests in the Aleppo Pine *Pinus halepensis* forests of Murcia, in southeast Spain, calculated that the median lifespan of nest structures built and used by Common Buzzards, Booted Eagles and Northern Goshawks was 12 years. The nests could be used by more than one species during that time, but it seems consistent with our experience of Buzzard nests in an area that lacked these competing species.

The same research group also sought specifically to investigate whether building new nests reduced productivity, because of the energy required for nest-building (Jiménez-Franco et al. 2014b). They could differentiate between birds setting up completely new territories, new arrivals in a territory that had been used by other Buzzards before, and those staying in the same territory. From 14 years of observations, 90 per cent of new arrivals in a previously used territory used the same nest as the previous owners, the others building new nests. That was very similar to the 85 per cent of those that stayed in the same territory using the same nest as the previous year, with the remaining 15 per cent evenly split between using an alternative nest and building a new nest (Figure 6.3). Essentially, building new nests occurred less than 10 per cent of the time if there were already existing structures. In Dorset, we did not record systematically how often a changed nest represented an alternative, because it was hard to spot nests in Ivy-covered trees. However, of 96 occasions when we recorded whether nests were new or changed, 75 involved reuse of the previous nest. Thus, the rate of building new nests may not have differed greatly from that in Spain.

Building new nests so infrequently might indicate that Common Buzzards do not want to waste energy that could be put into egg-laying and chick-rearing. Surprisingly, reproductive output was actually slightly higher for new nests in Murcia, but it wasn't a significant difference, which was probably due to the very small number (four) of newly built nests; thus, it could have been due to chance. On the other hand, perhaps the increase in productivity resulted from vigorous new males entering existing territories and starting a new nest after replacing a senescent male. Having said that, there is evidence that in general the first nesting attempt is not as successful as later attempts in long-lived birds (e.g., McCleery et al. 2008), and that also held true for one study of Common Buzzards (Krüger & Lindström 2001).

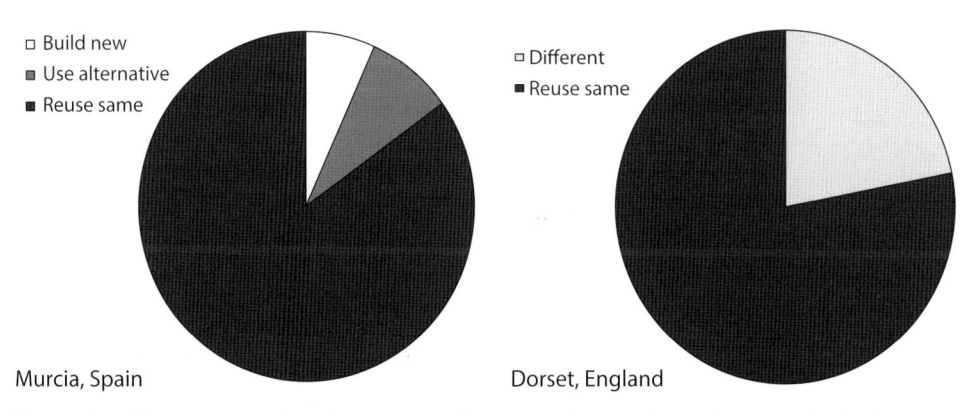

Figure 6.3. *The proportion of nests that were reused by resident Common Buzzard pairs was similar for 100 in southern Spain (Jiménez-Franco* et al. *2014b) and 96 in southern England.*

There are other possible explanations for having a number of nests at a site. For parasites with short life-cycles, or bacterial pathogens sensitive to UV light from the sun, changing most years between a number of alternative nests could provide the necessary degree of sanitation. Changing between years could also serve to confuse climbing predators, such as martens (*Martes* spp.), which might remember from a previous year that whitewash and pellets below a nest one year provide cues to a tasty meal of eggs the next year. Clearly, there are many factors involved, so discussions of the reasons for alternative nests by Buzzards and other large territorial raptors are likely to run and run. Nevertheless, having a variety of alternatives does give a female a choice of where to lay her eggs, with many subtle past and present experiences playing on her choice each year.

Decoration

The tendency of Common Buzzards to decorate several alternative nests during courtship makes the fieldwork of finding nests more difficult until the young are big enough to produce the tell-tale whitewash of uric acid under the nest, or are screaming expectantly for their parents to come and feed them. The decoration of nests with fresh green vegetation makes the rim of the nest look more broken than unmaintained nests that have become flat-topped during the winter winds (Prytherch 2013). Just as they show no preference for tree species to build in, so Buzzards are not selective about what vegetation they use to decorate nests. Typically, it is the same as the surrounding trees, but it can also include ferns, rushes, heather and moss (Tubbs 1974, Fryer 1986). There doesn't seem to be much evidence for Common Buzzards decorating the nest with objects that aren't plant material, whereas other species such as kites seem to collect all sorts of junk. Although, as an exception, at a Buzzard nest photographed by Julian Driver in mid-Wales the eggs were on scraps of white paper (Dare 2015).

The greenery typical of active Common Buzzard nests is described as decoration because, although it can be woven into the nest, pieces are often just rested on the rim or line the nest-cup (Plates 7 and 21; see colour section), and sometimes they fall to the ground. These pieces of vegetation are therefore not a structural necessity. Regarding the lining of the nest-cup, some have suggested that it cushions the eggs, but with the observation that sharp sprigs of Holly may be used (Tubbs 1974, Fryer 1986) it seems strange the female would consider it good 'cushioning' for her as well. A number of other explanations for nest-cup lining have been proposed for hawks, comparable tree-nesting raptors, including a role in insulation during incubation by Eurasian Sparrowhawks (Newton 1986), after Brüll (1964) noted that using lining to change the shape of the cup helped regulate Northern Goshawk incubation temperature. A sanitation role was suggested by Schnell (1958) who, with other authors, noted that both sexes engage in nest-lining activities through the incubation and rearing periods; vegetation contains many antibacterial agents that may reduce risk, for example, of *Clostridium* infections, which can be fatal for raptors (Kenward 2006).

Tubbs (1974), Newton (1979) and Blezard (1933) have all regarded nest decoration as marking territorial ownership. Fryer (1986), too, thought this the most likely explanation, especially as he had observed Buzzards putting greenery on cliff sites not suitable for nests, that is, decorating the general area. However, he also noted that green linings were not a necessity as he had found active cliff nests with no greenery whatsoever. The continued addition of greenery during chick-rearing (Picozzi & Weir 1974, Tubbs 1974, Dare 1998) may be because it is part of the pair-bonding, although it could also be that the male always feels motivated to bring something to the nest. On this basis, if he has no food to offer then vegetation is brought to pacify a female that tends to fiercely defend her young (Christine Hubert pers. comm.). Having said that, Dare (2015) considered that it was predominantly the female that added green foliage to the nest when the chicks were young. Perhaps that was a difference in behaviour, or maybe it is simply because the male doesn't get an opportunity as the female is ever-present and defensively guarding the offspring. In fact, the addition of green materials can continue into the autumn (Fabrizio Sergio pers. comm.), again suggesting it is some sort of signal, rather than necessarily to do with chick-rearing.

An elegant study of Black Kites showed how nest decoration can act as an honest signal about the occupier. Sergio *et al.* (2011) made some great empirical observations and then manipulated the nest decoration to see what the kites were choosing and what it said about the owners of the nest. Kites are easier to study than Buzzards in this respect, through a predilection for plastic objects, especially white plastic, which makes the decoration very obvious. This allowed the fieldworkers to provide pairs with decorative materials that were not common in the environment, which facilitated observation of use and effects. Birds in the prime of their life, aged 7–15 years old, when experience and authority is greatest and brood size highest, had the most naturally decorated nests. If nests were artificially augmented with plastic decorations, it was the youngest and oldest birds that were most likely to take the decorations off the nest. This was consistent with a theory that they didn't want to encourage conflict with other kites by 'lying' about their vigour. Those with more decorations did have higher reproductive success, which is logical given their age, and if their decorations were artificially increased, then these birds seemingly became bolder and more likely to attack trespassers. While we cannot apply this directly to Common Buzzards (individuals and populations, let alone species, can have different behaviours), it gives some hard evidence that nests and their decorations can be used as signals of fitness to conspecifics.

It is sometimes suggested that obvious nests may be cues about the suitability of nesting for conspecifics. These are part of the 'public information' by which individuals are able to assess the quality of an area by observing phenomena such as breeding success or density of conspecifics (Danchin *et al.* 2001). Kurt Burnham *et al.* (2009) even went so far as to say that nests were 'ecological magnets' for Gyrfalcons *Falco rusticolus*. Combining this concept with their own observations, María Jiménez-Franco *et al.* (2014b) have suggested that preserving existing nest structures and associated trees may help to attract raptors searching for new territories. Testing that possibility could be important for management, because there would be scope on one hand for encouraging nesting in new areas by decorating artificial nests; on the other hand it might also be possible to discourage nesting in problematic locations by destroying any existing nests during the winter, when there is no chance of harming nestlings.

There is also the question of whether building more than one nest is a true adaptation, or 'runaway selection'. The tail of a male peafowl *Pavo* spp. is the classic example of runaway selection, where the evolution of the character may be based on female choice rather than the survival of the individual with the trait. The theory proposes that progressively longer and more decorative tails are selectively chosen by peahens, because the sons who inherit that trait are more likely to be chosen by the better females in the next generation, therefore securing more offspring. Of course, the long, heavy tail increases energetic costs, and could be a hindrance when trying to avoid predators, so the selection has 'run away', or diverged from natural selection, and instead the trait is driven by sexual selection. In the same way, perhaps the number of nests could be something a female chooses that does not necessarily indicate a better male or a more productive territory. Zahavi's 'handicap theory' maintains that these apparent sexual selection traits may actually be a true indication of value, in that if the male peafowl can survive with the handicap of carrying around that great tail, he must be strong, a trait that a female will want for her offspring (Zahavi 1975). We know this happens in other birds, such as bowerbirds (family Ptilonorhynchidae), whose males build over-the-top decorative bowers to woo females. So, in the same way, perhaps a female Common Buzzard will choose a male that can make and decorate more nests as an indication that he has time on his hands after feeding himself. Or perhaps she is looking at the history of occupancy, and if it looks as though the male's territory has been used for many years, it's likely to be a good one. Decorating each nest will draw attention to them all.

The size of nest is very variable. Larger nests have to be more stable to survive winter winds and may be better anchored by their mass, whereas small nests may be more at risk of being blown out over the winter. Some nests become more than 1m across and nearly 1m deep, although never more than 2m in the New Forest, according to Tubbs (1967, 1974), perhaps because too much height and mass can also create instability. Prytherch (2016) gives more details of 104 nests he systematically quantified. Measuring the load-bearing platform, he found nests were only exceptionally 80–90cm across the top, more commonly 50–70cm across, and provided a mean surface area of $2,212cm^2$ (range $1,244–4,196cm^2$). He measured when the chicks were around 32 days of age, and most had a nest-cup of less than 5cm depth, maybe because of the addition of fresh greenery throughout the nesting. The mean depth of the whole nest was 28.5cm (range 13–55cm). He noted a trend of larger broods being found in nests with greater surface area and shallower cups. Like Prytherch, we very occasionally had new nests that were rather flimsy, that collapsed the following winter or fell apart even before the nestling left the nest.

Nesting can start before winter is over. For example, in the UK Dare has seen a pair starting to visit the nest in mid-January, but more usually males start carrying nest material in mid-February to mid-March (Tubbs 1974, Dare 1998, 2015). Nest-building itself seems to be a 'furtive business' (Dare 1998) so there is not much information about the actual process of constructing the nest except that it can happen very quickly, within a few days (Tubbs 1974). Nowadays, we can record some of the secretive activity of birds with cameras, but these are unlikely to help much with providing information on the nest-building process, because cameras are not normally mounted until the nest is properly built and the eggs laid; otherwise, there is a real risk of abandonment.

We know that nest-building and maintenance continues throughout incubation and chick-rearing. Nevertheless, the platform needs to be sufficiently sturdy to support both adults and a brood before the female starts to lay. Beyond that point, other factors may influence when laying occurs, as we shall see in the next chapter. Once she lays, there is less need for a courtship performance, and a greater need to care for the resulting clutch. So, at that point the dramatic aerial displays with both partners virtually stop, and females become far less conspicuous as they spend nearly all their time incubating their precious eggs.

Conclusions

1. Common Buzzards perform dramatic aerial displays, including roller-coaster flights, great stoops to the nest and carrying foliage for nesting, which are partly to court their mates but also to demonstrate their active defence of the territory to others.
2. Courtship, like territorial defence, is all about bluster, but it is likely that there are some true indicators of fitness in the display, as has been shown for skydancing in other species. Notably, the start of copulation is associated with courtship feeding.
3. Common Buzzards typically form long-term partnerships and only change mates after the death of one of the pair, although there are proven cases of bigamous male and female Buzzards.
4. Nesting typically exploits a raised site to avoid nest predators, generally trees and cliffs, but if elevated platforms are absent and predators scarce then Buzzards may nest on the ground.
5. Selection of trees for nesting seems to depend much more on physical structure, including whether there are suitable stable branches on which to build a platform, than on the species of tree.
6. Cliffs can be used for nesting. There seems to be no clear preference between trees and cliffs, but tree nests are more common due to the higher availability of trees across the whole geographic distribution.
7. Typically, a pair will decorate multiple nests in woodland with fresh green vegetation, even if only one nest is properly maintained for chick-rearing.
8. The function of the nest decoration is unknown, but may well be a signal of the occupancy and current vigour of the territory owners for the breeding season. Fresh decoration is added throughout the breeding season, until the youngsters fledge.

CHAPTER 7

Incubating and chick-rearing

From earlier chapters, we know what Common Buzzards need in order to settle in areas and construct nests, and the ways in which they go about defending these resources. Now, we will take a look at their production of young. Breeding brings a whole new set of challenges. The last chapter ended with nest-building and we now continue the story with egg-laying, chick-rearing and an exploration of factors that affect the success of nests and the lifetime reproductive success of individuals.

Finding nests could be difficult and long-winded, and that was the first stage of our Dorset fieldwork during the breeding season each year. Next, we wanted to climb up to nests to record brood size and the age of the hatched chicks, so that we could return at a later date to attach radio-tags to the fledglings when they were almost fully grown. It was exciting but arduous work. Having to carry a ladder, climbing equipment and tagging kit for long distances, or through thick rhododendrons, could be an exercise in itself, before gazing up a trunk that narrowed into the sky like Jack's giant beanstalk. Nests looked high when psyching yourself up to climb, but they felt even higher looking down from the nest while trying some awkward manoeuvre that required careful balance. Some nests were more difficult than others to reach. The worst were in tall pines or Beech trees with few branches, especially if the trunks were too thick to reach around. Others might be easy enough to climb up to, but then to reach out and over the nest to deal with the chicks was a different matter. Some required a scramble through Ivy covered in the chicks' faeces, only to have your head poke out at the edge of the nest, laden with half-eaten carcasses being enjoyed by wasps. But by hook or by crook (or harness and carabiner, to be precise) our team was defeated by fewer than five nests in 15 years of climbing (Figure 7.1).

Once at the nest on the first visit, we would count the clutch or take a wing measurement to age the chicks. When returning at a later date to tag the chicks, at the highest nests the

Incubating and chick-rearing

Figure 7.1. *A ladder helped surveyors to conserve their energy when climbing to nests; however, if coming down with chicks in the bag, it was faster for the climber (here, Maarit Pahkala) to abseil all the way (© Robert Kenward).*

climber usually remained secured at the nest and lowered the chicks by rope to the belay-partner on the ground. Although we had suitably sized bird bags, a pair of trousers with the legs tied below the knee provided a great carrier for a nest of one or two young. The climber waited while the belayer measured and tagged the chicks on the ground, before they were pulled back up to the nest to resettle them before leaving.

Eggs

It is difficult to tell exactly when eggs are laid, because Common Buzzard nests are particularly vulnerable to disturbance during laying and while incubating very young chicks. In their field guide to surveying and monitoring raptors, Jon Hardey *et al.* (2009) warn 'care must be exercised to avoid excessive disturbance during laying and early incubation. Incubating birds should not be flushed from their nest unless there is a clear need to record clutch size.' In later years of our Dorset study, we therefore left the first visit to confirm nest activity until chicks were likely to have feathers and not just down protecting them, even though we lost data on clutch size. Moreover, although we climbed to check on apparent failure at known sites, we likely missed a few new nest-sites that failed at an early stage of incubation. If we made checks early in the season, for example because it was easier to find some nests in deciduous woodland before the leaves were on the trees, then our visits were in warm, dry weather and for 15 minutes at most. If eggs are lost, then it is extremely rare for Common Buzzards to repeat the laying for that year. Exceptions to that include two cases from Picozzi & Weir (1974) and one case from Holdsworth (1971). However, replacement clutches were never recorded in Dare's extensive study, or by Sergio et al. (2002). Therefore, egg-cooling can mean that a whole year's breeding is lost. Although remote cameras, including those on drones, are likely to be used increasingly to check for eggs and young, to reduce the necessity to climb to nests, they need to be used with caution. It is only really safe to set up a camera to focus on an existing nest in the winter, hoping that the birds will reuse the nest during the next breeding season. Otherwise, camera installation should wait until chicks are a week old.

A study in Denmark investigated the distance from the nest at which a person would provoke an incubating Common Buzzard to leave, as a by-product of their annual surveys (Sunde *et al.* 2009). From 213 walk-ins, there were no records of Buzzards flushing from the nest when the intruder was more than 200m away. At closer distances, there was much variation, with 15 per cent staying securely on the nest even if the intruder walked up to the nest tree; otherwise, birds left at any distance within 200m of the nest. About 40 per cent had left before the person was within 100m and 60 per cent by the time the intruder reached 50m (Figure 7.2). The fieldwork was conducted when there were no leaves on the trees to hide the fieldworker, which would also have been when birds were on eggs rather than chicks. No variables (such as nest height, or distance to anthropogenic features) affected the distance at which birds left. We also don't know the possible influence of each Buzzard's hunger or stage of laying. Nevertheless it's useful to know typical distances at which humans may disturb incubating Common Buzzards.

The incubation period is 33–38 days. Incubation is almost exclusively carried out by the female, who sits progressively tighter on the nest as hatching approaches (Tubbs 1974, Dare 1998). She only leaves to eat food brought in by the male, making begging noises as she greets him (Dare 1998). While the female is off the nest and eating, the male will incubate the eggs. Once the eggs are laid, the female doesn't soar with the male as she did before laying, especially when there are young chicks in early June (Tubbs 1974). The male can

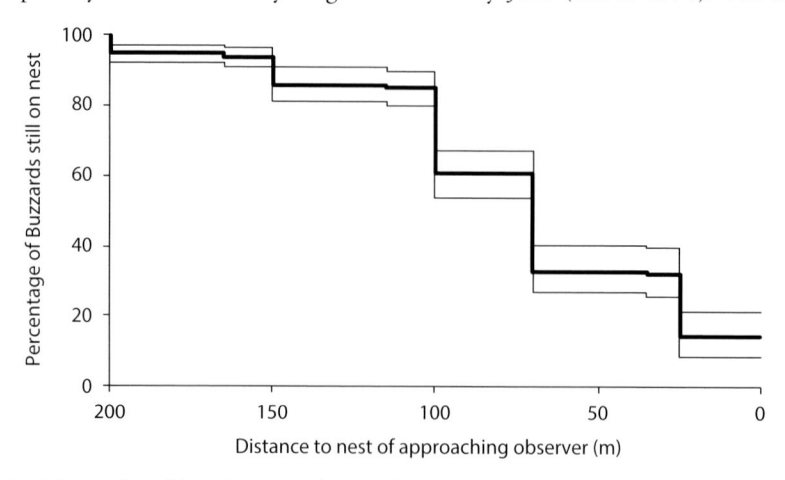

Figure 7.2. *Mean and confidence limits for observer's distance from nest when the female Buzzard left it. Data from Sunde* et al. *(2009), with kind permission from Ardea.*

frequently be present at the nest during incubation, as seen with the early use of a nest camera (Hubert & Carlier 1992). The female's strongest urge is to incubate, and sitting tight is the best thing she can do if there is no apparent danger. In fact, the male often acts as a sentinel when she is on the nest and unable to get a comprehensive scan of the area. It's different once the eggs hatch, as she then becomes more aggressive and dominant to the male at the nest. Although there isn't the extent of sexual dimorphism as found in many other raptors, she is still bigger than the male.

Different studies make different assumptions about incubation times and wing growth. This results in errors of a few days, but methods are sufficiently accurate to compare for any geographic or temporal trends. Laying dates are very variable but confined to an amazingly consistent window of about four weeks within areas of the UK (Austin & Houston 1997), although the dates of that window will vary across more extensive latitudes. In Europe, Buzzards usually lay their eggs between late February and the end of May, with a tendency for laying to be on average about ten days later for every 10^0 of latitude between the Tenerife population in the Canary Islands and populations in Finland in the north of the species' distribution (Figure 7.3).

The variation between individuals within an area is always apparent, no matter how abundant the prey. In our study area, certain nests were always earlier than others, suggesting that it depended either on some males being consistently better providers of food, or possibly that the laying date was genetically determined for individual females. More recently, Nayden Chakarov *et al.* (2013) found that there was consistency of laying date among females, based on the timing of 1,430 broods (a wonderful sample size to have). There was also consistency for some males, but not as strong as the females. While the difference was small (only a few days) the intermediate-plumaged morphs in their study laid earlier than either the pale or dark morphs. However, despite the Buzzards' laying dates

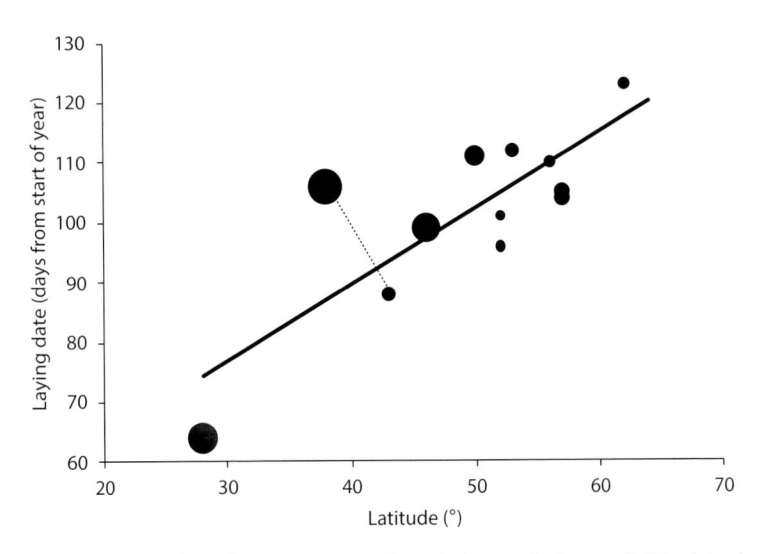

Figure 7.3. *The mean laying date of Common Buzzard populations varies from early March in the far south of its European distribution to late April in the far north. Buzzards at the site with the highest average altitude (represented by the largest circle size) of 900m in the south of Spain laid 18 days later than a paired site at 150m (represented by the dotted line and small circle for much lower altitude) and 5° to the north. Data sources in Appendix 2.*

being so diverse, they found that the gene-set often associated with timing of breeding in other birds was relatively invariable in Buzzards, suggesting that the genes played a very limited role in determining the laying date. Quite understandably, they reasoned that a species subject to fluctuating food conditions would be better off responding to the environment than being hardwired genetically. Nevertheless, there is still a chance that an unanalysed part of the genome or epigenetic effects could be playing a role. Rooney *et al.* (2015) found that supplementary feeding, which artificially improved the environment by providing abundant food such that it was not a limiting factor, did not affect the laying date. Thus, the date may be pre-set to an extent in female Common Buzzards, and thus differ from its strong dependence on temperature and food in Northern Goshawk (Kenward 2006), a less migratory species. Nevertheless, across many species the phenology has shown a trend for migration and breeding cycles to get earlier each year. This is thought to be due to global warming selecting for individuals that can respond earlier in the year, rather than individuals changing behaviour. Common Buzzards are no exception, in that the laying dates appear to be creeping earlier; the average date of laying in Finland advanced by ten days between 1979 and 2004 (Lehikoinen *et al.* 2009), equivalent to Finland moving 10° of latitude southward!

Since the early 1970s, ornithologists have investigated whether sex allocation in many species can be determined by environmental conditions (Trivers & Willard 1973). In particular, there has been much discussion of whether adaptive sex allocation, where it would make evolutionary sense to produce more of a particular sex in certain conditions, occurs in birds (Komdeur & Pen 2002). Therefore, Chakarov *et al.* (2015) sought to investigate this possibility using their well-monitored population of Common Buzzards in Germany. They were able to assess territory quality according to the number of years of occupancy since their study began, assuming that the best territories would be occupied in more years, if not all years, whereas the poorest territories would only produce in years with abundant food. Their hypothesis had several predictions. First, in harsh years it is better to produce the less costly sex, as that is more likely to survive and contribute to offspring in the future. Essentially, males are considered less costly because they are smaller and therefore need less food to rear, although that may be marginal in Common Buzzards given the small sexual size dimorphism. Moreover, as females take more food to rear, it would be better if they hatched earlier in the season, giving the parents more time to feed them, so they would expect the sex ratio to change during the season. In this way, although the sex ratios might be equal overall, differences might be expected between years, depending on how kind the environmental conditions were. The overall expectation was more males in harsh years and later in the season.

As expected, roughly equal numbers of males and females were hatched overall (50.7 per cent male). The most significant and safest finding was in keeping with the hypothesis, in that better-quality territories tended to hatch a higher proportion of the more costly young, the females. Beyond that, the relationships were more complex, especially when the morph of the birds was taken into account. In harsh years, light and dark females produced more males, agreeing with the hypothesis. In contrast, 194 attempts from intermediate females produced the opposite effect, namely more females in harsh years. The intermediate morphs are thought to be the fitter morph, so perhaps they were not so close to a threshold, and so could produce as many females in harsher years. Over a lifetime, light-morph fathers produced more daughters, but again this was a weak relationship and only apparent when

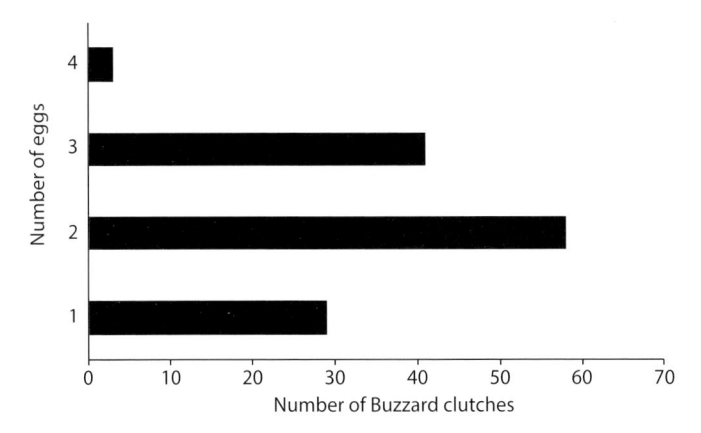

Figure 7.4. *Buzzards in Dorset mostly laid two or three eggs, although a large minority laid single eggs and a very few had clutches of four.*

brood size was greater than one. Perhaps more data will produce clearer results, but at least we know that one well-studied site definitely showed a relatively even sex ratio.

Typically, Common Buzzards lay either two or three eggs (Figure 7.4), although it is not uncommon to find clutches with only one, and at times there are clutches of four. A particularly high occurrence of large clutches came from Picozzi & Weir (1974), who recorded up to a quarter of nests with four eggs. Others have observed five or even six eggs in nests (Mebs 1964, Holdsworth 1971, Tubbs 1974, Björklund *et al.* 2013). As an experiment, where Picozzi & Weir knew one nest was going to fail, they took the eggs and added them to another nest to make a clutch of five. The result was that all the young fledged and appeared healthy (Picozzi & Weir 1974). More recent evidence from nest cameras, and the genetic analyses revealing that more than two birds can contribute to the eggs in a nest, raises the further question of whether broods larger than four are indicators that more than one female has laid the eggs, as was the case for the six eggs found in a nest in Northern Ireland (Rooney 2013, see Chapter 6).

Common Buzzard eggs are approximately 55mm long and 44mm wide, having a dull white base colour, often with variable amounts of red-brown mottling. The mass of the egg shell is about 5g, and the total egg mass is around 60g, roughly 5–8 per cent of the female's spring mass. Not all eggs will hatch a chick, and every year we would find a few addled eggs (Plate 21; see colour section), which sometimes contained part-grown fledglings. Dare estimated their occurrence to be 4.5 per cent of eggs in his Dartmoor study and 6 per cent in North Wales. Even rarer is finding a whole clutch of addled eggs, but that can happen if the eggs cool severely due to disturbance of the female from incubation. Most of the time though, all eggs hatch.

Chicks

The male may incubate the eggs for some of the time, but once there are young chicks he attends the nest far less (Hubert & Carlier 1992). Chicks in the same nest are of notably different sizes and stages of feather development, because the eggs hatch one at a time, usually with 1–3 days between each. From the first chip in the shell until the chick emerges can take up to 48 hours (Austin 1992). At first, the down is damp and stuck to the skin, but over a period of a few hours it dries out to provide the characteristic fluffy white down that will help insulate the chick.

After about a week, a second, greyer coat of thicker down replaces the hatching down (Plate 7; see colour section). It takes nearly two weeks until the quills of the primary and tail feathers appear. At first, these blood-filled quills are grey, almost fleshy looking, soft thin spines, which after a few days split at their points to reveal the feather tips. When the primary feathers start opening, at two weeks of age, the mantle and scapular feathers start to cover the back and wing stubs with a lush chocolate-brown. Then, at three weeks the flanks and ear-coverts start feathering, although it takes yet another week before a nestling starts to look properly feather-covered. Thus, at nearly a month old a young Buzzard nestling can have a relatively well-feathered body, but still looks quite comical because the head remains white and downy. When at last the crown feathers start to adorn the head in the fifth week, we know that nestlings are turning into fledglings and they might try to leave the nest as we climb to them, although if undisturbed they will remain in the nest for another ten days or so.

The flight feathers are the most crucial for enabling a tree-nesting bird to leave the nest and escape climbing predators, and they are the first proper feathers to start growing. Wing length is measured as the maximum wing chord from the carpal joint to the tip of the longest (5th) primary. Wing growth can be represented as an S-shaped curve, starting slowly as the flesh and bone of the forearm grow, but then increasing rapidly at about 8mm per day once feathers extend beyond the flesh after 35mm. The growth does not slow appreciably until shortly before the feather shafts harden to complete growth when the young are out of the nest. It is this growth pattern that researchers have used to calculate the hatch date of nestlings.

Austin & Houston (1997) give the following feather growth formula using just the length of the longest primary feather (not the conventional wing length): AGE = 12.75 + (0.125 × length of longest primary in mm). This equation estimates that feathers grow 8mm per day, and that 12.75 days must then be added to one-eighth of the feather length to estimate age. We visited broods twice if they were to be radio-tagged, the first time to estimate when to make the second visit when the chicks would be old enough to fit the tags. Consequently, we had data on the change in average wing length for 89 broods. The intervals for the change in length were from four days, for birds close to fledging when first checked, up to 35 days, for birds trapped for tags just after fledging. The growth rate of wing length (feather and bone combined) was indeed very close to 8mm per day (Figure 7.5). However, for the broods first visited just after hatching, when their wings were 25–30mm, we found that on average only two days needed to be added to the multiple of (total) wing length to predict ages of older chicks, so we used AGE = 2 + (0.125 × wing length in mm). Hardey et al. (2009) have produced a useful growth curve from data in Bijlsma (1999), which likewise shows wing length growing 7–8mm per day.

Based on wing length growth at a relatively constant 8mm per day, the intervals of hatching of chicks can be estimated, as shown in Figure 7.6. Hatching on the same day was likely when the female delayed the start of incubation after laying the first egg. Two-thirds of intervals between consecutive chicks were two days or less, which suggested that quite often incubation did not start from the first egg. However, wing length differences also

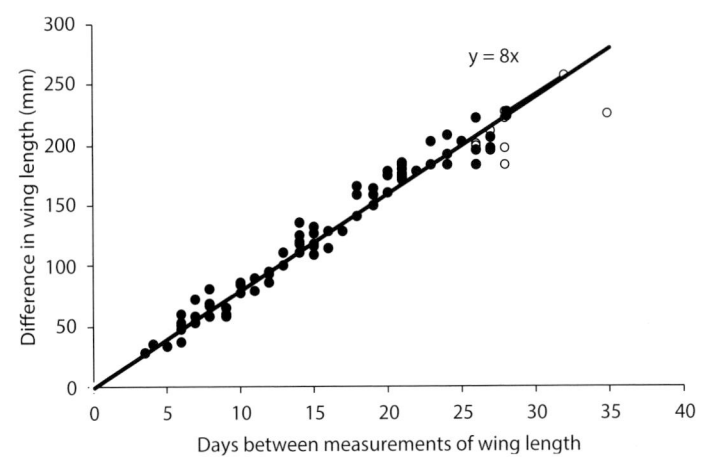

Figure 7.5. *Growth in wing length of nestling Buzzards in our Dorset study averaged 8mm per day, as shown by the line with that slope. However, a few days must be added when estimating age from wing length if wings are shorter than 35mm (see white dots), as shown by below-average growth for birds first measured closest to hatching.*

suggested hatching intervals of three, four, five or even seven days. Some chicks were often noticeably smaller than their siblings from an early age and therefore probably represented a considerable delay in completing that clutch, albeit perhaps also suffering from a decreased rate of wing growth (see later). Moreover, if any intermediate-aged chicks died, a greater difference between oldest and youngest chick would be expected than from consecutive chicks. Now that we know more than two females can lay in the same nest, maybe the biggest intervals are the result of a second female laying a few days later, but alas we did not have genetic fingerprinting available to us to check for that at the time.

Although mass will increase according to the amount and quality of what the chicks eat, feathers have a similar growth rate, whatever the food supply. This relatively constant feather growth results in daily growth bars that are a product of the periodic daytime feeding and night-time fasting. If you look very carefully, this can be seen on most bird feathers as the shading lightens and darkens along the feather, representing each day. These regular growth

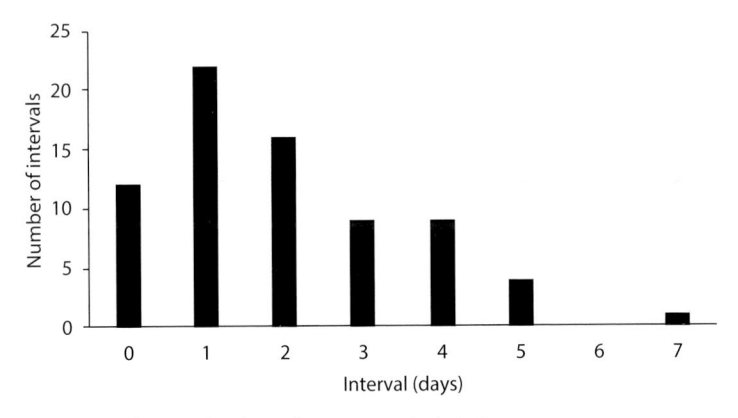

Figure 7.6. *The intervals between hatching of consecutive chicks in Dorset.*

bars are quite different from the more obvious stress bars that are caused by food shortages (Chapter 2).

When Common Buzzards hatch, they generally weigh between 40 and 50g. However, they gain mass very fast as they start to feed. On average the nestlings we measured increased to about 300g in mass during their first 10 days, at the upper limit of the range of 20–30g per day noted by Austin (1992). Then, from the 11th to 30th day, growth of chicks in Dorset was somewhat slower, at about 20g per day (Figure 7.7). During this period, the average mass of broods diverged appreciably, especially after 20 days, depending on the proportion of males, which ultimately weighed about 150g less than females. Due to much greater variability of mass than for wing length, and the difficulty of assigning a sex to chicks until they were close to fledging, mass was not useful for ageing chicks. At an individual level, a lot can also depend on when the nestling last ate something, and therefore how much undigested food there is in the digestive tract, especially the crop. Of course, the crops are relative in size to the nestling, and for the first few weeks the chicks are only able to take on a small amount of raw meat dissected from a carcass by a parent, usually the female.

Chicks spend their first week predominantly sleeping, with occasional feeds. After a few weeks they start to spend more time sitting up, but still spend much of their time sleeping and in the third week may start to hold their wings out in a threatening posture. By then, they can swallow smaller animals such as shrews or small voles whole and will produce castings of fur (Dare 2015). In the fourth week, once the feet become more effective and the chicks build their strength, they will grab food as soon as it arrives and start pulling chunks of meat, fur and feather off to swallow ravenously. For a raptor eating vertebrates, the chicks are able to handle prey independently at a younger age than most, owing to the fact that they are so much larger than their staple food, small mammals. Nestlings of Common Buzzards from a sample of five nests could eat more than 50 per cent of their prey items without parental help by only 12 days of age. This is very young compared with Northern Goshawks, a similar-sized species whose young become independent at a similar age, which

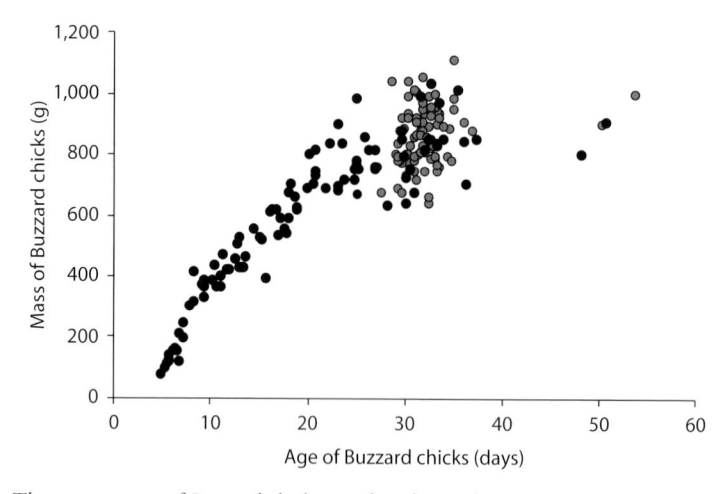

Figure 7.7. *The average mass of Buzzard chicks in a brood according to the mean age of the young, showing initial measurements in black and subsequent measurements when revisiting chicks more than 27 days old in grey, with a few young also trapped after fledging.*

in the seven nests observed were not eating more than 50 per cent of their prey unassisted until 29 days old (Sonerud *et al.* 2014).

Rapid growth of chicks, combined with the asynchronous hatching, can mean that siblings' mass may differ by 60–90g from quite early on. When they weigh only a few hundred grams, this difference is significant in a dispute over food. For the first few weeks after hatching, age will have the biggest effect on dominance, but after 20 days the sexes really start to separate in terms of mass (Biljsma 1999). A female of 20 days old is likely to be as heavy as a 24-day-old male. Or at extreme divergence, a female of 24 days is as big as any male more than 32 days old, because the male will have reached a mass that can only increase after leaving the nest and developing flight muscle. From his observations, Prytherch (2016) thought that the older dominant chick held the centre of the nest, pushing the young siblings towards the edge. The older nestlings also have the advantage of larger and better-coordinated feet. However, age and gender are not the only factors that determine dominance. Sometimes, the 'character', particularly the aggressiveness of younger chicks, will determine the pecking order.

Prey delivery to nests on Dartmoor varied between 0.1 and 1.1 items an hour, dependent on prey size Dare (2015). Small items such as voles were delivered straight to these nests in Devon, but larger Rabbits or birds were taken to plucking stations. In Dorset, we didn't come across many plucking stations, but we also did not spend long searching around nests. However, Dare observed that if the nest is on a steep hillside, then these plucking posts are always uphill from the nest and within 30m.

Younger chicks typically have to wait for older chicks to be satiated before they get to eat anything of the parent's deliveries. When the oldest chick is big enough, it will swallow a vole whole, so then the younger chicks will have to hope that parents bring the next food parcel before the oldest gets hungry again. In the early stages, a parent can choose who to feed, and it has long been thought among ornithologists that those chicks that call the most, or have the wider gapes, attract the parents to feed them more. What has only been appreciated recently is that there might be other cues to impel the adult to feed a particular chick, just as Eurasian Scops Owls seem influenced by the UV reflectance of ceres on underweight chicks (Chapter 2). However, whether the female has a choice when the nestlings are older is debatable, given that food is often dropped for the nestlings to fight over. At this later stage there is 'scramble competition' for food rather than the behaviour of a 'fair-minded parent'.

If food is really in short supply, then the oldest chicks will start attacking the weakest (usually the youngest) chick; a behaviour known as cainism, after the Cain and Abel story of the Christian Bible. Although some feel it unfair, from an anthropomorphic point of view, that the biggest chick eats all the food it can before younger chicks get anything, it is the way natural selection works. In 'selfish gene' terms (Dawkins 1976), it is better to have one juvenile get through a food shortage, taking forward half the genes they have in common, than to share the food between two, causing both to die and resulting in no genes going forward. Furthermore, in evolutionary terms, if food is really short it is better to eliminate the youngest chick or use it as a food source, than for it to continue to be a drain on parental resources. This is the chick with the least chance of surviving a food shortage, and into which least parental investment has been made. First, the victim loses feathers from the head and starts to look bald, and if the attacks persist then the skin is broken and the skull starts to show. If attacks endure for too long then the chick will die and often be eaten (Dare

1998). Despite that common conclusion, it is surprising how some young with serious-looking wounds to the back of the head and neck can recover and fledge if food becomes abundant again. Even if there are no external wounds, there is evidence that younger chicks may suffer from having older siblings. A study of Upland Buzzards in Mongolia by Reuven Yosef *et al.* (2013) found that the amount of corticosterone (a hormonal measure of stress) correlated with the brood's pecking order: those youngsters at the bottom had higher levels, implying that they were more stressed. Interestingly, the stress was affected more by being the youngest in a larger brood, when presumably food was not so short, than by being in a smaller brood when there was not so much food about.

Like the eggs, which have no thermoregulation and require constant incubation by the female, so small chicks with little feathering must also be sheltered from inclement weather almost constantly. Therefore, the female shelters and feeds the chicks, as well as maintaining hygiene in the nest, for at least the first two weeks, and hardly leaves the nest (Dare 2015). This makes her totally reliant on the male for food delivery. During this time, she is also very careful to tuck her claws under her feet as she moves around the nest, so there is less chance of damaging the precious, fragile chicks.

We know less about the male, because he is not present on the nest much, but it seems fair to suggest that he spends his time hunting for provisions, occasionally suspending this activity to defend the nest-site. His own need for food must be assuaged too, and he often eats the head and sometimes the forelimbs of prey before delivering them to the nest. It is not until nearly the third week after the chicks hatch that the female starts to leave the nest, other than for a quick feed on what the male has brought in. At first, she starts hunting close to the nest, gradually moving further away as chicks require increased amounts of food towards the point of fledging. As the food demands grow dramatically, the male has to catch enough to feed the female and the growing chicks as well as himself. Dare (1998) recorded that he may hunt for up to 16.5 hours per day, which is almost all the daylight hours in Devon in summer.

Raptors bringing food to the nest can sometimes result in some extraordinary situations. Very rarely indeed, it seems that Common Buzzards can end up in eagle nests and be adopted. Ivan Literak and Jakub Mraz (2011) have reported that they found a young Common Buzzard in the nest of a pair of White-tailed Eagles *Haliaeetus albicilla* in the Czech Republic. No adult Buzzard was seen at the nest, although a Buzzard had been seen soaring above the nest with the pair of eagles. The eagle chicks were around three weeks old and the Buzzard approximately two weeks old. The Buzzard appeared to fledge successfully, because it was seen perching within 150m of the nest when about five weeks old. The article cites three other occasions when this has been observed (Palko 1997, Fenyosi & Stix 1998, Horvarth 2009), and also notes that Red-tailed Hawk nestlings have been found in nests of Bald Eagles *Haliaeetus leucocephalus* (Stefanek *et al.* 1992, Watson *et al.* 1993, Watson & Cunningham 1996). More recently, evidence has been provided from nest cameras at White-tailed Eagle nests for the presumed mechanism by which Buzzards were brought to the nest as prey items (Neumann & Schwarz 2017) (Chapter 9). The females, with their maternal instincts, then responded as though the Buzzard chicks were their own. Presumably, with plenty of food there is little competition, so the Buzzards are able to survive living with some considerably bigger and more powerful 'step-chicks'.

For our studies, it was best to attach radio-tags for long-term tracking when chicks were about four weeks old, because then they had grown sufficiently to ensure a good fit, but their flight feathers were not developed enough for them to be likely to leave the nest or fly any distance if they did. When arriving at the nest, we soon noticed how differently individual chicks reacted. Some were placid, hardly moving while we measured their wings to see if they were ready for tagging, although others resisted by raising their wings, often moving into a 'fallen angel' position, with their talons stretched out in defence (Plate 16; see colour section). They often grabbed at our hands, and occasionally jumped towards us very aggressively. We were dazzled by how fast they were, with the feet sometimes locking onto a hand before we'd registered the movement. Once we had controlled their feet, the most aggressive chicks continued to peck at hands, which could be painful and sometimes drew blood, but not nearly as painful as extracting talons locked with their ratchet system.

A chick's capacity for aggression depends on several circumstances. There is not much a young buzzard can do before its feet become effective, so aggression is not really apparent in the first few weeks. Once the feet are well developed, aggression increases, but it may be dampened if there has been recent rain. This reaction to damp conditions can be used to calm near-fledged birds artificially, by first sprinkling water on them with a pump-pressurised water-pistol from below, before reaching the nest (Kenward *et al.* 2001c). We had the impression that those nestlings with particularly golden heads (as opposed to the more usual brown heads) were more aggressive. Unfortunately, this was not something we could prove conclusively because we did not record the 'goldenness' in an objective manner, and scoring aggression was not easy, either. Our primary concern was to spend as little time disturbing the birds while ensuring that the tags were fitted safely. Some parents were aggravated by our presence and spent the period circling overhead, sometimes calling. Adults never physically attacked us, but other researchers have reported being struck by Common Buzzards defending their nest (Chapter 11).

Both parents attend the nest progressively less from the third week on. Nevertheless, they remain very defensive of the chicks, probably because they have invested so much in rearing them to that fledging stage. We found this very convenient when studying Common Buzzards that had attained breeding age after being released in Sussex. We could only visit very infrequently to check survival and breeding. A visit when all females were likely to be incubating would reveal whether radio-tagged females were on eggs. This was because the radio-tags had posture-sensing switches, which sent a slow signal pulse when birds were perched but a fast pulse when their bodies were horizontal; this usually helped us detect when birds were flying. The signal from a flying Buzzard varied in volume as the bird and its tag's antenna moved, whereas a steady, fast signal from a female in spring meant that she was on a nest, and one could walk straight to it. This did not help us find whether males were breeding, because they incubate so little. Nevertheless, a visit that actually disturbed radio-tagged males at a time when nestlings would have been well grown tended to result in the upset parent returning to his nest, giving alarm calls. He would typically be joined there by the female, as a further indication of where we needed to search for the nest to measure and mark the young.

Once the chicks are sufficiently developed, with wings strong enough to at least act as parachutes and slow their descent if they miss their footing, they start 'branching': walking along the bough, away from the nest. Very soon, they then take their first short flights from branch to branch and enter their next phase of development as fledglings.

Fledglings

The exact timing of branching and flying varies. Using the radio-tags that were fitted to nestlings, we always knew where the birds were, and one year we had diploma-student Tony Tyack check on our tagged nestlings to find out when they left the nest and what happened shortly after that (Tyack *et al.* 1998). First flights from the nest tree occurred 43–54 days after hatching. Some of the variation depended on whether there were easy branches for the birds to move along, away from the nest. Some nests were built at the junction of the trunk with a few rather flimsy branches, so the birds were obliged to wait until their flight feathers were sufficiently long for them to jump and glide to other branches. Despite a difference in age averaging about five days between the oldest and youngest in a brood of three siblings, they typically all left the nest on the same day. Probably, the youngest were being inveigled away slightly early; if they stayed in the nest they risked not getting enough food from the parents, which tended to be mugged by the older chicks before they got back to the nest.

Although the fledglings are well feathered when they leave the nest, maybe with just a little down showing on the crown, the flight feathers are still 'in blood' as they continue to grow from the soft basal sheaths. The blood in the base of the feather supplies nutrients to the growing feather until it is fully formed and the shaft can harden at its origin. The longest wings we recorded in the nest were just under 300mm, and the average final wing was nearer 400mm (maximum 432mm). There is therefore typically at least 12 days of further feather-growth after the bird leaves the nest.

These under-size flight feathers, combined with undeveloped muscles and lack of experience, means that birds that have just left the nest are relatively clumsy, lolloping fliers that are easily distinguished from their agile parents. Distance from the nest tree increased dramatically after 65 days (Figure 7.8), when feathers were fully grown and hardened. The time spent flying and calling increased with age as the juveniles moved further away from the nest, usually in one sibling group, although older birds tended to be further from the

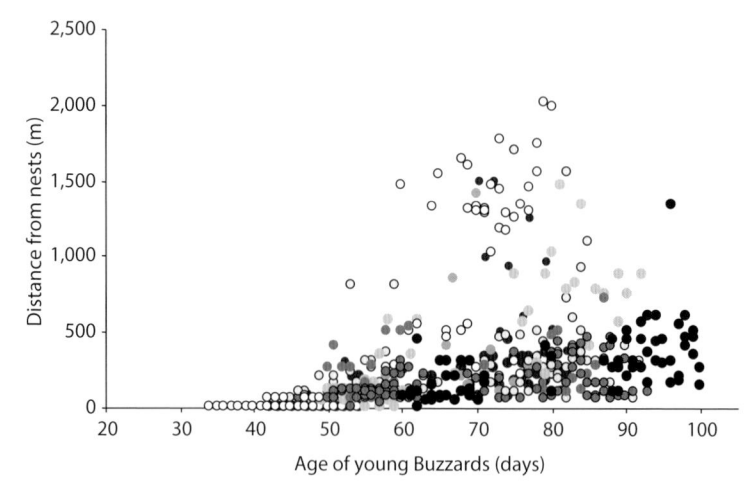

Figure 7.8. *Distance from nest with age of 26 young Dorset Buzzards radio-tracked in 13 broods (shown with different-coloured symbols) before they dispersed in 1991. Few went more than 300m from the nest until the main flight feathers hardened 65 days after hatching. Data from Tyack* et al. *(1998), with kind permission from John Wiley & Sons.*

nest than the younger ones. After the 65th day, calling reduced markedly and the broods became much less cohesive. Distance from the nest then became much more variable and showed no further consistent relationship with age, up to when intensive tracking stopped at 100 days. After the 65th day, Buzzards were never recorded at the nest, whereas distances away from the nest were greater for those broods whose parents were more often seen, which tended to be where there was more open grass and arable habitat, rather than woodland (Tyack *et al.* 1998).

The adults continued to deliver food, initially to the nest where the radio-tagged fledglings came back to eat, but as chicks became stronger fliers the parents carrying food were met before they got to the nest. There is an interesting difference from Northern Goshawks in the use of the nest, and in brood cohesiveness as a whole. In our studies, Common Buzzards stopped using the nest quite rapidly, and were not recorded there after 65 days (Figure 7.8). In contrast, Goshawks continued to sleep on the nest for about a week after their first flight (Kenward 2006), and about 20 per cent of females (10 per cent of males) were recorded visiting the nest after their 65th day (Kenward *et al.* 1993a). Buzzard young were far more likely to be close to their siblings after fledging than were Goshawks of any age, and were generally found all together (albeit, away from the nest) before their feathers hardened (Walls & Kenward 1994). Perhaps young Buzzards move away from the nest because they are most at risk of being detected there by stronger raptors, including Goshawks.

It is unclear, and again probably very variable depending on the circumstances, when fledglings start to make their first kills, but for most there is a period of two to four weeks when they are fed entirely by the parents, and mostly that extends to between six and eight weeks. Tubbs (1974) observed food being taken up on a soar with the juveniles following, and even a few food passes, but the latter (thought to be a way of training young birds to stoop on prey) are not nearly as common in buzzard species as with falcons.

Parents continue to look out for and protect their young during this fledgling period. Therefore, anxious nesting Common Buzzards will still scream at people if they are walking too close to the young. Juveniles are most easily recognised at this stage, because the call is discordant, more squeaky and piercing than an adult's. It could be that the discordance of fledging Buzzard calls is to make them more audible for parents trying to care from them away from the nest. In the days before telemetry was used to find lost hawks, falconers used bells to hear where errant birds had gone, and some considered it best to use two bells with different pitches. The idea was that the two clashing frequencies would be more noticeable. It's possible that the discordance of the call has evolved because parents will respond to it more often, or from further away. Young Buzzards often appear naïve, willing to perch on exposed branches screaming for food, or be more intent chasing the adult around for food. Inevitably, the young follow the adults out along the paths they take to the hunting grounds. Some will get into trouble at this stage, for example, with their poor flight they can hit wires and break wings, or end up drowning in water bodies. We will deal with mortalities in more detail in Chapter 9. For now, we will look at what factors affect the overall success of a nesting attempt to the point of fledging.

Breeding success

Thankfully for fieldworkers researching Buzzards, nesting success can be recorded with few nest visits. There is an assumption of only one brood per season; there isn't time over most of the Common Buzzard's distribution to have two broods considering that it takes more than three months to incubate eggs and then rear chicks to full growth of their feathers, let alone to independence. So apart from rare cases when Buzzards laid again after the early loss of a clutch, we have found only one report of double-brooding within the whole *Buteo* group, and these are for Galapagos Hawks *Buteo galapagoensis* – where the weather is relatively warm for most of the year. However, Marta López-Darias (2007) gave an interesting report of a pair of Common Buzzards on Fuerteventura (Canary Islands). The pair laid unusually early and three chicks were seen at about one week of age, but a week later only one chick remained. Three weeks after that, when the original chick was at least four weeks old, there was a new egg in the nest, followed by another, and incubation started, with the initial chick being pushed to the brim of the nest. This original youngster fledged at about nine weeks old and was not seen again in the territory. The new eggs were incubated for a few weeks, but did not hatch and were abandoned. There was no evidence that any other Buzzards were involved other than the territorial pair, whose distinctive plumage made them individually recognisable.

Thus, the second brood was not successful and the absence of post-fledging observations of the first chick suggests that it died, too. We consider this apparent double brood as a very rare exception. In general, therefore, studies of Buzzards assume that there is no double brooding; if it does occur, it would not affect any of the results significantly. Given that assumption, a second assumption is that larger broods represent greater breeding success. More work needs to be conducted on this assumption for Common Buzzards, but there is evidence from experimental manipulation of Great Tit brood size that, under equal environmental conditions, chicks from eggs laid into larger clutches are more likely to be recruited into the next breeding generation (Both *et al.* 1998). The conclusion of the Great Tit study was that either the early maternal care (including the female's physical condition when laying) or her genetic composition increases long-term success. It is therefore fair to assume that larger clutches represent better breeding success, in Great Tits at least.

To measure breeding success, it is standard practice to record clutch size, or the number of chicks large enough for ringing (brood size) or, more usefully, the number of fledglings leaving the nest, together with the number of nests that fail to produce young. The Mayfield technique (Mayfield 1961, 1975) can be used to estimate the number of nests not found before they failed, by using the known failure rate from observations at various stages of the breeding cycle to be extrapolated to nests that were never found, assuming that the failure rate remains constant during the breeding cycle. Some researchers have spent a lot of time in the field and got to know individual birds and their territorial boundaries, even if no nest was found, and so can give good estimates. For example, Goszczyński (1997) recorded 1.67 fledglings per active territory, including 1.78 young per nesting pair and 2.34 fledglings per successful nest (that raised at least one young to fledging). As it is much easier to find successful nests than unsuccessful nests (Sim *et al.* 2001), the brood size per successful nest is likely to be the most accurate estimate. Though the other estimates have their own value, to go back earlier than clutches as a basis for productivity estimates is an invitation to errors and possible bias, unless one has birds of all ages radio-tagged (e.g., Northern Goshawks in Kenward 2006).

Most of our Buzzards in Dorset didn't breed until after their radios had stopped, and it proved impossible to catch adults consistently. However, a process of checking all known nest areas carefully provided quite good information on clutch size at the start of the study. The checking of known nest-sites was followed by searching other woodland systematically when nestlings were likely to be calling and spattering the ground with faecal whitewash, which made data on nestlings more reliable. During chick-rearing, we also routinely climbed to freshly decorated alternatives if the prime suspected nest lacked a brood, and occasionally recorded evidence of early failure to further improve estimates of productivity for Buzzards that laid eggs in our study. Of course, we had occasions where we knew there were territorial pairs making a lot of a fuss and no nest was found, and we strongly suspected that there were others not being as overtly territorial in the 120km^2 we were investigating. A small percentage of the audible pairs may have built an inconspicuous new nest, laid eggs and then failed early without the nest being detected, but most are likely to have been among the notional 21–41 per cent of territorial non-laying pairs recorded by other authors (see Chapter 10).

In Dorset, we estimated the loss of about 5 per cent of possible productivity as eggs that failed to hatch because they were addled, which was very similar to that reported by Dare (2015) in neighbouring Devon. Another 13 per cent of chicks were lost during rearing, hence 18 per cent of 239 eggs we recorded during 1990–1996 failed to develop to fledglings (after 1996 we focused on checking nests while chicks were being reared). There was a tendency for hatched clutches of two to be most likely to lose a chick (Figure 7.9), with each egg in such clutches having a 21 per cent risk of failure to rear a chick, close to double that from clutches of one egg (10 per cent) or 3–4 eggs (12 per cent). As there was a strong tendency for large broods to be fed mostly on Rabbits (Hodder 2001), this may have reflected a tendency for birds to lay clutches of 3–4 eggs and rear young from most of them

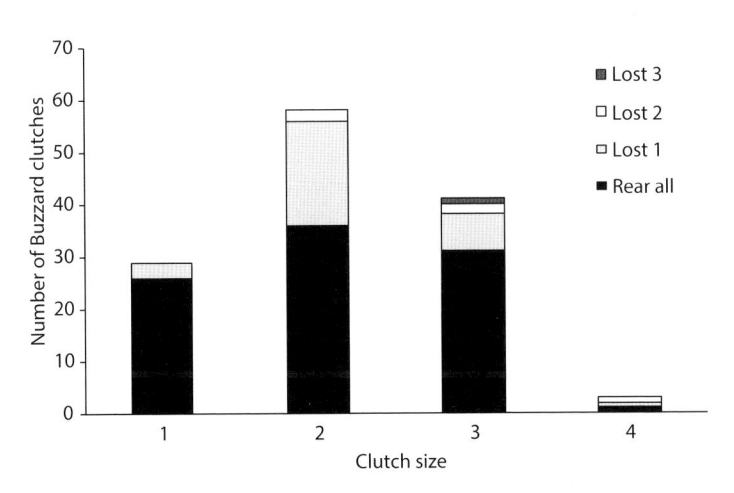

Figure 7.9. *The numbers of clutches of 1–4 eggs that lost 1–3 young before fledging in Dorset, showing relatively highest losses from broods of two eggs. Loss of all eggs may be under-represented by failure to find nests that failed during incubation.*

where Rabbits were available, but otherwise not to lay more than 1–2 eggs and often rear only one chick.

Most multi-year studies of Common Buzzards find that breeding success is highly variable between pairs and between years. Here, we explore what influences that success. We present not only the factors, such as food supply and weather, which have rather clear influences on annual variability, but also other factors such as competition and the performance of individuals, which have been more difficult to quantify in the past.

Food

It stands to reason that if there isn't enough food for the chicks then they won't survive, so it is no surprise that food availability is the most common factor found to affect breeding success. It is not uncommon for chicks to die in the nest and, if food is short, a dead chick is likely to be eaten by others in the nest. So, how much do we know about the influence of food on nesting success?

Fortunately, vole cycles and the historic effects of myxomatosis on Rabbit populations have provided sufficient natural experiments for a quantitative assessment, and more recently there have even been some deliberate experimental manipulations. Some projects have specifically looked for relationships between breeding success and quantitative measures of rodent abundance. The most rigorous studies assessed the small mammal abundance independently, for example, by snap or live trapping. In Norway, Spidsø & Selås (1988) found the number of fledglings in peak vole years was 2.6 per nest, whereas in low vole years there were only 0.3 fledglings. This is an enormous difference, given the ability of Buzzards to switch to alternative prey (Chapter 2), so it suggests the alternative prey was not very abundant. In Poland, Goszczyński (1997, 2001) used food remains at the nest as an index of prey abundance. Although we know there are difficulties with analysing nest remains (Chapter 2), the left-overs are actually a good measure of one end of the scale, because they indicate whether there was surplus food. Large broods occurred only in years where rodent remains were abundant, indicating their importance for good breeding success in that region. In Scotland, the number of foraging Common Buzzards increased over heather moorland in years when small mammals were particularly abundant, thought to have been attracted in by bountiful prey, but otherwise Buzzards were more dependent on Rabbits in the valleys (Thirgood *et al.* 2003, Francksen *et al.* 2017). Lõhmus (2003) calculated that two poor vole years in Estonia gave approximately the same number of fledged Buzzards as a single good vole-year. Others have found strong correlations between vole fluctuations and breeding success: Mebs (1964) in Germany, and Šotnár & Obuch (2009) in Slovakia. Recent studies have shown how some agricultural crops have higher vole numbers that can benefit Common Buzzards (Heroldová *et al.* 2007, Panek & Hušek 2014).

At times, the evidence seems not to support vole-driven productivity. For example, George Swan (2011) surveyed for voles to establish their abundance in three Scottish habitat types. Surprisingly, fewer Buzzards fledged when the vole index was higher. This was despite voles being the highest recorded prey species from prey remains. However, it was apparent that the most successful nests had more Rabbits, so voles couldn't be looked at in isolation. In fact, Rabbit abundance has often been linked with breeding success in other studies. Working in the West Midlands, which was then on the edge of the UK's Buzzard distribution, Innes Sim *et al.* (2000, 2001) found a huge increase in Buzzard numbers between 1983–1996. They considered one of the main factors to be Rabbit abundance, because investigations during 1994–1996 showed that large clutches and high numbers of fledged young were associated with high Rabbit abundance close to the nest. In our own study, Kathy Hodder counted the number of active Rabbit burrows as an indicator of Rabbit availability. Although there was a positive correlation between nest productivity and active burrows, it was never significant. However, late brood size was significantly correlated to Rabbit remains at the nest, and brood sizes were significantly lower when nest remains contained more birds, suggesting that provisioning with Rabbits was important for breeding success. Individual ability to catch Rabbits may have trumped the spatial variation of this prey for determining the success of Buzzard pairs.

Just as vole abundance tends to vary between years, so too can Rabbit numbers, and in a particularly spectacular way following 1954, when myxomatosis drastically reduced Rabbit populations. Nearly one-third of nests failed in Devon immediately post-myxomatosis, compared with only around one-tenth in the 1960s and virtually no failures in the 1990s (Dare 1957). Dare (1998) also considered smaller brood sizes to be a result of a reduced Rabbit population. On the other hand, while numbers of Buzzards on the Welsh island of Skomer have fluctuated with the Rabbit numbers, breeding success was not linked very strongly to the variation (Davis & Saunders 1965). In this case, the Buzzards were thought to be switching to abundant seabird nestlings, which were within easy reach on open nests provided the Buzzards could brave the protective parent seabirds.

More evidence of the importance of food comes from proxy measurements. Holdsworth (1971) gives a very detailed account of the number of young Common Buzzards fledged over 20 years (1947–1967) in Yorkshire. Without any data on small mammal availability, he looked at the number of Common Kestrel nestlings ringed in the whole of the UK, and noted that the year-to-year changes were similar for Buzzards in the years after myxomatosis. Kestrels primarily eat small rodents, and so Holdsworth proposed that the correlation demonstrated the Buzzard's dependence on voles during 1956–1966. Similarly, at one of their Scottish sites, Swann & Etheridge (1995) found that over a 12-year period Buzzard breeding success was highly correlated with the four-year vole-cycle. Their other site, only 50km away, had higher breeding success, and this was attributed to the larger prey items, predominantly Rabbits and Woodpigeons. Others have been more speculative. For example, Tubbs had an incredibly long dataset running from 1961–1982, which showed a dramatic decline in breeding success after 1973. There were 35 per cent fewer breeding attempts, the number of successful nests were down 44 per cent and fledged young dropped 40 per cent. He considered various possible factors but the only hypothesis that seemed to fit was an association with reduced Jackdaw numbers after the 1960s, followed by a decline of small rodent abundance, potentially in both cases as a result of increased grazing pressure (Tubbs & Tubbs 1985).

Supplementary feeding experiments by Rooney *et al.* (2015) have attempted to quantify what the effect of food abundance may be at various stages of the breeding cycle. Without any manipulation, nests in bog scrub and semi-natural grassland (containing fewer Rabbits, the primary prey), bred later and had lower productivity compared with those on more productive agricultural land (good habitat). If food was artificially increased before laying, clutch size increased by an average of 0.6 more eggs, after accounting for habitat influences. The relationship was not quite significant, but then the sample size was not that large, so perhaps a larger sample might have provided the power to really distinguish this. However, there were more telling results for those nests that were fed through the chick-rearing stage. Without supplementary feeding, nests in lower-quality territories had lower hatching success, smaller early broods and fewer fledging from the nest. If those in poor habitat were fed supplementary food during chick-rearing then they could do as well as those in better habitats. It was notable that territory quality had a huge effect on the breeding, and that the experiments show quality equates to the amount of food.

In more southerly areas of the British Isles, and further south in Europe, an abundance of different food sources probably acts to reduce food-based annual variation in Buzzard productivity. For example, there was little productivity variation between years in the Italian pre-Alps (Sergio *et al.* 2002), where the predominant food was birds, although reptiles formed a sizable contribution, too. Perhaps small-mammal fluctuations were insufficient to offset the switching between birds and reptiles which maintained constant food abundance. Even though food is fundamental, success can also be variable due to other factors, potentially masking a food effect. One of those phenomena is the weather.

Weather

Most studies of Common Buzzards have not specifically looked at weather and its effect on breeding success. Nevertheless, weather has been proposed as the reason for nest failures where other factors look less likely. For example, nesting success in Wales was reduced (to 86 per cent from the average 91 per cent) in 1979 after a hard winter (Newton *et al.* 1982) and Davis & Saunders (1965) thought that a particularly severe winter could have put Buzzards into such poor condition that they couldn't nest in 1963. To really see the effect of weather, multiple years of weather and other factors need to be recorded.

Thankfully, we have a few long-term studies to help us. In Germany, Kostrzewa & Kostrzewa (1990) looked specifically at weather and the effect on the breeding success of Common Buzzards, as well as Common Kestrels and Northern Goshawks, over a decade. They monitored 53–75 Buzzard territories per year (508 in total) from 1979 to 1988, and compared their success with monthly averages of temperature and rainfall, plus a crude vole index (high, medium and low). Weather during certain periods appeared to be critical, more so than the vole index. Complete failure was most correlated with high rainfall in May, whereas the number of fledglings was also negatively related to the rainfall in April and May. Rainfall in June had no effect. They also described some particularly poor years where they found freshly dead young, completely soaked through, presumably dying from starvation and exposure. This clearly gives a potential mechanism and also explains why rain in June is not so influential, because by then the chicks have better developed waterproof feathers. They concluded that the hatching and the following two weeks were the most critical in terms of susceptibility to weather, something with which other authors agree. Goszczyński (2001) found that losses were greatest during incubation and the first two weeks after hatching, although he did not find correlations with either mean temperature or precipitation from April to June. This may be because the critical period for the majority of nests was until the end of May, as would be the case in Dorset (Figure 7.10). More recently, Prytherch (2013) used his 20-year study near Bristol to show that mean productivity was lower with higher May rainfall. However, Prytherch also had reduced productivity with

Figure 7.10. *Four-week-old chicks, close to fledging in Dorset, in good June weather as experienced by most broods in our study. The nearest chick has a leg-mounted transmitter, and in the front right is a Rabbit leg (© Robert Kenward).*

high temperatures in April and June. Unlike the high May rainfall, it's not easy to see a direct mechanism for high temperatures reducing success in the UK, where temperatures did not reach dangerous levels during our study years, at least. Perhaps there is some undiscovered relationship with temperatures and food supply, such as hot dry weather affecting grass growth and hence Rabbit breeding, or reducing the availability of molluscs and worms, which thrushes and European Blackbirds need to produce abundant young as another food source for Buzzards.

Weather impacts may not be confined to the exposure of broods to poor conditions that cause chilling. For example, Truszkowski (1976) found that long periods of rain reduced prey delivery to the nest. If this is an important mechanism for influencing productivity, then it might be expected to be most critical at the end of the nesting period, when the chicks are bigger and therefore the demands higher. However, by that stage the female will be hunting for prey too, so maybe there is some compensation, and the young may have accumulated appreciable body reserves.

Snow cover, too, can affect food supply. Initially, you might think that snow would represent a hard winter, and therefore fewer small mammals, exacerbated by the fact that they are also obscured by the snow and consequently more difficult for Buzzards to catch. As expected, Selås (2001) found that Common Buzzard brood size in Norway was negatively correlated with maximum snow depth in April, although not quite significantly. The food supply, in terms of vole numbers, was by far the most significant factor. Although Buzzards are only just arriving and starting to nest in April, vole vulnerability could affect the condition of the female and therefore how many eggs she laid. However, the overall importance of snow may be reduced because it tends to be the snow-free years that are worse for vole survival over winter and overall abundance in spring. Goszczyński (2001) likewise found that prey abundance was the key, and despite looking for effects of temperature and precipitation, he found no weather effect on productivity.

Martínez et al. (2013) investigated how an extreme weather event during the winter might affect forest-nesting raptors. Murcia, in south-east Spain, was hit by very unseasonal weather in January 2007, when persistent winds exceeding 100km/h combined with heavy snowfall, affecting much of the Aleppo Pine forests in which Buzzards and other raptors nest. Opportunistically, Martínez et al. could make a comparison due to a long-term dataset that started in 1998 and continued for three seasons following the storm damage. There was no change to the number of territorial pairs. In fact, the number of Buzzard territories continued its gradual rise even the year after the storm. Buzzards didn't appear to be building significantly more nests, as we might expect if nests had been destroyed by the wind. The nests that were lost were old unmaintained nests, and Buzzards only need one nest in which to raise young. There was more of an effect on Northern Goshawks, which abandoned nests if the canopy cover surrounding the nest was reduced by more than 30 per cent. There was also no significant difference in reproductive output. In conclusion, despite a severe weather event in winter, Common Buzzards seemed to cope.

Although this extreme weather outside the breeding season did not seem to have much effect, there seems plenty of evidence that weather can act in several ways to affect breeding success through all stages of the breeding cycle: by reducing prey abundance and hunting opportunities, causing chilling of eggs and wetting or cooling of the nestlings. Indeed, two German studies investigating long-term nesting success of individuals found weather to be a better predictor than vole indices (Kostrzewa & Kostrzewa 1990, Krüger 2002a). They considered that result to be partly due to the vole index being too crude, because food supply seems a simpler mechanism, therefore intuitively more likely. In fact, weather may have a more pervasive influence on productivity, having an effect on food abundance and the ability of parents to hunt, as well as increasing energy requirements of the chicks to maintain heat, and reducing their ability to conserve that heat when wet.

Competition

Until recently, there had been no evidence that 'interference competition' between Common Buzzards had an effect on breeding success, despite concerted efforts to look for that effect (Sim et al. 2001, Krüger 2004b). In our area, productivity was greater (though not significantly) where Buzzard nest densities were higher (Hodder 2001). However, Prytherch's study in Somerset has now obtained some interesting relationships that suggest Buzzard productivity reduces if breeding density becomes high enough (Prytherch 2013). At the beginning of the main study in 1988 there were only 19 pairs, with a mean distance of 1km between nearest nests, and they produced 1.89 chicks per pair, or a brood size of 2.12 if only those nests that produced young were considered. In 2007 there were 92 nests, at a mean distance of 0.6km apart, producing only 0.87 chicks per pair, or 1.53 per successful pair. The trend in productivity, while fluctuating, showed a steady decline over the 20 years and the number of successful nests dropped from near 90 per cent to only half being successful. Overall, the study area produced more chicks over time, but for individual nests, brood size reduced. The article also assesses other studies with sufficient years of data and shows that

▲ **1a.** Pale morph.

▲ **1b.** Intermediate morph.

▲ **1c.** Dark morph.

▲ **2a.** Yellow cere on wild chick.

▲ **2b.** Blue cere on domestic chick (© Sean Walls).

◀ **3.** New dark feathers mix with faded older feathers during the moult.

▲ **4.** Greater wing-coverts on juvenile Buzzards (left) tend to have paler fringes and hence be more distinct from the primaries than is the case with adults (right).

▲ **5.** A trained buzzard worming (© Sean Walls).

▲ **6.** A vole for young Buzzards.

▲ **7.** A surfeit of rabbit for two Buzzard chicks (© Jason Fathers).

▲ **8.** Pellets below a Buzzard roost (© Robert Kenward).

▶ **9.** Grouse feathers on a wall used for plucking under a Buzzard nest (© Robert Kenward).

▲ **10.** A captured vole gives worms scope to escape (© Gerhard Kornelis).

▲ **11.** It is Buzzards, such as the one pictured here on the ground, that access the food most confidently at a feeding station for Red Kites.

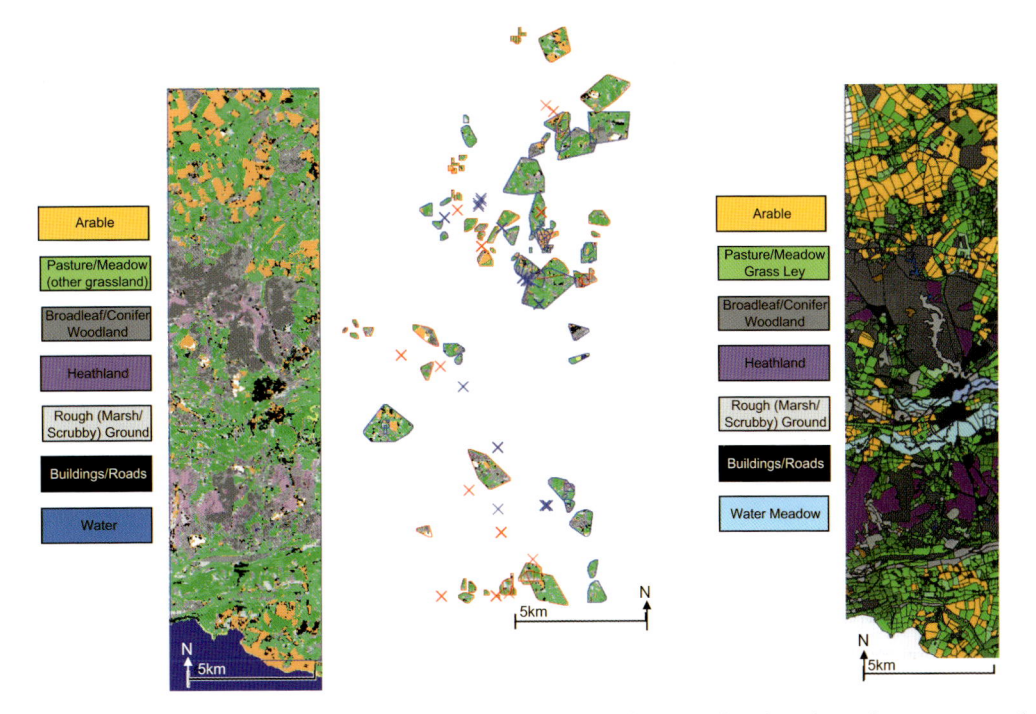

12. Two land-cover maps of the study area in Dorset (left, from satellite-based Land Cover Map of Great Britain; right, survey-based field map from Hodder 2001), showing also (centre) land-cover in range cores and natal nests (X) of adult male (blue) and female (red) Buzzards.

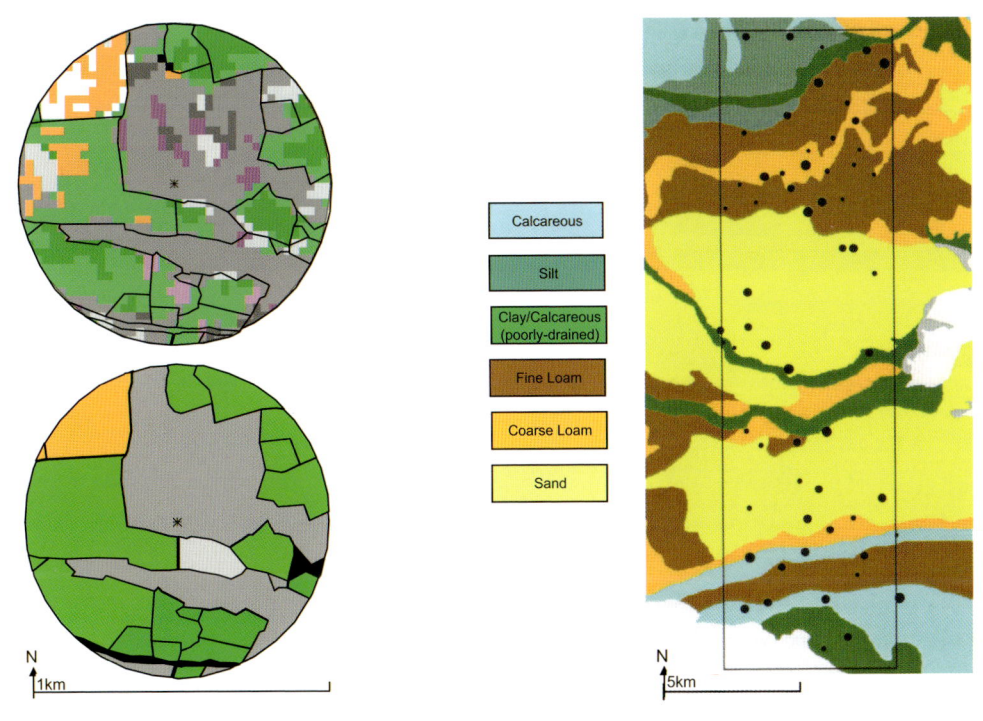

13. Land-cover classifications (for keys, see Plate 12) in 500m radius from same nest on the Land Cover Map of Great Britain (top) and field survey map (below).

14. Nest locations (black dots; size shows productivity) on geology-based soil map of Great Britain (from Hodder 2001).

15. 'Whitewash' from faecal urates under a Buzzard nest with well-grown young (© Robert Kenward).

16. The 'full-angel' (left) and 'fallen angel' (right) postures.

17. The pale morph bird is 'mantling' to protect food from the dark morph.

▲ **18.** High nests in a Larch (left) and an oak (centre) in Dorset, and relatively low in a young Sycamore (right) (© Sean Walls, Jason Fathers).

▼ **19.** The nest in a single Dorset field-boundary oak is completely obscured by foliage (© Robert Kenward).

20. Nest 5m up in a Beech, with a higher alternative nest in the distance (▶), and the view across moorland (▼) from this Scottish site (© Robert Kenward).

21. Ten-day-old chick with unhatched egg and fresh green nest decoration.

22. Common Buzzard found hidden under a large log in 1994. A post-mortem showed that this bird had been shot (© Sean Walls).

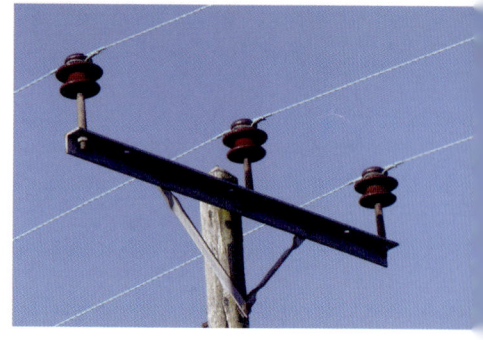

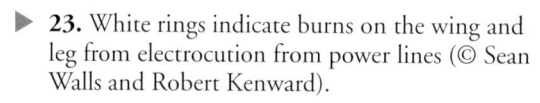

23. White rings indicate burns on the wing and leg from electrocution from power lines (© Sean Walls and Robert Kenward).

24. A young Goshawk (left) hesitates about directly attacking a Common Buzzard.

they have reduced productivity with increasing density, but only after the 1980s since when Buzzards have been allowed to recolonise areas and build high densities in the absence of other large raptors. It seems highly likely that earlier studies that found no effect of density on breeding probably failed to reach a threshold density above which it could be detected. It's also intriguing that once density increased above 0.75 pairs per km^2 in 1996, productivity fluctuated more, which could have reflected competition amplifying the effects of variation in food.

So there is evidence that Common Buzzard breeding success is affected by other Buzzards, as well as other species. One particular species that has had a large share of the influence on nesting success in so many ways is, of course, ourselves.

Human disturbance

Human activities affect breeding both deliberately and accidentally. In later chapters we will discuss the indirect effects of anthropogenic practices such as pesticide use, and how human ways have affected Common Buzzard populations over time. Here, we will concentrate specifically on the direct effects of human activities at individual nests. The most recent extensive analysis in the UK (Sim *et al.* 2000), which gave a total nest failure rate of 15 per cent, found that 2.5 per cent had been taken by egg-collectors and 5 per cent 'disappeared', with most other failures due to damaged or unhatched eggs, or injured or predated young (Figure 7.11).

There is no doubt that in the past (and to a lesser extent now) some people have aimed to reduce nest productivity by killing adult breeders, destroying the nest or removing eggs. Looking at nest recording cards between 1937–1969, Tubbs (1972) found that 'human predation' was the most common cause of nest failure. Failures in 49 of 146 cards (34 per cent) were recorded as human predation, and this was only if there was direct evidence, whereas cooled eggs or disappearance could also have been caused by humans. Picozzi & Weir (1974) estimated that in Scotland an average of 35 per cent of nests were probably deliberately disturbed every year. Even though Swann & Etheridge (1995) found only three of the 71 failures recorded at one of their study sites in Scotland had direct evidence of

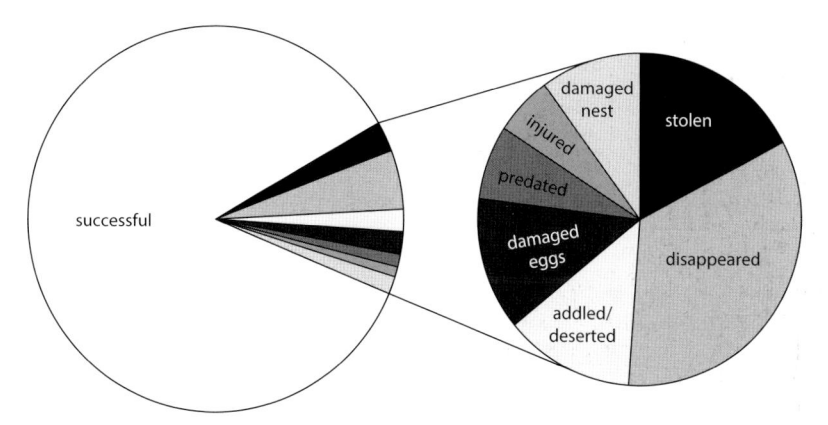

Figure 7.11. *Records of 952 Buzzard nests in the West Midlands and Wales 1950–1995 (with most data for 1970–1995) as compiled by Sim* et al. *(2000), with kind permission from* Bird Study.

interference by humans, they still suspected that certain shooting estates in the Moray district were affecting nest occupancy. Not only did they find eight poisoned Buzzards on occupied territories just before they were due to lay eggs, they were also surprised that the number of nests was so low compared with a similar site (Glen Urquhart), despite high breeding success in Moray, which suggested plenty of food. In Poland, Goszczyński (1997) considered 30 per cent of breeding failure in his study area was due to human disturbance, including nest destruction.

Although not intending to reduce Buzzard numbers, egg-collectors inevitably directly reduced productivity at nests in the past (Tubbs 1972, 1974, Picozzi & Weir 1974, Tubbs & Tubbs 1985, Sergio *et al.* 2002). However, there is no evidence of this illegal pursuit being a particular problem for UK Buzzards now. Unfortunately, those increasingly interested in countryside recreations, including birdwatchers whose last wish is to affect breeding, can disturb the nest at a critical time and result in cooling of eggs or chicks (Davis & Saunders 1965, Picozzi & Weir 1974, Dare 1998). This unwitting damage may become more prevalent as Buzzards become more abundant.

Woodland management can result in nest trees being felled, or disturbance nearby leading to chilling of eggs or young chicks (Picozzi & Weir 1974, Sim *et al.* 2000, Löhmus 2005). However, we had a case where trees were felled to within five trees of the nest of an incubating Buzzard. That year, the nest had higher productivity than any previous years, we think because the felled area provided rich small mammal pickings not usually available so close to the nest. The following year, the pair nested in a line of trees within the same woodland plot. Talking to a forester on another estate, she described how working near Buzzard nests is not normally a problem unless you spend time looking at the nest. Working close to a nest with a chainsaw, a female would sit tight on the eggs, but if you downed tools and looked up to the nest then she would be off. Nevertheless, felling one tree in a garden near another incubating female resulted in the loss of the brood. Scope for disturbance may well reflect previous experience of each female.

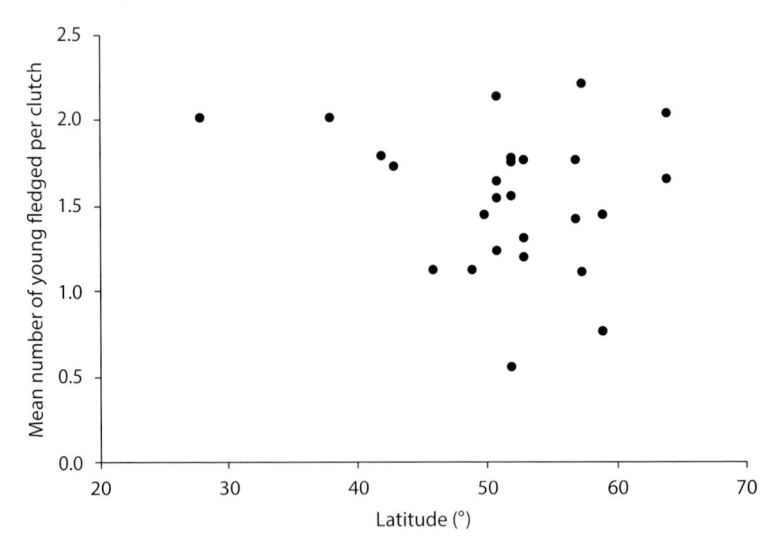

Figure 7.12. *The average productivity of Common Buzzard pairs (mean number of young fledged per clutch) in multi-year studies shows no consistent trend with latitude from Tenerife to Finland. Data sources in Appendix 2.*

Productivity and geography

It is clear that many factors influence the productivity of Common Buzzards across their geographic distribution. Food, climate, competition, predation and human activities may all play a role, with the result that productivity, defined as the number of young per clutch laid, shows no trend with latitude in the European and Atlantic part of their distribution, from Tenerife in the south-west to Finland in the north-east (Figure 7.12). The observed 1.71 young per clutch in our Dorset study lies pretty much in the centre of the range of values, from 0.55 young in a Welsh population at a time of relatively poor food supply and when only 65 per cent of nests reared young, to around two young per pair at all latitudes in other multi-year studies.

Individual characteristics

Having explored the external influences, we can now look at how much difference there is between individual Buzzards and areas across their geographic distribution. Looking at the fledging success of 13 Common Buzzard territories on Dartmoor, Dare (2015) showed that five territories (35 per cent) produced 65 per cent of the fledged young in 17 years. The least successful four territories (31 per cent) produced only 12 per cent of the fledged young. The individuals holding those territories undoubtedly changed over those years, because the study was conducted in two periods separated by 21 years, and not all territories were occupied every year. This result therefore reflects variation in territory quality, although some individual variation (perhaps including ability to secure a better-quality territory) cannot be ruled out.

There are various ways in which individual characteristics might affect breeding success. The number of eggs laid might be genetically determined. It is also likely to be condition-dependent, in which case the hunting ability of the male, and hence his delivery of food to the female (perhaps, with genetic influence), is a likely contributory factor. Once there are young, again the ability to deliver sufficient food for them to survive will vary between different males in the early stages, and between females once the youngsters are old enough to leave. It is difficult to assess the ability of individuals to hunt quantitatively, because faster prey delivery will also depend on the availability of prey, but there is evidence that hunting ability changes in terms of the species that are delivered to the nest, and the ease with which they are caught – for example, large Rabbits will be harder than small Rabbits and Woodpigeons are probably much more difficult than voles. Lastly, it is likely that there is variation between individuals in their defence of the young in the nest, which may start with building the nest in a less conspicuous place, but also depends on how they react to predators that do find the nest (Chapter 5).

Age may underlie the individual's ability in any one year. Krüger & Lindström (2001) showed that reproductive output was dependent on experience and senescence. The average number of fledglings increasing with each successive breeding of an individual until the fifth to seventh attempt, and then declined again. Similarly, Prytherch (2013) found that the success of the first breeding year was significantly less than the average of the following four years. Lifespan showed a similar pattern, with the youngest parents likely to disappear after their initial attempts, and the oldest birds also less likely to survive until the following year.

So, there is evidence that age has an effect both on the likelihood of surviving to breed the following year, and on success within a particular year. Let's now look at the breeding success of an individual Common Buzzard over its lifetime.

Lifetime reproductive success

Breeding success in a season is easily defined as the number of young fledged, but from an evolutionary perspective the success of an individual is not merely the number of young raised in one season or even during its lifetime, but the success of those young that survive through to breeding. Marking with visual tags, rings or long-life radios has the potential to provide data on this, but the effort required has so far proved prohibitive. A commitment of up to 40 years would be needed to follow through ten cohorts to 20 years of age, when breeding potential declines, and then to follow their offspring until the majority are breeding, say another ten years.

We have known for a long time, from a study on Eurasian Sparrowhawks in southern Scotland by Ian Newton (Newton 1985), that a small proportion of a raptor population can contribute a disproportionately large proportion of genes to following generations. Only 15 per cent of females produced 50 per cent of all fledged young, and 42 per cent of females fledged 75 per cent of young. The most influential factor affecting lifetime reproductive success (LRS) in Sparrowhawks was longevity: those that live longer have more offspring, especially when there are year-to-year variations in food supply and only a few young are produced each year.

Thankfully, we also have a study that has specifically investigated individual variation in Common Buzzard LRS. Krüger (2002b) used data collected from his German study population over 12 years (1989–2000) to investigate the combined effects of food, weather, competition, habitat and the individual Buzzard on LRS. Individuals could be identified from their plumage markings. Any two birds that were confusingly similar were excluded. Breeding lifespan was defined as the time of first nesting until the bird was not found in the breeding population for two or more years (some missed breeding for one year). LRS varied between 0–22 chicks, with a typically very uneven spread within which a small percentage of birds produced a large percentage of the young (Figure 7.13): 17 per cent of birds produced 50 per cent of young (Krüger & Lindström 2001), very similar to the Sparrowhawk population studied by Newton. There was no particular difference between males and females, and as with other species, there was a strong correlation between breeding lifespan and LRS. Average breeding duration was relatively short, at 2.8 years for females and 2.2 years for males, compared with a few that bred for more than 10 years. The breeding lifespan was mainly associated with weather in the first year an individual bred (Krüger 2002b), which also would have reduced the number of chicks fledged by first breeders. The greater LRS of those surviving longer than two to three years would have been enhanced not only by their larger number of broods, but also because reproductive success increased to a peak between attempts five and seven. Surprisingly, habitat around the nest played a relatively weak role. Lastly, for females, plumage morph was a factor. The most successful female birds were those of intermediate morph. Intermediate males were also the most successful, but much of the variation could be explained by other factors, and once these were entered into multivariate tests, then the pure effect of male morph was not significant.

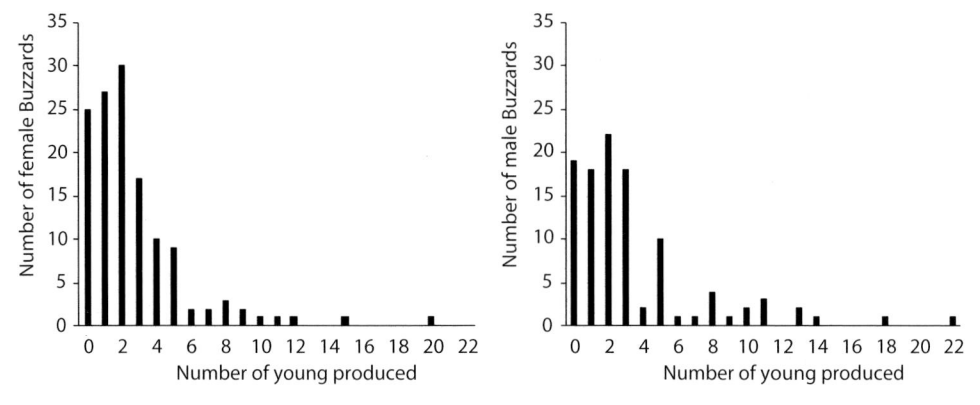

Figure 7.13. *The lifetime reproductive success (number of young produced) by known individual female (left) and male (right) Buzzards during ten years in a Westphalian study. Data from Krüger, O. ('Dissecting common buzzard lifespan and lifetime reproductive success: The relative importance of food, competition, weather, habitat and individual attributes') in* Oecologia *(2002) with permission from Springer Verlag.*

Having described the courting and nesting phase that produces young Buzzards, we have been able to examine two aspects of Buzzard populations that are potentially important for conservation. One is the genetic significance of lifetime reproductive success, which means that factors selecting against birds early in life are likely to strongly influence the genetics of following generations, if 50 per cent of a generation are produced by only 17 per cent of those that lay eggs in any year. The other important aspect is the productivity of breeding pairs, in terms of young fledged per clutch, because that is not only relatively straightforward to estimate, but is also a fundamental component of modelling populations, as we shall see in Chapter 10. Next, it is time to explore what happens to young Buzzards as they leave their nest, start to make a life of their own and ultimately begin their own breeding.

Conclusions

1. Common Buzzards typically lay from one to three eggs, between late February in the most southwestern part of their Eurasian distribution, to late May in the far north. In a local area, there can be about a month between the first and last pairs laying.
2. In our Dorset study, about 5 per cent of eggs incubated to term failed to hatch and another 13 per cent of eggs hatched but failed to fledge chicks. With insufficient food, older chicks may attack younger chicks and sometimes eat them.
3. Very rarely, more than two females contribute to the same brood, and produce up to six fledglings in the right conditions. How often smaller broods are the result of more than two individuals is currently unknown.
4. An increase in wing length of chicks of 8mm per day is remarkably constant and can be used to age nestlings and thus predict hatching dates. An increase in mass of about 20–30g per day occurs until chicks are about 30 days old.

5. Fledglings start branching and making short flights between six and eight weeks after hatching, during which time they can be very vocal.
6. Breeding success for a pair of Common Buzzards is better when there is ample food and dry weather during the period when chicks do not have sufficient feathers for waterproofing. It also benefits from lower surrounding nest density, fewer predatory raptors and less human disturbance.
7. Breeding success of pairs increases over the first five to seven years that they reproduce, after which it decreases until senescence prevents further breeding.
8. An individual Buzzard may produce more than 20 chicks in its lifetime. However, lifetime reproductive success is very variable; although some pairs are known to have bred for over ten years, a German study found that the average breeding span there was between two and three years.
9. Because most Buzzards do not have a long reproductive life, some of the variation in the number of young raised over a lifespan will be due to chance, through local and temporal variation in weather and food abundance. There is also evidence that there may be genetic differences in defence against predators and other factors that increase an individual's ability to raise many young during its lifetime.

CHAPTER 8
Dispersal and migration

Like looking for nests, searching for a radio-tagged Common Buzzard that had vanished from our Dorset study area required perseverance, sometimes spanning several months, before the jubilation at finding the wanderer. We dedicated most of our time and resources to the dispersal and settling behaviour of Common Buzzards, because this was where we expected to contribute most to raptor biology. We were working in a time before the remote download of GPS data was possible, so tracking involved being on the road in a Land Rover for several days a week during the non-breeding period, repeatedly stopping to raise the mast on the vehicle to take bearings, or jumping out to climb off-road high points to listen for missing birds. It was very satisfying to know that, for the first time after decades of using rings or wing-tags to estimate Buzzard movements, there was less scope for bias to cast doubt on the accuracy of the data we were recording, because we could find every bird in the study area when we wanted to. Further afield, the radio-tracking took us up hill and down vale, across moorlands, along rivers and to many wonderful places that were completely new to us. When we could not find a number of birds after ground-based tracking, we took to the sky in a Cessna light aircraft. Planned survey flights could be frustrating as we waited for autumn fog to clear before flying, but at other times were exciting when, after many hours of hissing white noise in the headphones, a momentary 'beep' promised contact with a bird a long distance from home. Sometimes, flying was more exciting than we bargained for, such as the time when the Cessna's engine started to cough due to icing between two cloud layers. We had to make an advisory landing at RAF Yeovilton, among hovering Harrier jump jets and greeted by on-site firefighters poised to take action that thankfully wasn't needed. When we heard radio-tag signals from the air, we had to find them as quickly as we could on the ground after our flight, to check that the signal was indeed one of our tagged Buzzards and to get a more accurate location. We then spent years fascinated by computer routines we'd programmed for the data, from which patterns in the movements of our tagged Buzzards gradually emerged.

So, where do Common Buzzards go when they leave the nest? At first, a young Buzzard will follow its parents to get food. If its flight has not developed sufficiently to keep up with a parent, it will scream repeatedly and elicit a feeding response from parents with freshly caught prey. But at some point the young bird reaches a stage at which it starts to explore what lies beyond the breeding territory.

Dispersal movements

Animal dispersal is a captivating phenomenon. For those who cherish ideas, it inspires many intriguing theories about philopatry (favouring one's homeland) versus moving away, and what cues individuals use to decide where to settle. It's also useful for wildlife managers to understand dispersal, because it enables predictions about how effectively a species is likely to colonise new areas, or how well a species reintroduced for conservation purposes is likely to reinhabit the old haunts.

Defining dispersal

In an article in 1980, Paul Greenwood differentiated natal and breeding dispersal in ways that were easy for ornithologists to define through marking and reidentifying individuals at nests (Greenwood 1980). 'Natal dispersal' was the movement of a young bird from its hatching nest to a place where it would breed. 'Breeding dispersal' was a movement from one nesting place to another. Of course, it was always understood that some birds might migrate or otherwise move around between leaving a nest and finding a place to breed, but it was difficult to fill in other movements without the ability to track birds away from the nest. Common Buzzards don't breed in their first year, and so they don't simply move from the nest where they hatched straight to a new area to build their own nest. Luckily, Buzzards are also big enough to carry radio-tags that can transmit for several years, enabling us to find the birds at all times of year and therefore describe additional dispersal stages before they settled to breed. Moreover, we could discover when Buzzards made long-distance movements, what habitats they were dispersing from and the nature of the habitats (and hence presumably the resources they required) where they eventually settled.

Technology at the end of the 1980s, when planning our work, was less advanced than today. We were using VHF radio-tags that could be detected from tens of kilometres away. However, to achieve our standard accuracy of 100m for locations required triangulation from 1–2km away. Thus, with 25 birds tagged each year, and the majority still detectable each subsequent year, it was very time-consuming to track all the tagged Buzzards that remained close to home. It was also a lot of work driving around, and sometimes flying, to search for Buzzards that had moved. However, it was well worth the effort to find how Common Buzzards really dispersed. GPS tags have the potential to provide a big saving in time and cost of travel to obtain locations from dispersing Buzzards, but sadly there has not been as much interest in tracking Common Buzzards as for other raptors. Therefore, the GPS tracks of Common Buzzards are still very few in number, and our VHF tracking data are still the most comprehensive available.

A fledgling Buzzard is initially very constrained by its flying ability. Time is needed to complete growth of its flight feathers; before that, the wing and tail surface area is smaller than an adult's. The resulting higher wing-loading means that the bird will have to flap harder to initiate flight; it will also lose more height during gliding and be unable to take full advantage of updraughts. Not only does that make movement more energetically expensive, but it also renders the young Buzzard less manoeuvrable if attacked. As a consequence, staying with parents who will provide defence is a sensible strategy, at least until the feathers are fully grown (Chapter 7).

It also takes time to build flight muscles and develop experience of how to use the air. In the first few weeks after leaving the nest, it is easy to differentiate young birds from adults, because they are so much clumsier in flight. Even later in the season, it's surprising how adults in a thermal will rise much quicker than their offspring who are trying to follow them. Prytherch (2009) notes that juveniles tend to hold their wings flatter than adults, sometimes raising them slightly but still looking more rakish. His diagram shows that even if the inner wing is raised, the outer remains flat, giving the wing a bent look, rather than the straighter V-shape of the adult. This is interesting if viewed from the perspective of a more nervous youngster, wanting to use as much of the updraught as possible for lift and therefore maximising the downward-facing area. However, the V-shape helps stabilise the Buzzard, as it does for vultures and other similar soaring birds; in time, the bird may learn that and then prefer to hold its wings higher to provide more stability. Some recent work by researchers at the Hebrew University in Israel found that there were also age differences between young and older Eurasian Griffon Vultures in their use of rising thermals (Harel *et al.* 2016). They used accelerometers to look at the detailed movement and found that older birds turned in larger circles in a moderate thermal, riding the outer reaches of the ascending air. The air does not rise so strongly away from the centre, but the birds did not bank so strongly and therefore could better take advantage of the rising air. Perhaps, young birds need the security of the strong air current near the centre of the thermal, whereas older birds have learnt to trust the smoother peripheral updraught in which they can devote more attention to looking around for signs of food and less to coping with turbulence. However, adults also conducted more 'thermal centring', where they turned into the strong updraught and rose suddenly. Overall, on the foraging trips that were measured, juveniles demonstrated less soaring-gliding efficiency, and the accelerometers showed they spent more energy than adults flapping. Common Buzzards, too, seem to use the centre of the thermal for rising suddenly before soaring around in the periphery again.

Let's return to the close of the last chapter, with our tagged Buzzards starting to explore more distantly after 65 days of age, when all feathers are fully grown and they had some experience of mastering the air (Tyack *et al.* 1998). After our tagged fledglings were 100 days old, there was no limit to where they could go, and we had to spend many hours with our receivers and Yagi antennae scanning from hilltops far and wide, listening for signals from Buzzards we had temporarily lost. The best time of day was between 10am–2pm, when the late-summer thermals had developed and Buzzards were more likely to be riding them high into the air.

If we were on a hilltop with our Land Rover mast elevated (Chapter 2, and Figure 4.1), we could hear a radio-tag on a Buzzard at 40km if it was flying high or perched high in a tree on another hilltop, and hence in line of sight. On the other hand, our detection range was considerably reduced if a Buzzard was perched low in a tree within woodland. We might receive a signal at 20km if the Buzzard was on a hilltop, or only a few hundred metres if it was on the other side of a hill or on the ground feeding. If the Buzzard had not moved from where it was last found, then this shorter range was no problem. On the other hand, after a

Figure 8.1. *A three-element Yagi antenna was mounted on each wing strut of a Cessna aircraft for tracking Buzzards with VHF radio-tags across the woods and fields of southern England in the 1990s (© Robert Kenward).*

big movement we needed all the detection distance we could get, and so times when Buzzards were flying were crucial to us for reconnecting with them.

With our limited movement on the ground, we could not always keep up with the Buzzards that we heard flying away, because they could cover any landscape in a straight line at considerable speed, exploiting winds and updraughts, and then drop to a perch in a distant valley. Luckily, we had one more tool available for when an individual Buzzard had deserted us. That was to beat them at their own game by flying in a single-engined Cessna 172 aircraft, with a Yagi antenna fastened securely to each wing strut (Figure 8.1). Flying over southern England, with our great friend Alan Morriss piloting the aircraft, allowed us to hear soaring birds that we knew were more than 80km away. These day-long flights, which were planned with tracks 40km apart, took us north to a line from Lincoln to Liverpool (Appendix 3). This was not quite as much fun as it may sound, as it involved spending 3–4 hours at a time (between refuelling stops) listening to a receiver attached to each antenna and programmed to cycle for a few seconds through the frequencies of up to 30 lost tags in the latter years of the study. The hiss in the earphones changed pitch slightly as the receiver switched through each programmed frequency in turn and only rarely rewarded us with a *bleep-bleep-bleep*; when we recorded the signal's strength and time it was heard along each flight track (see figure 7.17 in Kenward 2001). This search technique enabled us to find most of the 'lost' birds, sometimes more than 100km from their nest-site.

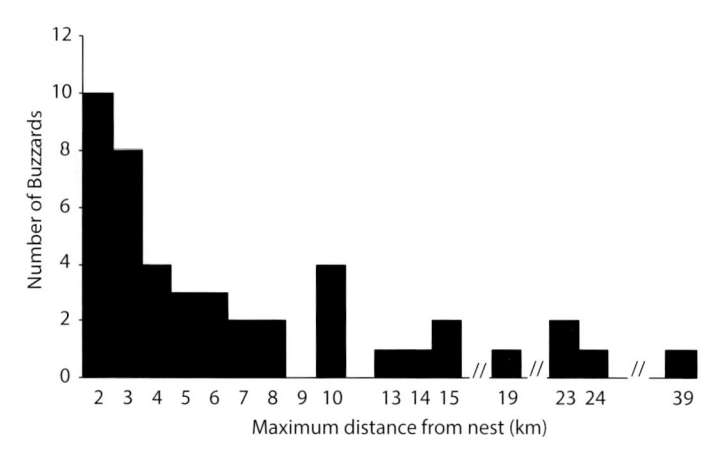

Figure 8.2. *The maximum distance from the nest recorded in excursions of 45 Buzzards tracked during the post-fledging period in Dorset.*

We also sought, and found at greater distances, Red Kites lost by English Nature and the RSPB; these had been released to increase their distribution in the UK.

Our tagged Buzzards did not simply disperse abruptly and move straight to a new area where they settled, as we had observed Northern Goshawks do (Kenward *et al.* 1993a). We found that juvenile Buzzards often made exploratory excursions before properly dispersing (Walls & Kenward 1995). These were similar to the '*véritable erraticism*' found by Nore *et al.* (1992), who radio-tracked some Buzzards in France, or the local wanderings of visually marked Buzzards in Scotland observed by Picozzi & Weir (1976). Once fully feathered, Buzzards excursed in very varied directions, which did not seem to relate at all to the final dispersal direction, and sometimes went more than 20km, not necessarily returning the same day (Figure 8.2). Although we recorded 2–4 excursions from about half of the birds, it is likely that they made many more excursions that we did not record, because we were checking each Buzzard only every 3–4 days (at most) and short-distance excursions could have been completed in hours.

This behaviour was consistent with young birds first building up knowledge of the area around their natal nest, perhaps as a standard with which to compare new areas. Excursions in some directions may have encountered unfavourable landscapes or inhospitable territory owners, in which case it would make sense to return to the familiar safety of the nest territory. Other wanderers may have spent some days exploring a promising area, and then, if prevailing winds were strong, not found it easy to return home. These different behaviours would explain the great variation in what we recorded. Some dispersed abruptly soon after their feathers hardened, within a month of leaving the nest. Others explored, returning to the nest for a while before eventually dispersing, sometimes not until after the first winter. There were then a small proportion of birds that never properly left, settling into a home range in or next to the natal territory (Figure 8.3).

More remarkable still, we found that in the following spring and summer, Buzzards that had left their hatching area but not started to breed tended to return close to where they had fledged. Indeed, this was the case for all birds that we knew to have wintered some distance away, even if the visits were short (Walls & Kenward 1995, 1998). The return visits included

The Common Buzzard

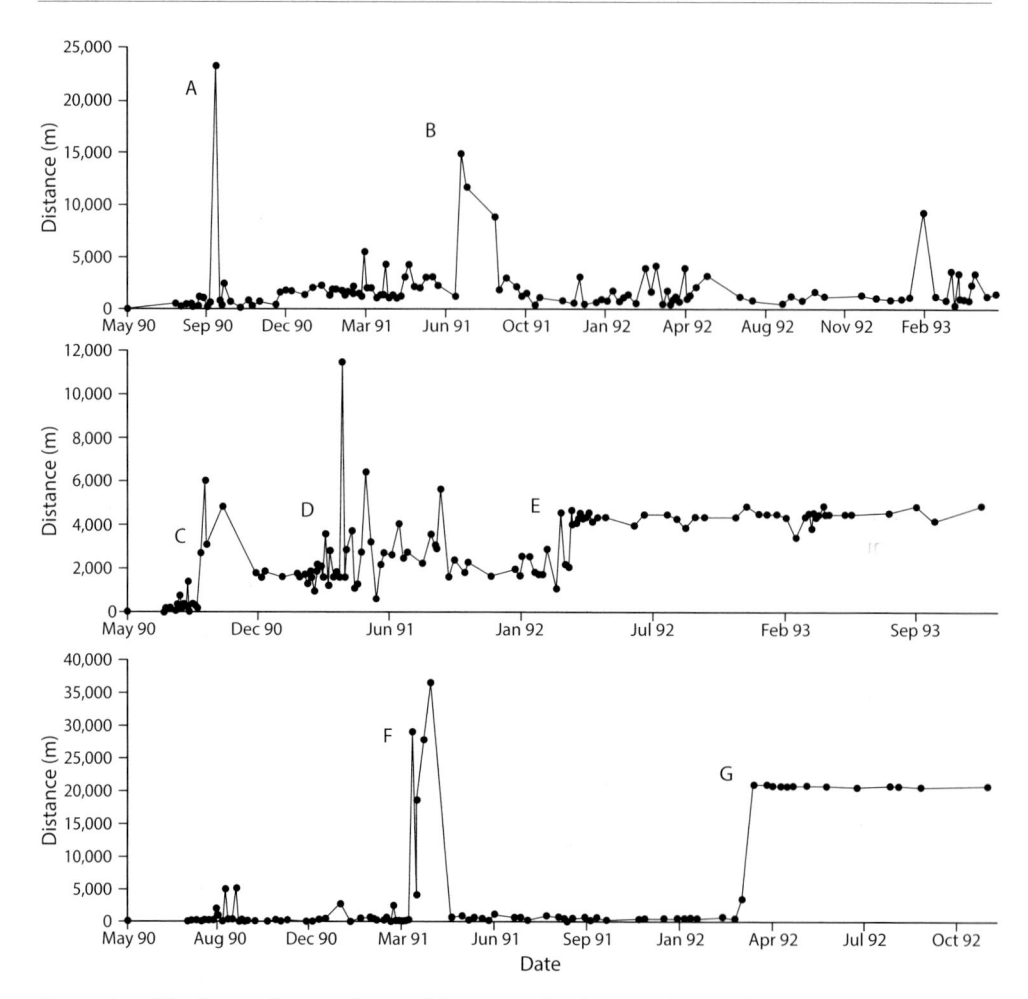

Figure 8.3. *The distance from natal nests of three young female Buzzards tracked in Dorset during 1990–93. The bird represented in the top graph made a 20km excursion (A), then settled 1–2km from the nest but spent its second summer 10–15km away (B) and then returned to the range near its origin. The young female in the middle graph dispersed to 2km from its nest in autumn (C), became very restless the following spring (D), before changing to a more stable range 4km from its origin the following spring (E). The female in the bottom graph moved considerably from her natal home range in her first spring (F) but then resettled for the following winter, leaving and promptly breeding 20km away the next spring (G).*

a few birds that we had recorded leaving in the autumn but had not found throughout the winter. Therefore, we could not simply define dispersal as when an individual left its parents' territory, because many quickly returned, sometimes to remain until the following breeding season but also sometimes for a brief visit before leaving again.

With these patterns of excursion from natal areas, and then from areas where birds had settled for winter, it was challenging to come up with a definition of dispersal that could satisfy the diversity of movements. It is no use having an improved technique for collecting dispersal data, to remove observational bias, unless the analysis, too, is as free as possible from bias in how the data are used. Ideally, one needs a definition that is easy to understand

and program into a computer for objective processing of the data, without any slippery subjective human decisions about each bird.

Defining when birds dispersed required us first to decide a minimum distance beyond which a Buzzard had to have settled before the movement could be considered dispersal. Ideally, that would have been beyond its parents' territory. Unfortunately, we did not have radio-tags on the breeding birds to work out their home range, nor did we have the time to spend making detailed observations of territorial boundaries, so we had to use the data we had collected from tracking juveniles. Eighty-eight percent of locations from 23 juvenile Buzzards tracked in July and August were found within 1km of their nest (see Chapter 7, Figure 7.8), with many of the remaining 12 per cent due to three young from one particular nest that spent a lot of time with their parents 1.5km from the nest. Furthermore, range diameters of Buzzards tracked in their first October were an average of 1.1km away. Therefore, it seemed reasonable to consider that birds settling more than 1km from their hatch nest had dispersed. For the initial analyses we also tried a more conservative 2km limit, but it made no difference to the results. In truth, some birds could have been out of their parents' territory before reaching 1km from the nest, especially in view of some recently recorded densities of Buzzard nests (Chapter 10).

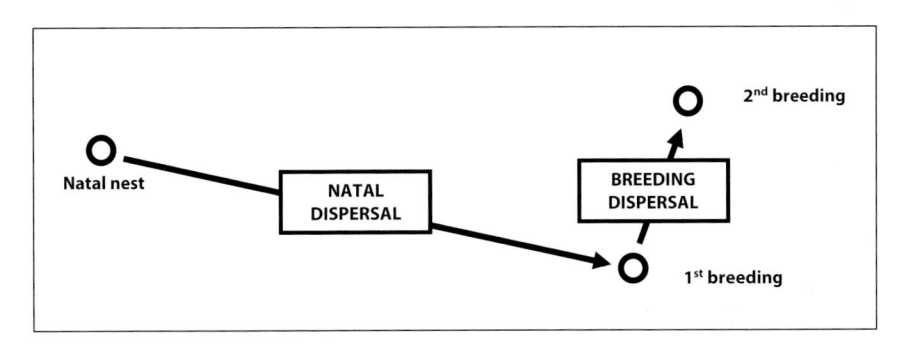

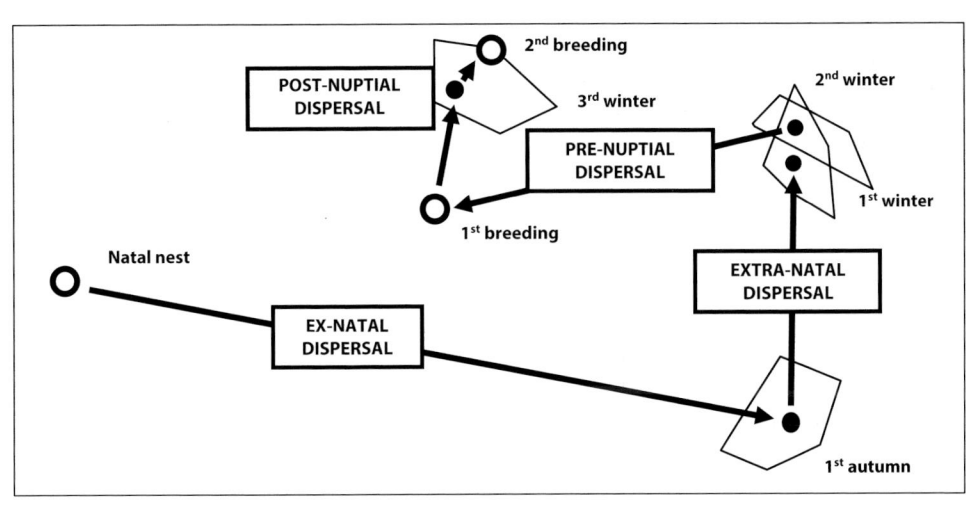

Figure 8.4. *Data from ringing and re-recording birds at nests (circles) enabled natal dispersal and breeding dispersal to be defined (top), but movements recorded by radio-tracking enabled transit between additional areas of settlement (polygons) to be defined and studied in detail.*

We then had to put a time requirement on the duration of absence, given that birds could not only return from excursions to live with their parents through the winter, but could also return from wintering areas to visit their natal areas in the spring, after they had dispersed. Therefore, we defined a new term, 'ex-natal dispersal' (from the Latin *ex* for 'out of' the natal area) for when an individual went more than 1km from their natal nest without returning for more than two consecutive detections there again until after January. Included among the dispersers were some birds that seemed to settle in one area, but then went on to settle in another area away from the nest, sometimes making the second movement after the first year (Figure 8.4). These 'extra-natal dispersal' movements (from the Latin *extra* for 'beyond' the natal area) were similar to ex-natal dispersal, but there was no nest involved. Therefore, rather than using a nest we had to use a different focal point, the centre of activity. Thus, 'extra-natal dispersal' was defined as when a bird moved more than 1km from its centre of activity, without returning. If the bird attempted to breed following the movement, then that would be considered 'pre-nuptial dispersal' (although it could also be 'extra-natal', see Figure 8.3G). A move after a breeding attempt could be either 'post-nuptial' if it did not result in another breeding attempt (Figure 8.4), or breeding dispersal if the bird was attempting to breed the following season. Based on that conceptual framework, we can describe the dispersal of Common Buzzards with consistency.

When Buzzards disperse

From an analysis of 77 Common Buzzards we tagged as nestlings during 1990–94 and tracked for more than a year (Walls & Kenward 1995), 56 per cent dispersed between July and December, aged 58–191 days, average 107 days old. This corresponds with observations of Nore *et al.* (1992), who found their radio-tagged Buzzards first left at the age of three months, in mid-August. Those birds often found a loose gathering of other Buzzards to feed with, sometimes staying several days in one place, especially during bad weather. In Speyside, Scotland, it seemed that the wing-tagged juveniles started to leave earlier than in Dorset, and 84 per cent of the Scottish Buzzards had apparently left their parents' territory by November (Figure 8.5), while unmarked birds travelling through Speyside were most abundant in September and October. Sometimes, all members of a single brood dispersed at the same time, while members of other broods departed sequentially and irregularly (Picozzi & Weir 1976). The disappearance of visually marked Buzzards did not necessarily mean they had permanently left; they may have been on an excursion or out of sight of the observers, so the difference in early leaving may be a product of the technology available. The dispersers we radio-tracked sometimes made excursions initially, but for the final dispersal movement they tended to leave suddenly and settle into a new area reasonably quickly. Although some birds made further travels after an initial dispersal movement, all surviving Buzzards seemed to settle for the winter by December.

While some previous studies observed a few young staying in the adults' territory into December (Tubbs 1974, Davis & Davis 1992), we were surprised that as many as 39 per cent of radio-tagged Buzzards were still foraging less than 1km from their natal nest in midwinter. Another 5 per cent of our tagged Buzzards left during the winter, but tended to make initial, relatively short movements and then continue to wander (Figure 8.6). In spring, almost all those that had not already dispersed then left, with their dispersal peaking

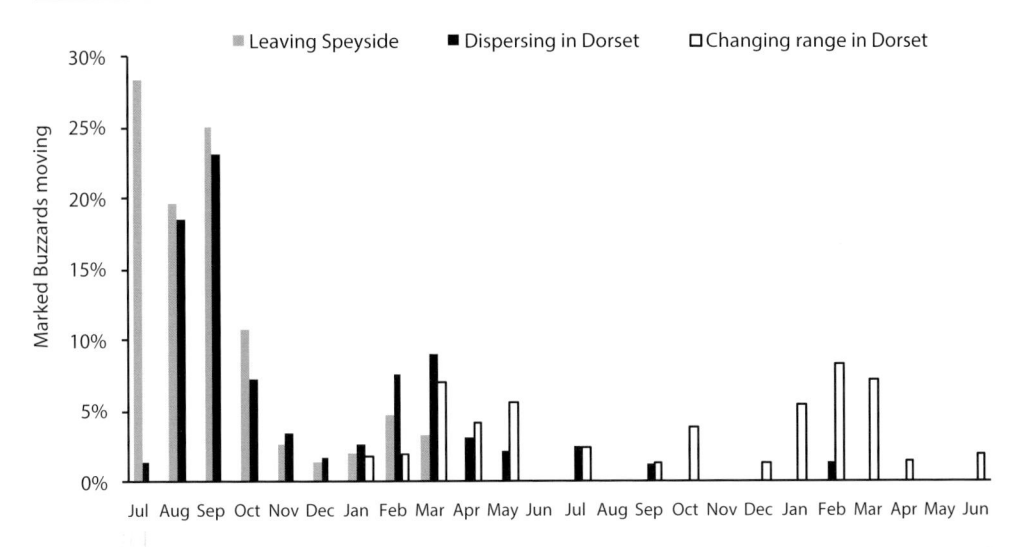

Figure 8.5. *The movements away from natal areas by visually tagged Buzzards in Speyside, and of radio-tagged birds in Dorset, together with subsequent changes of home range (i.e., 'extra-natal dispersal'). After initial ex-natal dispersal in the first autumn, further ex-natal dispersal (and peaks of extra-natal dispersal from initial home ranges) occurred in the first and second springs, after which there was little further movement. Data from Picozzi & Weir (1976) and Walls and Kenward (1998).*

around March (Figure 8.5) (Walls & Kenward 1998). This matched observations of Picozzi & Weir (1976), who reported a peak of unmarked birds in their study area during March–April. While the remaining birds were making their ex-natal dispersal in spring, some of those that had already dispersed started to move again, making more exploratory travels and some making an extra-natal dispersal to start a new home range elsewhere. About 15 per cent of Dorset birds remained in home ranges close to their natal nest during their second summer. This may in part have reflected favourable conditions in Dorset, because Picozzi and Weir (1976) recorded only 16 per cent still in natal territories by their first November, and only 2 per cent by the following March.

The spring wandering also revealed the philopatric tendencies of juvenile Buzzards: those that had dispersed tended to return back to their hatching area during the following breeding season (Walls & Kenward 1998). This continued as they got older, and we recorded some birds returning to their natal area in their third year. Moreover, older birds tended to return earlier (Figure 8.7) but did not return when they started breeding. Common Buzzards do not breed in their first year, so there may not have been any adverse consequences of returning to the natal area after others had started breeding, but the fact that older birds, ready to breed, arrived earlier suggests that they were looking at the site for breeding opportunities.

We could not find six of our tagged Buzzards during winter, despite searches from the air over southern England and Wales, but they then returned the following spring. Our best assumption was that a few birds dispersed further than our searches. At least one may have gone to France during the winter. That bird twice arrived in spring from the east, where the English Channel is narrowest. Although no direct evidence existed from our tracking, ring recoveries have shown Common Buzzards do cross the Channel between the UK and the

The Common Buzzard

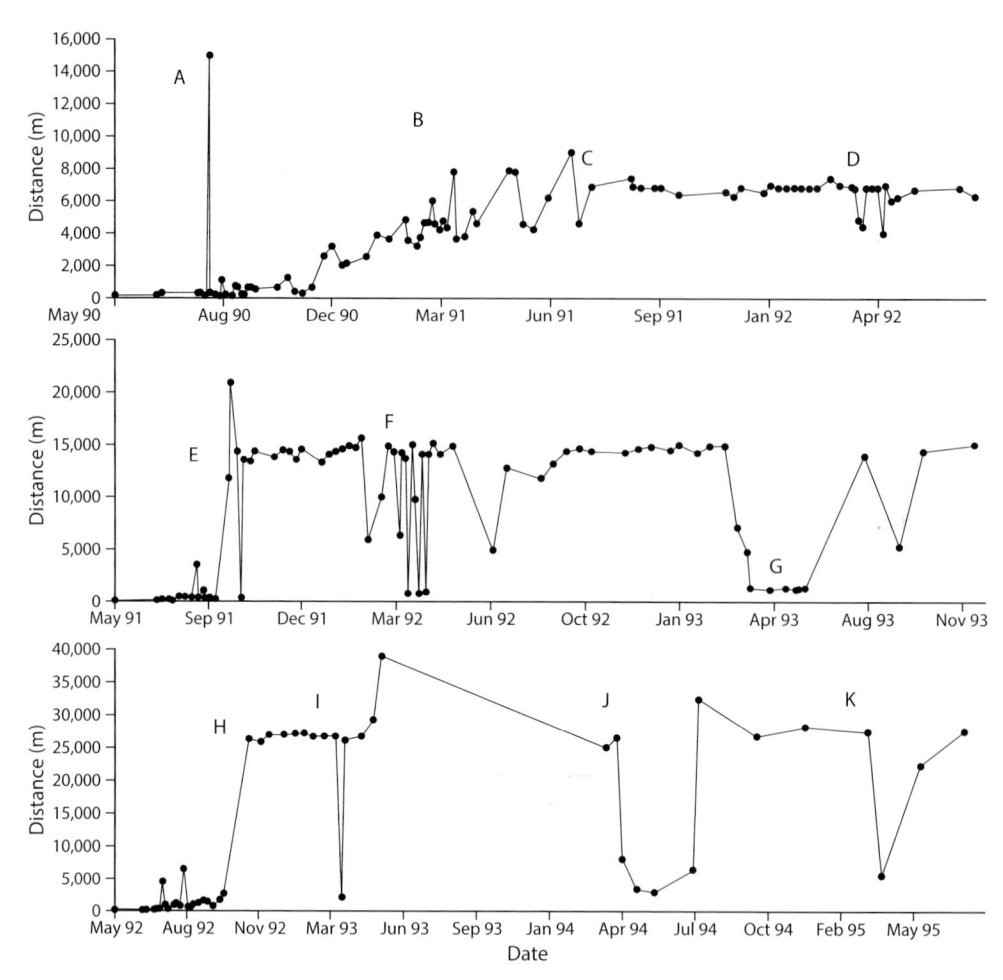

Figure 8.6. *Two male Buzzards showed very different behaviour, with one remaining near its nest after an initial 15km excursion (A) until drifting away in mid-winter (B) and after continuing to explore through the breeding season (C) settling into a range from which it made its way back towards its natal nest the next spring (D). The second male left abruptly in September (E) and made many separate philopatric movements the following spring (F) before spending its third summer close to its origin (G) but then returning to its wintering area. A female settled initially about 27km away, but then made philopatric visits towards its natal site in each of the next three springs (I, J, K). None of these birds were breeding.*

Continent. In 1962, Tubbs received a ring recovery of a New Forest bird found dead in France in its first autumn. A one-year-old Scottish bird was found in Germany in January in 1986, and in 1994 a Buzzard ringed as an adult in Belgium was found dead ten months later in Wales (Robinson *et al.* 2015).

The returns of our radio-tagged Buzzards were always for a brief period; none stayed to breed, and they tended to go back out to their wintering areas before the next of our regular checks. All our distant settlers that remained alive were recorded returning to their natal area in the first summer, but those that had settled within 20km were much less often found again near the natal nest, probably because they could return within a few hours and

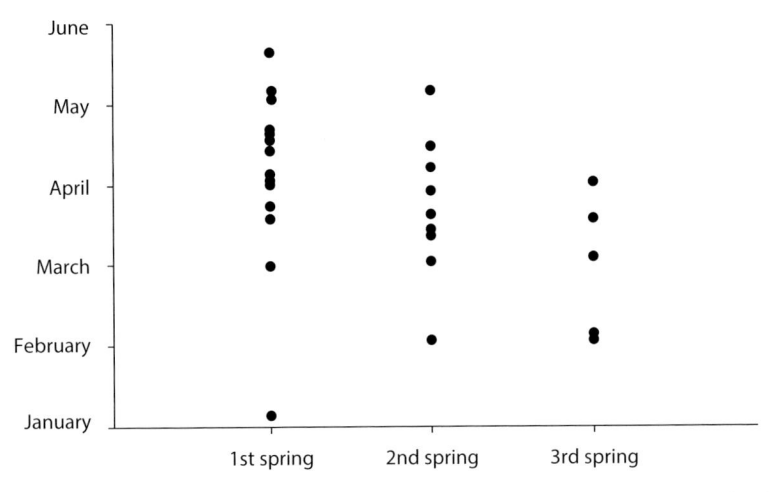

Figure 8.7. *The dates at which 27 radio-tagged Buzzards were detected returning to within half of their dispersal distances from their natal nest in the first three years of their life, with a median of 9 April after one year, 19 March after two years and 7 March when they were three years old. Data from Walls & Kenward (1998) with kind permission from John Wiley & Sons.*

therefore visit undetected between our tracking sessions. In some cases they could have observed their natal areas by soaring high within their wintering areas.

Others have suspected this type of philopatric behaviour from visually marked Common Buzzards. From their 215 Buzzards wing-tagged in Wales (Appendix 3), Davis & Davis (1992) found some birds appearing in spring after they had not been seen all winter, and there is a record of a colour-ringed Scottish Buzzard reappearing briefly close to its birthplace the year after hatching (Tubbs 1974). Our observations therefore confirm not only the ubiquity of this philopatric behaviour, but also the absence of settling close to 'home', which raises the interesting question of why, if eventually settling close to where they were raised is so rare, birds should tend to revisit the natal area at all.

Although dispersal sometimes involved a little meandering, most Buzzards we tracked seemed to find a new place to settle within a few weeks, and then stayed there for the winter. Most would then remain in that new 'home', perhaps adjusting slightly as the seasons moved on. From this new home range, or territory, excursions could be recorded, including those philopatric visits back to their hatching area, sometimes as part of a wider wandering before returning to their new home. However, between 35–45 per cent of Buzzards that had initially dispersed to one focal area then made another dispersal of more than 1km before settling again (see Figure 8.5 on timing of movements). This extra-natal dispersal was much shorter than the ex-natal dispersal they had already made. The subsequent short moves seemed to be a 'fine-tuning' exercise after having chosen areas that sufficed during the initial dispersal; perhaps the moves were made to include some other resource needed in the next season or phase of their life. Presumably, once they had spent time in an area, and built up some knowledge of the resources and other Buzzards, it was better to stay put and use that knowledge to make a small adjustment, rather than risk trying to settle afresh in an entirely new area, although continued excursions suggested that further resettlement (as opposed to local readjustment) remained a strategy of last resort.

Why Buzzards disperse

If there really is an advantage to acquired local knowledge, then we must address the question of why Buzzards disperse at all. From an evolutionary perspective, ex-natal dispersal is easily explained. If the young birds stayed close to their parents and siblings, it is likely they would compete for food and, eventually, nesting resources. If their combined production of breeding offspring was greater than for the single original pair, a colonial system might be favoured, as is the case for some kites and harriers. With potential advantages to both territoriality and coloniality, apart from communal defence (which is more frequently found in herbivores), the deciding factor may well be food supply. If hunting tends to make food less available, either through depletion or disturbance, there is an advantage in being territorial. Eurasian Sparrowhawks and Northern Goshawks extensively feed on birds, which are easily disturbed and reproduce slowly, and these hawks are highly territorial. Although Buzzards may congregate when worming in winter, they are strongly territorial in the breeding season, when young birds can be important prey in the absence of mammals.

For all these species, there is also the possibility of mating with siblings, leading to inbreeding and an associated chance of weaker offspring due to lack of genetic diversity. On the other hand, an advantage of a more colonial system occurs where the offspring actually help to promote genes into the next generation, most commonly by helping with feeding or defence of the related young. This too has, on occasion, seemingly been detected by cameras placed at Buzzard nests (Chapter 6). So, if the younger Buzzards are not usually actively helping to raise younger relatives, and thereby mitigating the poor evolutionary strategy of staying at home, is there something that actually motivates the young Buzzards to explore and disperse?

Young Buzzards do not always leave due to food shortage, because some of those that stay continue to get food from parents through the winter. Dare (2015) recorded a youngster getting a European Blackbird from its parent in early February, and we've certainly heard begging youngsters well into the winter months. Also, there has been no evidence of adults driving their young from the territory during the autumn. So it seems likely that the youngster is intrinsically motivated to leave, presumably to search for a better situation. This was clearly the case for Northern Goshawks, which dispersed early in areas with less food but, even if fed artificially in such areas, eventually left at the same age as birds in areas naturally rich in food (Kenward *et al.* 1993). However, whereas all Goshawks dispersed during their first autumn, many Buzzards remained until the following spring. Parental territoriality would likely have made foraging harder for them during the breeding season, although a small proportion managed to hang on and 'adjust' into a neighbouring territory through the summer.

After the second spring, it was very rare for Buzzards to disperse any distance again. However, our sample size was small because while our radio-tags had some advantages over the visual markers used in previous studies, unfortunately their batteries rarely lasted through the lifespan of a Buzzard. In comparison, the wing-tags used by Davis & Davis (1992) allowed longer-term observations, up to 13 years in some cases. However, these authors too found that birds were rather sedentary after the first year, and as the years passed those birds that they could monitor did not disperse again. Breeding started in the third summer and most were breeding in their fourth summer. Visually marked Buzzards were never seen outside their territory after they had started breeding.

We can now return to the second question of the previous section. Why, having dispersed, do Common Buzzards then make return visits to the natal territory in spring, especially given that they were self-motivated to leave? Perhaps this is an evolutionary strategy, based on the self-evident fact that an area in which an animal was bred is suitable for breeding, whereas the area to which it dispersed for winter has not yet proved satisfactory for breeding, only for wintering. This could be considered 'proto-migratory' behaviour, leading rapidly to short-distance migration (e.g., from low to high altitudes) if birds that stayed in the natal area for winter tended not to survive. The tendency for older (and hence more mature) birds to return earlier would be entirely consistent with being the first resettlers in areas emptied by overwinter mortality.

Radio-tags have given us more consistent information faster than previous methods, and with fewer assumptions. As well as the wing-tags used by Picozzi & Weir (1976), ring returns also helped to build an early picture of Common Buzzard movements. In one analysis, Mead (1973) looked at 111 recoveries from 1,562 Common Buzzards ringed in the UK. The 7 per cent ring return is a reasonably good recovery rate, but took a tremendous marking effort over 60 years to build up an impression of population movements averaged across the UK. In about five years of study, VHF radio-tags gave much more detailed information than both approaches, albeit just for central southern England. In time, GPS tags with solar panels (for long life) may well give much more detailed information on the actual dispersal routes as they happen, as well as more accurate departure dates. In the meantime, let us now turn to what we found about where Common Buzzards disperse from and to, and what we can infer from that about drivers of dispersal.

Whether to disperse

The discovery of much variation in the timing of ex-natal dispersal, and in how much extra-natal dispersal occurs, raises the question as to whether variation in individual personality (and hence perhaps genetics) or environmental variables play an appreciable role in promoting dispersal. Along with the radio locations, which informed us when individual Common Buzzards dispersed, we also had a digital land-cover map compiled from satellite imagery (Chapter 4), which was our best way to assess land cover used for foraging around each natal nest, and directly comparable to the land cover where birds settled.

To understand the effects of individual versus environmental variation on dispersal, we first asked whether brood size had an effect on the tendency of radio-tagged young Buzzards to disperse. As a simple analysis, there was no effect; dispersal occurred from all brood sizes. We soon realised that this was because, within a brood, some might disperse in the first autumn whereas others would not. It was only when we broke it down into who was left in the nest territory at the time of dispersal that we could understand the pattern properly. If Buzzards had siblings that were still around, they were far more likely to disperse. Thus, in a brood of three, the first to leave had two siblings it was 'dispersing from', the second to leave would have another sibling still present, and sometimes the last bird, with no siblings remaining, did not leave (Figure 8.8). In fact, the presence of other siblings was by far the strongest factor associated with dispersal. It was unusual for more than one young to remain in the parents' territory after autumn. Only once in our study did two young remain after November, and in that case one gradually left later in the winter, and continued to drift,

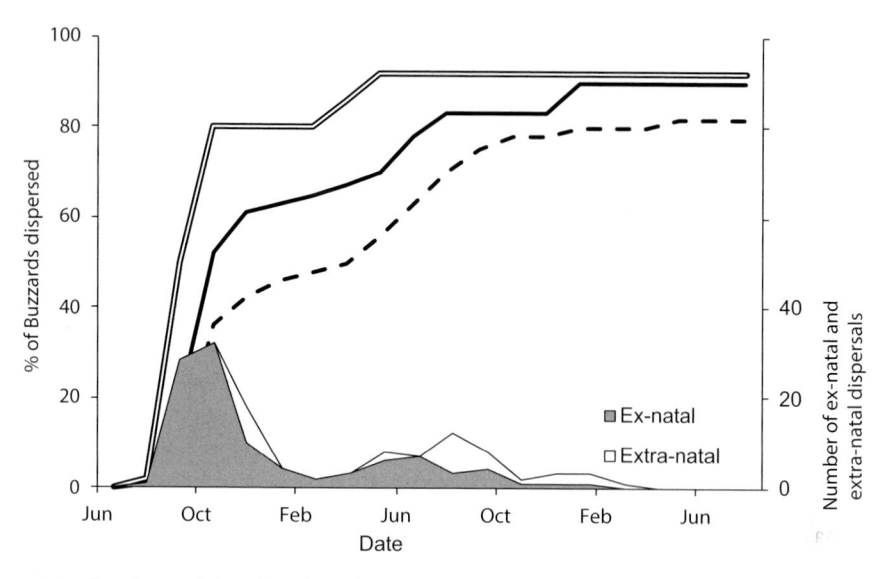

Figure 8.8. *One Buzzard from broods of three chicks (double line) tended to disperse early in autumn and almost always to have left by the following spring. The second bird from such broods or the second of a brood of two chicks (thick black line) left later, and a few birds without any siblings (dotted line) remained through to the following breeding season. Ex-natal movements of siblings left alone coincided with extra-natal movements of Buzzards that had already dispersed. Data from Kenward et al. (2001a) with kind permission from John Wiley & Sons.*

slowly changing its range and never quite settling. This pattern of only one remaining near the nest makes perfect sense if family members should avoid competing against each other for resources. Although we lacked data for robust statistical tests, there was a tendency for the older and larger (possibly female) birds from a brood to remain over the winter. Perhaps the most dominant in the brood is able to tolerate the parents best, whereas the weaker birds experience competition from both the parents and other siblings as a strong motivation to leave. Tags that can transmit some indication of stress may eventually reveal more about mechanisms. The physical size and ubiquity of Common Buzzards could make them a good model species to carry tags for such an investigation.

Whether a Buzzard dispersed from its parents' territory was also correlated with the nearby land cover (Kenward *et al.* 2001d). Buzzards avoid foraging in lowland heath (Chapter 4), probably because that land-cover type provides limited food. Correspondingly, juveniles were far more likely to disperse if the nest was surrounded by a lot of lowland heath. On the other hand, they were far more likely to stay at home if there was plenty of arable or short grassland nearby, habitats that would benefit their worming diet during the winter. It was interesting that the presence of fine loam soils had a higher predictive value in the models than any of the vegetative land-cover types sitting on top of the soils. This may have been indicative of a substrate that earthworms as well as Rabbits found best to burrow in, and of vegetation that favoured both staple foods of Buzzards.

By the following spring, there were no longer other brood-mates with non-dispersed Buzzards, so the number of siblings in the original nest was no longer relevant to dispersal. Instead, dispersal tended to occur when the increasingly territorial parents displayed in preparation for the new breeding season. However, there was no actual evidence for parents driving young away, in our study or others (Tubbs 1974, Nore *et al.* 1992), although only about 5 per cent of radio-tagged Buzzards remained in their parents' territory after May.

Following the ex-natal dispersal, Buzzards that had dispersed were more likely to move again if there was not enough seasonally long grass (representing meadows that were good for worming in winter and for small mammals to feed on grass seeds in early summer), or if there was too much coniferous woodland or arable land nearby (Kenward *et al.* 2001d). Plantations of non-native conifers planted in unnatural rows of monoculture are renowned for a lack of fauna and therefore wouldn't provide much food for Buzzards. It was interesting to see that the arable land favoured by Buzzards in the autumn and winter became a land cover associated with dispersal during the spring. It is likely that earthworms would have bred and become more numerous with increasing warmth in the summer, but their availability to Buzzards decreased. Destruction of worm burrows by tillage may have reduced worm survival overwinter and, once the crops start growing, it may also be less easy to see the worms. Then, as summer progressed, the earth dried, which meant the worms kept away from the surface to avoid desiccation. The vegetation growing up during the spring should have provided more food, with insects and small mammals feeding on the growing crops. However, people demand the cheapest food possible, which typically means spraying with herbicides that remove food for insects, or with pesticides that kill insects directly, and often push the crop right up to the fence line. 'Weed' seeds and insects are important elements of food chains for other Buzzard prey, so without them there are fewer rodents and breeding birds, which Buzzards need to survive the summer and feed their young. So, although the availability of arable land has probably improved feeding for Common Buzzards early in the winter, it is not ideal during spring and summer. That said, some crops may be better than others (see Chapter 4).

Our work agreed with that of others, in that almost all Buzzards eventually dispersed, but also showed that whether they dispersed quickly or waited until the following breeding season depended on food availability and who they had to share those resources with. We were also able to use the satellite mapping (Chapter 4) to look at land cover where these Buzzards settled.

Landscapes for settling

From the analysis above, we knew that young Common Buzzards were more likely to disperse from land cover likely to hold poor food resources, and away from competition. Therefore, we would expect them to settle in areas where there is good food and where competition is sufficiently low for them to survive and breed. So, we were surprised to find that when we examined where our radio-tagged Buzzards settled, and specifically compared the difference between dispersed and non-dispersed juveniles in their first winter, we found no particular difference in average range size or land-cover content between the two groups (Walls *et al.* 1999).

However, if we restricted the analysis to the dispersed juveniles, excluding those remaining in natal territories, those that had dispersed furthest had smaller home range cores. This indicated that the settled youngsters that dispersed furthest didn't have to go so far on a daily basis to get all the food they needed, and were therefore probably in areas with better resources (Chapter 4). Perhaps the richer feeding grounds pulled birds further away.

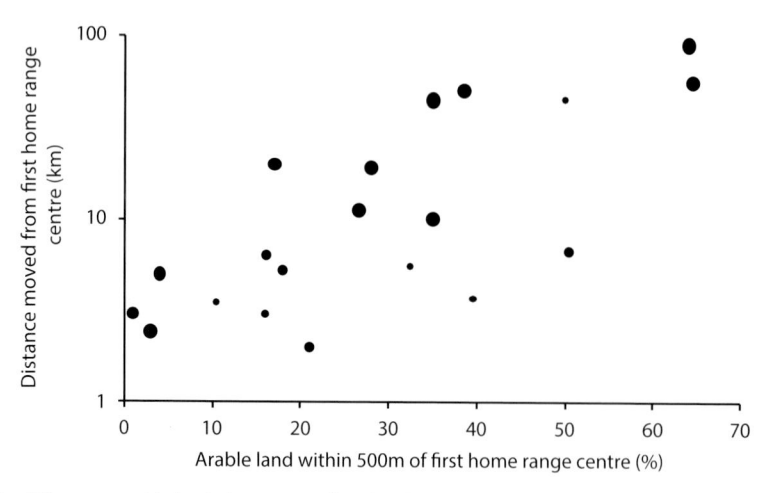

Figure 8.9. *The more arable land there was within first home ranges in winter, the further Buzzards moved to new settlement areas, with those that had first settled close to their nests (shown by small dots) tending to move relatively shorter distances than those that had travelled further from natal nests to their first home range (large dots). Data from Kenward* et al. *(2001a) with kind permission from John Wiley & Sons.*

There was some evidence for this, in that the further the Buzzards dispersed, the more arable land they had within their range, and we know that this was one land cover associated also with staying at home in autumn. It therefore seemed that those hatched in poor areas had moved to areas similar to those used by Buzzards from nests in better areas, likely because arable land provided abundant earthworms in autumn. However, once they had established their first home ranges after dispersal, extra-natal movements were longest for Buzzards that had settled in areas with most arable land (Figure 8.9). We gained the impression that arable areas can be a 'poisoned chalice' for young Buzzards (Chapter 4).

Where others have been able to observe dispersal movements, these movements appeared to be similarly motivated by food. In the colder north of the UK, Picozzi & Weir (1976) found that most visually marked Buzzards dispersing more than 20km had moved to lower ground nearer the coast or inland valleys, where there would have been fewer frost days and worming would have lasted for longer in the winter. In France, Nore et al. (1992) found that in some years juveniles dispersed to areas with small mammals (especially voles) but in one year they flocked to areas where there was an abundance of field crickets (subfamily Gryllinae) to feast on. So, the areas in which Buzzards settled could vary from year to year, depending on annual variation in food supplies, and sometimes reaching high densities to exploit good food sources.

Although Common Buzzards do not show the cohesion associated with colonially nesting birds, they will nevertheless gather at important resources, whether that is abundant food or helpful thermals. We certainly found Buzzards spending the winter in good worming fields with many others, breaking down the territorial boundaries by arriving in numbers the despotic territory holder could not rebuff (Chapter 5). The large groups of Buzzards demonstrate that dispersed Buzzards did not necessarily get rid of all competitors, but at least they weren't competing appreciably with their own family. On the other hand, when we were conducting our study we had areas to the east that still had relatively few nests. Therefore, our tagged birds could dramatically reduce the competition if they went far enough. There was a weak tendency for our Buzzards to have moved eastwards, especially by their second year (Walls & Kenward 1998, Kenward *et al.* 2001d), although whether that was to do with reduced competition, or the prevailing winds (Walls *et al.* 2005), is difficult to assess.

Dispersal distance

Discovering the distances at which individual Buzzards settled was really fascinating. We know, from the migratory movements of Steppe Buzzards, that it would be perfectly conceivable for UK Common Buzzards to reach South Africa, or to go east to Mongolia and beyond. But our birds did not. That was just as well, because we had limited ability to follow radio-tagged birds in the early 1990s. Even if tagged birds had only flown across the English Channel, we couldn't follow. Thankfully, there seemed to be a strong pull to stay as close to the birthplace as possible, demonstrated not only by the philopatric returns of dispersed birds in the spring, but also by the relatively short distances travelled from natal nests. A distribution of Buzzard dispersal distances is typical of radio-tracked animals, with many dispersing relatively short distances and fewer dispersing the longer distances (Figure 8.10).

Only one-third of the Common Buzzards that were tagged dispersed further than 15km from their natal nests. However, although there were fewer and fewer Buzzards settling further away, there was no absolute cut-off and a small minority were recorded settling more than 100km away. Moreover, all the birds whose radio signals were lost (about 7 per cent of the total tagged) were birds that had already moved from their natal ranges. Thus, the lost birds could have dispersed even further, and not made a philopatric movement. That would be consistent with the long distances of a few recovered rings for Common Buzzard.

Median values for our radio-tracked dispersal distances in the first year were 13km (Walls & Kenward 1998), which did not differ significantly overall from the 15km recorded among BTO ring recoveries. However, our data did not include the birds that had not dispersed at all in their first winter. The birds that dispersed late tended to disperse very short distances, so that by their third year (and hence completion of ex-natal dispersal) their

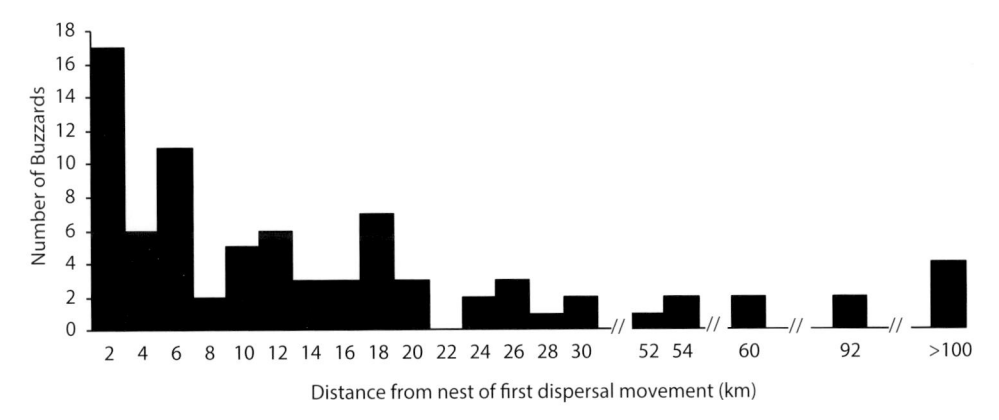

Figure 8.10. *Distances travelled from the nest during ex-natal dispersal by Buzzards in Dorset.*

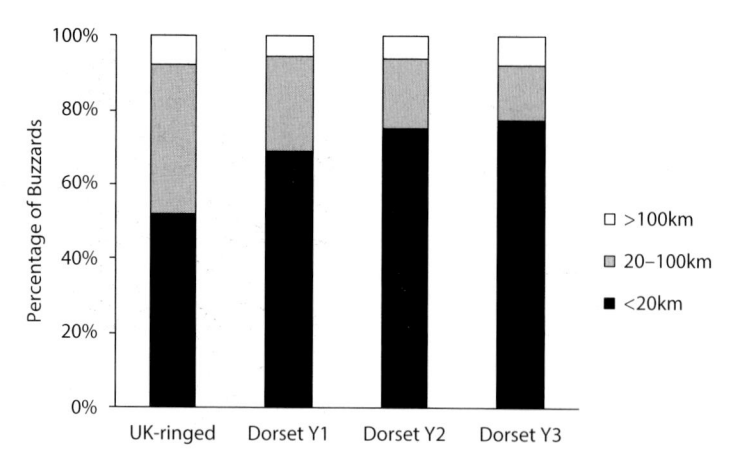

Figure 8.11. *The percentage of Buzzards that were recorded less than 20km from natal nests, between 20km and 100km, and greater distances, among 254 BTO ring recoveries for the UK (UK-ringed), and after one (Y1, n=74), two (Y2, n= 64) and three (Y3, n = 53) years of tracking birds radio-tagged in Dorset. The Dorset birds were very sedentary.*

median distance was close to 10km. Less than a quarter of surviving birds were then more than 20km from their natal nests (Figure 8.11).

The overall picture from ring recoveries in the UK probably conceals some regional variation. From colour rings used in Speyside, Picozzi & Weir (1976) found evidence that 24 per cent of young dispersed more than 50km from their natal nest, compared with only 15 per cent in Dorset. However, although there is a possibility that these Buzzards ringed in Scotland were most likely to be recorded in distant areas, with more people to observe them than in the sparsely inhabited Scottish Highlands from which the birds originated, it is also possible that the harsher conditions at the northerly latitude caused birds to move further. Another study that has lasted long enough to build up ring recoveries nearer to our study, in southern England, found that 22 of 24 recoveries were less than 10km from the natal nest, whether they were recovered in their first year or later (Prytherch & Roberts 2012), fitting the pattern we recorded in Dorset.

Species that don't have such a bias towards short-distance dispersal are typically those requiring resources with a fragmented distribution, for example waterbirds that prefer estuaries (Paradis *et al.* 1998). They cannot have a smooth distribution of dispersal distances, because as soon as they get to the edge of their estuary, there is then a considerable distance to the next available estuary. Typically, these species form many subpopulations, with sufficient exchange between the patches to maintain a single larger meta-population, rather than producing genetic isolation. Our relatively smooth distribution is another indication of how Buzzards are extremely adaptable to most environments in the UK: thus, there are few areas they would actively avoid, other than in very harsh winter conditions. It also means that in the current climatic conditions, continuous contact is possible between neighbouring populations and likely a smooth graduation in phenotypes.

Another similarity with most other birds is that female Common Buzzards dispersed further than males (Walls & Kenward 1995), which is commonly thought to improve the chances of keeping a healthy genetic diversity by avoiding inbreeding (Greenwood *et al.* 1978, Greenwood 1980), because the females settling in an area will have likely originated from further away than the males and are therefore less likely to be closely related. While bigger broods tended to have at least one youngster that stayed at home, those youngsters that did disperse from bigger broods did not seem to disperse as far away, and it would be reasonable to assume that bigger broods meant more food around the nest. We therefore also used the satellite land-cover map to look at the effect of land cover on dispersal distance. Buzzards dispersed further if there was not much short grassland (which is good for worming, small mammals and Rabbits), or if there was much heathland (not so good for feeding on worms or Rabbits) near the nest. Moreover, those individuals that had made the furthest ex-natal dispersal were likely to move further on the second movement (Figure 8.9). These are very similar results to those for the timing of dispersal. Thus, the factors associated with dispersing early seemed to also 'prime' the dispersers to go far, which started us wondering how that could happen.

Pre-nuptial movements for 14 Buzzards, from where they had temporarily settled to where they started breeding (Fig. 8.4), were typically less than 3km (Figure 8.12). The only example greater than 6km was a bird that delayed ex-natal dispersal until its second spring and immediately bred successfully 20km from its origin (Figure 8.3G). There were only seven post-nuptial records, between breeding attempts, and none was greater than 2km. There were no land-cover correlations with these, perhaps because our study was in one of the UK's biodiversity hotspots. We purposely chose our study site to cover lots of different land types, with conifer forests, large arable fields, pasture, expanses of heathland, deciduous woodlands etc. This was a great place to live and to study the effects of land cover on Buzzard behaviour. However, it was a mosaic of relatively small patches, with both suitable and unsuitable areas not far from any point. The effects of land cover on nuptial movements might have been greater in relatively treeless parts of eastern England. In time, it would be good to see dispersal studied in the east of England, where woodlands are spread more sparsely.

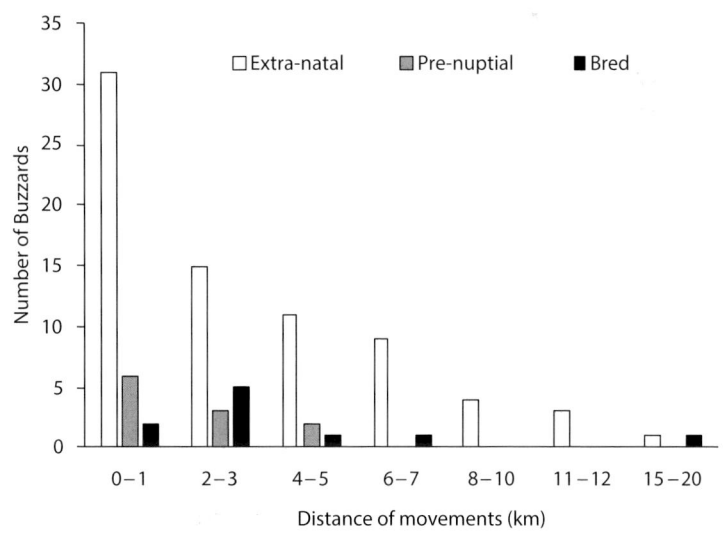

Figure 8.12. *The number of Buzzards that made extra-natal movements (white columns), changing the area where they first settled, separated from pre-nuptial movements that led to pairing (grey columns) or successful first breeding (black columns) in their second and third years. Data from Kenward* et al. *(2001a) with kind permission from John Wiley & Sons.*

Land cover was not the only environmental factor affecting dispersal distance. Unexpectedly, weather gave us more insight than habitat into how far a Buzzard went before it stopped dispersing, and may help to explain why birds dispersed especially far when they were in particular habitats.

We have long known, from reviews by Newton (1979), Elkins (1983) and Alerstam (1990), that birds on migration can be affected by weather. This is especially the case for soaring species with large, broad wings. However, although there has been a lot of research into the effects of weather on the dispersal patterns of airborne plant seeds or insects, which people expect to be blown about, at the time of our work there had been little study of how weather may affect the dispersal of birds, primarily because dispersal had been so difficult to study in detail. From our radio-tracking in the early 1990s we had dispersal dates, which gave us an exciting opportunity to see if we could discover the effects of weather on the Buzzard dispersal process. We knew to within five days when our tagged Buzzards had dispersed, and the British Atmospheric Data Centre provided weather from stations local to where the birds were dispersing from. To be consistent with our checking frequency, we divided the weather data into the same resolution of five days.

Weather preceding the dispersal didn't follow any particular pattern, except that juvenile Buzzards were more likely to leave when the wind changed to a more southerly direction, from the more common westerly winds, and when minimum temperatures had been lower. This could be seen as representing the arrival of warmer air coming from the landmass of France to the south (as opposed to off the Atlantic), or the wind being in a better direction for assisting dispersal. Lower minimum temperatures during the autumn are consistent with high pressure systems, which are associated with clear skies at night, which allow cooling, followed by warmer days because no clouds are blocking the Sun's rays. The greater temperature differential between night and day would then create good thermals during the day. On the other hand, poor weather did not seem to initiate dispersal, as is the case when migration serves to avoid the worst of the winter conditions in a breeding area (Elkins 1983). So, young Common Buzzards that are resident in the UK throughout the winter seem to use opportunities of dry, high pressure weather to disperse, rather than being driven by bad weather (Walls *et al.* 2005).

There were further intriguing findings when we looked at the distance Buzzards dispersed according to the weather at the time of departure. We measured dispersal distance between the natal nest and where the bird was settled in December. This relatively long period was chosen because some birds were lost when they first dispersed and were only located again after extensive searching. However, whether dispersed or not, by December we had usually found the lost birds. Moreover, by that month all birds tended to have settled for the winter, with no further tendency to complicate distance records through long excursions from their home ranges.

The biggest influence on dispersal distance was maximum daily temperature in the five days after leaving: Buzzards dispersed further away if they left during warmer, thermal-forming weather (Figure 8.13, left). Knowing that the early dispersers tended to go further, and that the weather gets cooler as autumn and winter progress, there was a chance that this effect was merely a temporal correlation. For example, changes in dispersal behaviour could also have been caused by hormonal changes induced by daylight hours, or changes in the territorial behaviour of other breeding pairs, rather than a change in the weather conditions. To disentangle the competing hypotheses, we artificially shifted the observed dispersal dates by small amounts and re-analysed, so that any correlations with short-term fluctuations in weather would be more likely to reduce than any correlations with gradual seasonal changes in daylight, territorial behaviour or temperature. We did this by adding or subtracting the same number of days to all the observed dispersal dates. What we found was that if we shifted dispersal dates, even by only one five-day period, the relationship between maximum temperature and dispersal distance reduced dramatically, and by ten days it had disappeared altogether. This was a strong indication that it was the short-term warm weather that increased dispersal distance, rather than other unrelated factors that happened to change through the season. Remembering that females disperse further than males, and there was further dispersal to the east, we then created a dispersal-distance model based on the sex of the individual, the wind direction and the maximum daily temperature when leaving in autumn, to predict the dispersal distances of individual Buzzards at that time of year. The formula from this model provided strong predictions of the much shorter distances Common Buzzards dispersed later in winter, a different period than used to calibrate the model (Figure 8.13, right).

These findings were also consistent with most of the dispersal distance being accomplished within those first five days, during the bout of warm weather, which was an indication that most dispersal was a sudden quick flight to another place, not a prolonged wandering before settling (Walls *et al.* 2005). It will be exciting to see what more can be discovered with more frequent locations from GPS tagging, which could confirm the fine details of dispersal and possibly also allow a more detailed analysis of the land cover in areas through which individual Buzzards travel without settling.

At the time when we made the analysis, there was precious little published evidence for weather affecting dispersal for any vertebrate species. One exception was Miguel Ferrer's work on the Spanish Imperial Eagle *Aquila adalberti* (Ferrer 1993). He found that the first departure of young eagles was with the prevailing wind (north-west for one study population and south-east for another), and the effect was independent of nest origin. The combination

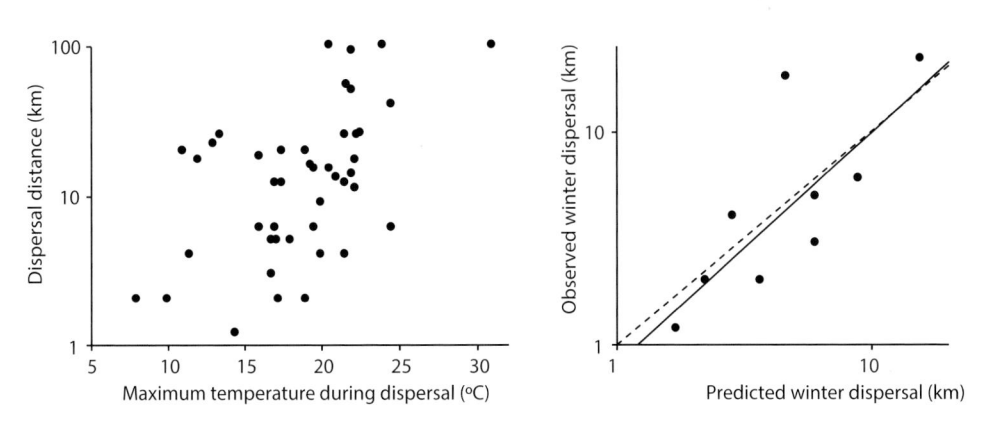

Figure 8.13. *The strongest relationship with dispersal distance of Buzzards was maximum temperature during the five days in which dispersal occurred (left). A model based on temperature, wind direction and gender of dispersing bird for 37 individuals in autumn (right) closely predicted dispersal distances (solid trend line) of nine birds in winter (dotted line for 1:1 matching), when temperature and travel distances were less than in autumn. Data from Walls* et al. *(2005) with kind permission from John Wiley & Sons.*

of his observations with our analyses makes a strong case for the dispersal outcomes in these species being more strongly based on environmental than genetic factors.

To understand why wind direction and temperature would have such an effect on dispersal distance in Buzzards, we need to return to their morphology. Large, broad wings and low wing loading may have been an adaptation to cover large distances on the wing as efficiently as possible, but they make it very difficult for Buzzards to fly into strong winds, whereas tail-winds can increase their cross-country speed. Buzzards can use gentler headwinds or side-winds to create lift for gliding, but their greater drag limits their speed, so high-pressure conditions with light winds are best for them. As for temperature, the dependence on thermals is apparent to anyone watching large soaring bids on migration. This reliance on temperature makes Buzzards very sensitive to weather for long migratory movements (Rudebeck 1950, Alerstam 1978) as well as for dispersal. They are best able to move when winds are not too strong, and when there are either thermal or orographic updraughts. In fact, the rate that birds can spiral up in a thermal is less dependent on the species (morphology) and more dependent on the velocity of the rising air. Spaar & Bruderer (1997) found no difference between bird species migrating through Israel, where the thermals are strong, although the Steppe Buzzards did have the fastest upward movement in the thermals. So, although Buzzards are designed to make the most of weakly rising air, a substantial thermal will boost most species' ability to migrate over long distances by providing lift. Between thermals, heavier birds sink faster than lighter birds, but Buzzards have a lower wing loading, which allows them to travel further between thermals.

In the Americas, Bohrer *et al.* (2011) made a study of how two large raptor species – Golden Eagle and Turkey Vulture – used thermal and orographic updraughts. The Golden Eagles appeared to use orographic updraughts, whereas the Turkey Vultures mainly used thermals. Both species travelled faster in warmer weather, but while Golden Eagles could make progress when the weather was colder, the Turkey Vultures hardly moved if there were insufficient thermals. In fact, later work has shown that Turkey Vultures employ 'contorted soaring' to use local small-scale updraughts very efficiently when there are few thermals, but this is for short-scale foraging movements, not dispersal or migration (Mallon *et al.* 2016). Buzzards may look more like eagles than vultures, but we note that Turkey Vultures hold their wings up in V-shape, just like the dihedral of a Buzzard described in Chapter 1. Thus, Buzzards appear to use the updraught of thermals more like vultures, and are to be expected to benefit from warmer weather for long-distance movement.

We have established that Buzzards appear motivated to leave in their first autumn by sibling pressure and land cover (as a presumed proxy for resources), and that they do so when weather conditions are good. Buzzards that leave when they are younger do so when there are more and stronger thermals generated by warmer temperatures in late summer and early autumn, so they can travel further in a short period of time and therefore settle in a more distant area. So the 'priming' effect of habitat around the nest (Kenward *et al.* 2001d) could be explained by weather, through nests with the poorest foraging around them causing birds to leave earliest and experience weather that takes them furthest. However, the tendency of long-distance ex-natal moves to be followed by long further movements would more likely reflect characteristics of the individual birds, perhaps based on genetics or stress prior to that later movement.

Migration

Migration can be a matter of survival for those whose breeding grounds experience winters harsh enough to reduce access to food. However, migration also has costs, such as flying through unknown areas with hazards (large water bodies, deserts, hunters and predators) and the potential that a migrant could return to find they have been usurped from their summer territory in their absence. Migration is therefore a dilemma, and Common Buzzards show a variety of migration behaviours, depending on the climate. Those in northernmost Scandinavia tend to migrate south and west; each autumn more than 10,000 are observed migrating through Falsterbo at the southwestern tip of Sweden. In contrast, in the UK and southern Europe there may be very little migratory behaviour. For intervening areas, there may be partial migration, where an appreciable proportion of individuals migrate, leaving others resident throughout the winter. For example, ringing suggests that in Denmark 57 per cent stay within 100km of their breeding sites, the rest migrating south-west into the countries bordering the North Sea and English Channel (Nielsen 1977). A small number of Common Buzzards get as far south as North Africa or the Middle East, but the vast majority stay within Europe and tend to move south-west into warmer areas.

As with dispersal, thermal convection also has a strong influence on flight altitude (Pennycuick 1972, Spaar 1995, Spaar et al. 2000, Shannon et al. 2002, Shamoun-Baranes et al. 2003), and so soaring heights of diurnally migrating birds generally increase until mid-afternoon, in phase with the rising temperature (Kerlinger 1989, Spaar 1995, Leshem & Yom-Tov 1996). Shamoun-Baranes et al. (2006) tracked the flight altitude of 447 Buzzards by radar in the Netherlands during the breeding season, while their flight behaviour was also recorded visually. Ninety-one percent of the flying time was spent gliding, with only 4 per cent as active flapping and 5 per cent foraging. The mean daily maximum flight altitude was 665m (208–1,592m), which gives us some idea of the elevation at which Buzzards can fly. They were generally gliding higher than all other species, demonstrating their wonderful ability to master lift. Migration is faster if it is higher, partly because longer glides can be used before the next thermal and sometimes because there are faster winds in useful directions at higher altitude (Shamoun-Baranes et al. 2006), so we would expect Buzzards sometimes to migrate as high as possible, and that can put them out of sight.

Although there are very few data on Common Buzzard migratory flight performance, Spaar & Bruderer (1997) have made detailed studies of migrating Steppe Buzzard flight, which is a good substitute given that morphologically the subspecies are almost identical. From tracks of 141 Steppe Buzzards migrating through Israel, 85 per cent of their time was spent soaring and gliding at an average of nearly 500m above ground, similar to the study in the Netherlands. Mean air speed when gliding was 16.3m/s (59km/h). Birds descended between thermals at an average of 1.7m/s, and then climbed faster on thermals at 2.4m/s. The rate of descent was reduced in higher wind speeds, due to increased orographic updraughts from the winds colliding with the escarpments that run north–south through Israel. The speed of travel over the ground, taking into account that some of the movement through air was up and down, was 30–40km/h, depending on wind strength and direction. Newton (2010) writes that Common Buzzards will lose about 1m of height for every 15m travelled between thermals, consistent with a 1m/s speed of descent. The shortest sea crossing at Gibraltar is 15km, so Buzzards need to gain 1km of altitude on a relatively windless day in order to make it across without flapping. That height is certainly attainable in reasonable conditions (Shamoun-Baranes *et al.* 2006), but conditions are not always perfect and so it is not surprising that Buzzards and other raptors are observed flying low and flapping hard when approaching the other side, presumably because they didn't attain sufficient height before the crossing to succeed in one glide, or the weather conditions increased the speed of descent.

The reliance on thermals to assist long-distance movements (dispersal or migration) means that one of the major barriers to their progress is large expanses of water. The water bodies on their routes in Europe neither have much thermal activity nor sufficient updraughts from waves (which are exploited by species such as shearwaters and albatrosses over the oceans). Besides, unlike waterbirds, Common Buzzards cannot land on water to rest without drowning; since they need to be able to reach sufficient height to soar and then glide over the water in one attempt, unless weather conditions are favourable, the risk of crossing is too high. This drives soaring birds to the shortest water crossings, forming bottlenecks of migrating raptors. Consequently, most observations of migrating Buzzards have been at these pinch-points, on either side of water bodies, where keen birders and surveyors can enjoy the spectacle of large numbers of migrating raptors. There are anecdotal observations of Buzzard flocks away from these crossings, such as those observed by Dare (2015) in south-west England. The largest movement he noticed was in late April when he saw a total of 71 Buzzards flying north-east in three large flocks at great altitude. Such a large number suggests migration, perhaps birds heading back to Scandinavia, rather than one of the more frequent collections of local transient birds, many immature, who have not managed to establish a territory and so go on exploratory excursions in spring.

For Buzzards in Europe, the Mediterranean Sea is the largest barrier for migration to Africa. The Black and Caspian Seas will also act as barriers, but information is less readily available along their shores. Studies from islands in the middle of the Mediterranean confirm that extremely few Buzzards will fly a long distance across water there. On the Maltese islands, Beaman & Galea (1974) watched migrants for between five and 10 hours per day during migratory periods in the years 1969 to 1973. Infrequently, they recorded a Common Buzzard on spring migration and up to seven during autumn migrations, but these were very rare compared to 467 European Honey-buzzards during the same period. On the island of Merrittimo, Agostini *et al.* (2005) set up a watch over the Channel of Sicily, between Italy and Tunisia. They started in the year 2000 and, despite this being a shorter crossing than via Malta, they struggled to see any Common Buzzards at all, eventually spotting six in 2004. This may reflect strong selection against this sea crossing. Many bird species, including Buzzards, move down the western European coast and then cross into Morocco near Gibraltar. Common Buzzards from eastern Europe are likely to go around the east of the Mediterranean and cross into Turkey, together with Steppe Buzzards, one of the most common raptor species counted crossing the Bosphorus and Dardanelles Strait (Nielsen & Christensen 1969, Panuccio *et al.* 2017). From Italy, migration around the Mediterranean Sea is a detour of at least 2,000km, compared with less than 200km to cross the Channel of Sicily. Once around the Mediterranean, hundreds of thousands of Steppe Buzzards move down the eastern side of Africa, many to the South African Cape.

Standardised counts at Gibraltar recorded only a few hundred Common Buzzards in the 1970s, and often fewer than 100 per year since 2000 (Martín *et al.* 2014). On the other side of the Mediterranean, Leshem & Yom-Tov (1996) analysed data from extensive counts in different locations in Israel and usually there were only a handful of Common Buzzards compared with thousands of Steppe Buzzards. Over 98 per cent of trapped buzzards migrating through the Gulf of Aqaba in Israel were thought to be Steppe Buzzards, which if representative means that fewer than 10,000 of the 300,000 buzzards observed per year are Common Buzzards (Yosef *et al.* 2002). So, most of the buzzards migrating through the Bosphorus are Steppe Buzzards coming from further north, and likely taking a more south-easterly flight-line down to Turkey, rather than heading due south until encountering the Mediterranean and then having to follow the coastline around.

Further north, at Falsterbo in Sweden, there are seldom favourable thermals to help cross the Baltic Sea, because it is too cold and exposed to winds during the autumn. Recent studies there showed that Buzzards avoided headwinds, migrated at higher altitudes, and were more sensitive to the wind strength and direction than Eurasian Sparrowhawks (Malmiga *et al.* 2014). The research team considered that this was primarily because birds such as Common Buzzards were far more dependent on soaring and needed to obtain a good height to get over the water without spending too much energy flapping. A secondary influence may be that, unlike Sparrowhawks, which preyed on weary, migrating passerines, Buzzards did not hunt while migrating, and therefore did not need to come close to the ground. Nevertheless, there is other evidence that buzzards feed on spring migration (discussed below). Panuccio *et al.* (2017) found that the number of Common/Steppe Buzzards crossing the Bosphorus from west to the east reduced if the winds were too strong and northerly, again demonstrating sensitivity to the wind.

Mountain ranges, too, can act as barriers. Radar has shown that some birds migrate at such a high altitude (thousands of metres) that they would be missed by human observers (Evans & Lathbury 1973, Beaman & Galea 1974), so there is a chance some Common Buzzards were crossing at height. However, despite their ability to reach great altitudes with good thermals, buzzards were not commonly seen crossing higher mountain ranges, probably because of the extreme cold, low oxygen levels and thin air. Thermals are less likely and the orographic updraught may not be helpful if broken up by craggy areas or if strong winds are funnelled along valleys. Bruderer & Jenni (1990) observed that some Common Buzzards occur in the Alps in early autumn, but that their rate of climb reduced by one-third over the autumn. Thus, it may become impossible for soar-and-glide raptors to cross the Alps later in the year, such that the mountain range becomes a seasonal barrier.

In some areas, buzzards exhibit altitudinal migration. For example, Dare (2015) observed Common Buzzards in the mountainous region of north Wales retreating to the lower valley during winter. In the particularly harsh weather of February 1956, in Dartmoor there was continuous snow and temperatures remained below -10°C. From what Dare could see, Buzzards stayed on territory for the first nine days and then most disappeared over the next four days. A rapid thaw then followed and most birds reappeared within two days, suggesting that they had remained close by. Mortality was low, except for one that was shot when raiding poultry (Panting 1955). In this instance, short-distance migration enabled most to survive this weather shock.

For those birds that migrate longer distances, the window of opportunity to go south opens when the young are properly independent and closes when inclement weather starts. As with the dispersal patterns described above, young Common Buzzards that migrate can also make exploratory movements, some of them in a northerly direction, before starting the longer journey south. Detailed data, from Buzzards satellite-tagged in Sweden, showed juveniles settling up to 200km from their natal nest before starting the southward migration (Strandberg *et al.* 2009a).

The travelling speeds recorded by Spaar and Bruderer in Israel were only short term, over a matter of minutes. Migration distances travelled in a day may be much more variable. It is not uncommon for Buzzards to have stopover periods, when they pause in the same place for days, or even weeks, at a time. Migration stopover by many species is considered to be a necessity to refuel or avoid moving too fast and encountering bad weather, but in the case of Common Buzzards it could also be to assess local conditions and the prevailing weather, because there is no point in going further than required. From studying autumn migrations of the 15 Swedish satellite-tracked Buzzards, Strandberg *et al.* (2009a) found that the birds spent from four to 41 days on migration, with average speeds of 16–133km per day over the whole migration, or 33–133km per day if stopover days were excluded. The maximum distance flown in one day was nearly 300km.

Stopover periods are extremely variable and appear much affected by weather conditions in the north. For example, Bylicka & Wikar (2007) studied Common Buzzard numbers in an area of Poland with severe weather conditions. Buzzard numbers there declined over the winter as the weather got worse. However, Wuczyński (2003) found that lower daily temperatures temporarily increased Buzzard numbers in an area of open farmland in south-west Poland, probably due to an influx from the north, or Estonia, where weather conditions were even worse. The same has been shown for Rough-legged Buzzards in Poland (Kasprzykowski & Cieśluk 2011). Del Marmol (1997) has also recorded increased numbers of buzzards during winter in Belgium. Some Common Buzzards seem only to go the minimum distance and then move on again when weather conditions deteriorate further. Some end up in areas with non-migratory resident breeders that they must avoid by not using high perches or being too close to their nests (Hohmann 1994, Schindler *et al.* 2012). Further south in Portugal, Lourenço (2009) found that Common Buzzards in rice fields were abundant before the start of a new calendar year but rapidly decreased subsequently. This suggested that Buzzards were still moving through on migration, but it is not clear whether food shortage, rather than weather, instigated these further movements within wintering areas.

A common theme through many Common Buzzard migration studies is that not necessarily all adult Buzzards migrate. At Falsterbo, there has been some excellent long-term record-keeping. The numbers leaving the peninsular to cross to Denmark have been observed every autumn since 1942. Larger numbers of Common Buzzards were counted in earlier years, with the maximum being 41,000 in 1950. More recently, numbers have fluctuated around 12,500 since 1973, although 32,692 were counted in 2017, the last year before publication (Figure 8.14). When analysing the counts between 1942 and 1997, Kjellén & Roos (2000) found that total numbers recorded were declining, but that the proportion of juveniles had increased. This may indicate that fewer established pairs were migrating whereas the juveniles, without territories to defend, were better off leaving the harsh winter behind. They also noted a shift in peak migration timing from mid-September to mid-October. Adults tended to be more abundant in the earlier crossings, and juveniles later on. From the Swedish birds with tags for tracking by satellite, there was no difference in departure date between juveniles and adults, but the adults tended to move more directly and so it could be that adults reach Falsterbo earlier and cross over the Øresunde more quickly. Thus, if a higher proportion of adults chose to stay behind and defend their territory

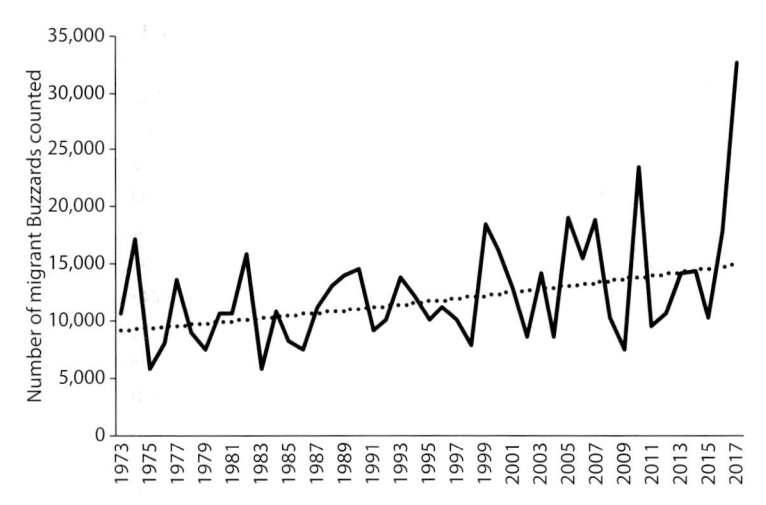

Figure 8.14. *Annual counts of Buzzards migrating over Falsterbo in southern Sweden fluctuate considerably; counts reached 41,000 in 1950 but, despite an increasing trend (dotted line), have only recently begun to approach that number again. Data from https://www.falsterbofagelstation.se, with kind permission from Lennart Karlsson.*

rather than migrating in the warmer winters, the relatively greater number of migrating juveniles would delay the peak. Between 1986 and 1991, Kjellén (1994) estimated that 60 per cent of Common Buzzards migrating through Falsterbo were adults. He compared this to the wintering population, in which the adults formed about 80 per cent of the population, four times more adults than juveniles. Those juvenile Buzzards that remained were found mainly on marginal habitat near the coast and there was a suggestion that juveniles were driven from the better inland territories. Another interpretation could be that without a territory to defend, juveniles gained most advantage from going to the coast where there was less persistent snow during the colder months. Looking at the last few decades, counts have tended to increase during the last 20 years, so perhaps we will again see numbers near the peak counts recorded 70 years ago.

There is good evidence that the tendency for birds to migrate is, at least in part, genetically controlled. For example, some Blackcap *Sylvia atricapilla* populations are partially migratory, i.e., some individuals migrate while others do not, depending on their parents. Experiments found that if Blackcaps of differing migratory behaviour were crossed, their offspring had intermediate behaviour, and if adults of similar behaviour were crossed then the behaviour became more enhanced (Berthold 1999). With Common Buzzards having similar variability in migratory behaviour, and therefore possibly a genetically determined component, it is possible to think that the migratory tendencies will be lost in warmer, more temperate regions, because there is little cost to staying and a higher risk of encountering physical hazards, or of being usurped, when absent after migrating. The migratory behaviour shown by Common Buzzards is complex and plastic. It is not a simple switch (migrate or not), but more dependent on environmental conditions. There is evidence from ring recoveries, that migration distances of European Common Buzzards have become shorter over the last 50 years, likely due to global warming (Martín *et al.* 2014). Moreover, the numbers of Common Buzzards wintering in areas of Europe is very weather dependent, demonstrating adaptive features in the migration rather than merely a binary decision of whether to migrate or not. As we shall see, even normally resident individuals can migrate to avoid the worst of weather. The ability to migrate if conditions change is yet another example of adaptability in the Common Buzzard.

In previous decades, ring recoveries of Swedish Common Buzzards suggested that the majority travelled south-west, hugging the coastline after crossing the Øresunde (Olsson 1958). The Swedish satellite-tracked Buzzards mainly went to Denmark, and the furthest south was to Belgium, so this was a comparatively short-distance migration. This contrasts with Olsson's earlier analyses of ringing data, which suggested rapid travel through Denmark and on down to France. The more recent shorter distances recorded from the tagged Buzzards are in keeping with the reduction in migration distances over time (Martín *et al.* 2014) and may also reflect a shift of wintering due to climate change, as there seems no good reason to expect that past ring recoveries from France were more likely than from Denmark. An analysis of recoveries of 62 Estonian Common Buzzards ringed as nestlings (Väli & Vainu 2015) showed that wintering birds were all found in Europe, generally having moved south-west to warmer areas, but none further south than France. Before these tracks had been recorded, there was much debate about whether the Estonian buzzards were Common Buzzards, Steppe Buzzards or even intermediate birds (Härms 1927). The observed migratory movements are typical for Common Buzzards in northern Europe, not Steppe Buzzards.

Like many species for which the migration of individuals has been tracked, the northward spring migration is faster than the southerly autumn migration, and adults tend to precede juveniles (Olsson 1958, Gorney & Yom-Tov 1994, Leshem & Yom-Tov 1996, Yosef *et al.* 2002, Strandberg *et al.* 2009b). The northward spring migrations of satellite-tagged Common Buzzards were faster and more direct than in autumn, with no stopovers for four adults, although the average speeds per day were only 51–148km per day. A comparison for one male showed he migrated at only 23km/day in autumn, but 148km/day in spring (Strandberg *et al.* 2009a). The most obvious explanation is the advantage of getting back to defend a territory and prepare for breeding as early as possible. Stopovers are not as common in the spring, but it is likely that buzzards feed along the way. A study that examined fat reserves of Steppe Buzzards migrating through Israel did not consider the fat recorded on them to be sufficient to fuel adults all the way to northern Europe without further feeding. Adults tended to be better hydrated, with more fat reserves than juveniles, which may allow them to get further and complete the journey faster (Gorney & Yom-Tov 1994). The Gibraltar crossing only spans a three-week spring period at the end of March, compared with 12 weeks in autumn (Evans 1974). The timing appears later on the eastern side of the Mediterranean. Yosef *et al.* (2002) observed about 300,000 predominantly Steppe Buzzards passing through the Gulf of Aqaba (and thus before they reach the Bosphorus), with numbers peaking in late April. Interestingly, the Swedish Common Buzzards tracked by satellite appeared to migrate earlier (mid-March) from wintering areas in northern Europe than the buzzards observed around the Mediterranean. This may be because they were Common Buzzards rather than the majority Steppe Buzzards seen further south. Past conditions might have selected for Steppe Buzzards to migrate later because they tend to breed further north, where spring and its resulting flush of young mammals comes later. Perhaps it is more crucial, when going to an area with a shifting food supply, not to arrive too early when conditions are harder, because a lack of food at that point will prove fatal. Such behaviour would be selected against.

We live in interesting times. Not only is our climate changing fast, but the human development that has accelerated global warming has also given us technology that allows us to track birds. Tracking devices allow us to assess the effects of climate change on the movements of many species, and perhaps can lead us to a greater understanding of how to mitigate against the change. The technology certainly gives us some fascinating stories, and we look forward to new discoveries about the movements of this adaptable and very successful species.

Conclusions

1. Electronic tracking devices have dramatically improved our understanding of bird dispersal, and this chapter reviews what we found from radio-tracking in the 1990s.
2. Results concurred with previous studies, showing that Common Buzzards in the UK can disperse over 100km from their hatching place, but that the majority stay within 10km, at least in lowland areas.
3. Young Common Buzzards often made excursions, sometimes of more than 20km, and then returned to their parents' territory.
4. About half of the juveniles dispersed from the territory in their first autumn. Others remained until the following breeding season when they started to explore again, and typically dispersed at that point. Most often, only one juvenile from a brood remained within the parental territory overwinter.
5. Buzzards that dispersed earlier from our area tended to go the furthest before settling.
6. Those that dispersed a long distance were recorded making visits back to their parents' territory in the following breeding season, and continued to do so earlier each year until they established their own territory and started breeding.
7. Those from areas with poor resources, such as lowland heath, surrounding the nest were more likely to leave early, whereas those surrounded by good winter feeding grounds, such as arable land and meadows, were more likely to stay in their parents' territory for the winter.
8. Initially, dispersers tended to settle in more arable areas, but after the first winter those with more arable land in their foraging areas tended to move further, to settle in areas with more woodland and grassland.
9. Dispersal distances were correlated with the weather conditions at the time when the individual Buzzard left. Birds dispersed farther if that period showed records of good thermal-forming weather and a tail-wind.
10. Whether Common Buzzards migrate seems to be very much weather-dependent, and therefore it is far more obligate in northern latitudes. On migration, they do not readily cross water and hence seldom leave Europe, unlike the closely related and more northerly nesting Steppe Buzzard.

CHAPTER 9
Longevity and survival

If you have the ability to radio-track more than 100 young birds from when they leave their nest, inevitably you find birds that have died. What surprised us when we started our study in Dorset in the 1990s was how few deaths we did record, compared with the mortality rates of raptors previously estimated from ringing. There were the expected early deaths as immature birds, still learning to master flight, encountered unnatural hazards such as vehicles, high-tension cables or artificial water bodies. In one case, although it was impossible to tell from the carcass, it seemed likely that a young, naïve Buzzard had landed in a field of maize from which it was unable to fly out, and ended up running around until it was too exhausted to go on any longer. Then there were Buzzards found underground, or in a slurry tank, or hidden under a large log, and post-mortems were able to reveal the true nature of deaths from shooting or poisoning. These were a small proportion of the surprisingly small number dying, so a tiny fraction of the population. We were able to track those that survived up to four years old, and it was sad when we no longer heard the radios where we were used to finding these well-established Buzzards, but we knew it was probably just because the battery had run out of energy. However, in many cases the story did not end there, because 20 years later we were still hearing of some of the birds we tagged, thanks to the BTO – a pioneer of what is now called 'citizen science'.

Long-lived Buzzards

How old can a Common Buzzard live? Before looking at what Buzzards have to contend with to survive, let's gain a perspective on how long they can live, provided they have good fortune and the genetics that can capitalise on that good fortune. Buzzards are relatively long-lived birds, at one time holding the record for the oldest-known ring recovery in the UK. We now know that they can reach more than 30 years old. A British recovery record

Figure 9.1. *A 24-year-old Buzzard from the Dorset study, identified from the BTO ring visible on its right leg, also showing the (shortened) antenna on its back-pack radio-tag (© Chris Ashurst).*

was set in 2009, when a Welsh Buzzard ringed as a nestling in 1983 was found dead only 6km away from its nest 25 years, 6 months and 26 days later. And the known longevity is still rising. Another record was set in 2013, with a bird that had survived at least 28 years, one month and 11 days before being found freshly dead 12km from its hatch nest in Cumbria. Then, in 2016 a bird in Wales was found still alive but sick at 30 years and five months old, 10km from where it was ringed as a nestling (Robinson *et al.* 2015). As this book goes to press, 29 years after ringing our first Dorset Buzzards, we'll be waiting to see if any ring recoveries from our study will be found older than that. In 2015, a 22-year-old Buzzard was found recently dead with one of our radio tags, albeit with a long-expired power cell, less than 3km from where it hatched. Another 24-year-old tagged Buzzard was seen alive, identified by photographs of the BTO ring, 10km from its hatch nest and still on our study site (Figure 9.1). This is exciting, because it's like news of old friends, and it's also useful to know that the Buzzards we fitted with backpack radio-tags were able to go on and live as long as Buzzards are known to live, thereby providing evidence that the methods we used were not adversely affecting survival. As with most species of raptor, or indeed most wild animal species, it is rare to die from old age.

Another way of looking at longevity is to estimate an average length of life, rather than the longest that birds can possibly live. We can calculate this as the time for half the birds to die, known as the median age, by estimating the proportion of birds that survive from one year to the next. Before our radio-tracking project, the only ways used to estimate the annual survival rate of Buzzards were from ringing data or, for breeding adults, the recognition of individual breeding birds that survived from year to year. For example, Rob Lensink (1997) used ring recovery data in the *Ecological Atlas of Dutch Raptors* (Biljsma 1994) when modelling the recolonisation of Common Buzzards in the UK. These data estimated that 47 per cent survived the first year and then 77 per cent survived each subsequent year, so the median age was apparently just less than one year. There are useful data to be gained from the 4–8 per cent of ringed Buzzards that die in places where people can find them, provided that large numbers are ringed and finders are sufficiently interested to submit the data to a ringing scheme. In years before it was illegal to kill raptors, many rings were returned by game-keepers, who were interested to know more about birds they had shot. With a majority of rings from such sources, about 64 per cent of ring recoveries from Finnish Northern Goshawks prior to their protection were from birds in their first year, which would suggest only 36 per cent of the ringed birds survived beyond their first year and hence a median life expectancy of less than a year for this species, as Lensink assumed for Buzzards (and which has been thought likely for many other raptors). However, a smaller proportion of rings from hawks that had died naturally were from first-year birds, so Erkki and Matti Haukioja recognised that ring recoveries from shooting might be biased towards young, inexperienced hawks, thereby underestimating survival of Northern Goshawks in their first year (Haukioja & Haukioja 1970).

Luckily, these days we have technology to make less biased estimates, and it was exciting to conduct research that led to a reassessment of survival in raptors. Radio-tagging in Sweden showed that some 65 per cent of female Northern Goshawks and just over 51 per cent of males (the smaller and more vulnerable sex) survived more than a year, which substantially affected thinking about their population structure (Kenward *et al.* 1999). Was the same true for Common Buzzards, and how would this combine with revelations on dispersal to build a bigger picture of population dynamics over a wider area?

To investigate survival, we selected the first eight years of our study in the 1990s, during which we radio-tagged 147 Common Buzzards as nestlings in 1990–94 and followed them consistently through until tags expired when they were four years old (Kenward *et al.* 2000). Using the radio-tags, we were able to establish if a particular bird was dead or alive, apart from a very small percentage that disappeared. This gave us information on survival through their early years, with little scope for bias. We found that the survival of radio-tagged nestlings to June of the next year was 73 per cent if only dead birds recovered with functioning radios were considered dead, or 66 per cent if we assumed that all birds with lost radio signals had also died. First year survival of our Buzzards, even if it was closer to the 66 per cent estimate, was much higher than the 47 per cent used by Lensink.

It was no surprise to find that the highest mortality of our tagged birds was just after the birds had left the nest, when they were still learning to fly and exploring their new surroundings. The most critical period was the first four months; 5–8 per cent died per month in their first July–October. During the second year, survival was unexpectedly very high, at 91–97 per cent through the entire year. Then in the third year it dropped slightly to 82–86 per cent, due to gradual attrition from summer onwards, but with a peak of deaths in March and April (Figure 9.2). Possibly, this was because they became potential breeders, and therefore probably needed to take more risks to gain breeding sites in a territory while hunting vertebrates to feed a female. Thus, using the most conservative estimates, about half of our birds (=0.66 × 0.91 × 0.82) survived for three years. Although we had radio-tags working on only 25 of our original 150 birds by May of their fourth year (because radio-tag

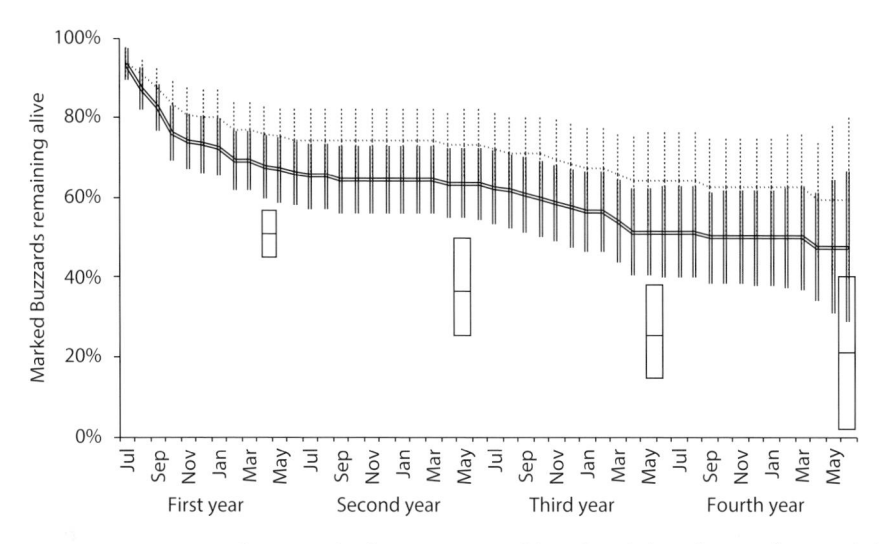

Figure 9.2. *Four-year survival estimates (with 95 per cent confidence limits) for radio-tagged Buzzards found dead (dotted line), assumed dead due to signal loss (hollow line), and UK ring recoveries (box plots). Data from Kenward et al. (2000) with kind permission from John Wiley & Sons.*

cells were exhausted), survival seemed to be very high again, at 92–95 per cent. We found no differences in survival between the sexes, or in whether birds had dispersed or not, but small differences would have been hard to detect without much larger samples.

We compared our results with 288 ring recoveries from Common Buzzards ringed in the UK between 1969 and 1989. It took quite a bit of time to extract all the data, by days of searching every paper sheet submitted to the BTO archive for G and H ring-sizes, which are used on Buzzards. Thankfully, today the BTO has embraced the digital technology that has followed, such that comparable data can now be mined by a computer in minutes. Stephen Freeman and Byron Morgan, from the University of Kent, helped by creating survival models from the ringing data, and estimated that 55 per cent survived their first year, 75 per cent survived the next two years and then 88 per cent survived the subsequent years (median age just over one year). To account for potential regional differences, the analysis was re-run using just data for the southern UK (including Wales and the Midlands) to estimate survival in the first (61 per cent) and second (79 per cent) year, but there were too few ring recoveries to calculate the survival of older birds. For all analyses it was important to separate recoveries into these age categories, because the first-year survival rate was so different, and strongly influenced the survival estimated as a product across years, thereby resulting in a lifetime survival estimate that was too low.

Looking at Figure 9.2, the divergence can be seen between the survival estimates for ringing and radio-tagging (with or without assuming death of lost birds) developed during the first two years of life. After that, the lines representing the three different categories run parallel. Thus, the adult survival that emerged from the analysis was very similar between the ringing and radio-tracking data. In the case of the radio-tagging estimates, the initial divergence in the radio-tracking estimates when lost birds were assumed dead occurred when the signals were lost after birds had dispersed appreciable distances. With these birds not being recorded again, despite our aerial surveys and philopatric spring return of dispersers know to be alive, it was reasonable to consider the lost birds dead. However, the possibility of one or two radio failures, or of distant dispersers not returning within detection range, would mean that this tended to slightly underestimate survival.

According to the ringing analysis it seemed that only 28 per cent survived through to the fourth summer, very close to Biljsma's 1994 estimate from ringing in the Netherlands. However, nearly twice as many of our radio-tagged Buzzards survived through the same period, even if we assumed that every lost radio signal was a dead bird. If the ring-based survival of 47 per cent in the first year and 77 per cent for each subsequent year is used, then by the 16th year there would be less than 1 per cent of the Buzzard fledglings left. We've already had records of five (3 per cent) of the 147 birds we radio-tagged before 1995 in Dorset that have survived close to or beyond 20 years.

There are several potential explanations for why survival estimates based on ringing data and those based on radio-tracking are different:

1. Putting radio-tags on Buzzards could have improved their survival. Of course, this seems unlikely, because we would expect birds carrying a radio to survive less well than those without a 3 per cent addition to their body mass.
2. Our location and time was unusual compared with previous studies. This explanation has more merit. We were based in southern England in a time where persecution seemed to be greatly reduced compared with previous studies and more northern areas of the UK (see Chapter 11). However, for this reason we checked with ringing data from the southern UK for comparisons, although we could not avoid using older ring recoveries so that there were sufficient data. An updated analysis of BTO data for the many Buzzards in the southern UK ringed during 1990–2000 could resolve this possibility.
3. Another very likely explanation for the differences in estimated survival was biases of recovering ringed Buzzards. Of course, our local situation could have been atypical, but we do know that there were biases in the ringing data when we compared causes of death between the two techniques, and it is likely those contributed to differences in the survival analyses between radio-tagged and ringed Common Buzzards.

The biases we found in causes of death were primarily to do with how likely a dead Buzzard was to be found and reported. People are relatively likely to find dead birds on a public road, next to buildings or under power lines that are regularly inspected. In contrast, they are less likely to find those that die away from paths, or whose bodies end up in vegetation so they are hidden from view. In broad terms, from 50 Buzzards we found dead by regular radio-tracking, 40 per cent died from natural causes (disease, starvation etc.), 36 per cent of deaths were associated with human artefacts (cars, trains, electric poles etc.) and 24 per cent were shot or poisoned. In contrast, from 160 ring recoveries where the cause of death was reported, 67 per cent of deaths were associated with human artefacts, nearly twice as frequent an occurrence, and only 16 per cent were attributed to natural causes (Figure 9.3).

A significantly lower proportion of radio-tagged Buzzards was recorded as killed by human artefacts compared with ringing results, and this was true for both first-year and older categories. There was also a tendency for more shot and poisoned juveniles to be found by radio-tagging than just by rings, whereas no radio-tagged adults were found suffering the same fate. Although these differences were not statistically significant, it was intriguing, and

□ Natural   ▦ Collision   ■ Shot/poisoned

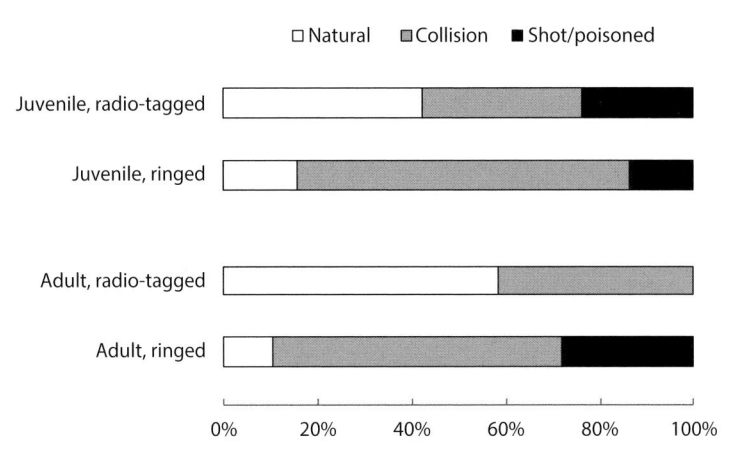

Figure 9.3. *The causes of death recorded from Buzzards monitored by radio-tagging (38 juveniles, 12 adults), and from BTO ring recoveries of Buzzards in the period 1969–89 in southern UK (103 juveniles, 57 adults). Deaths were assigned to shooting or poisoning, or to collision with human artefacts, or to natural causes, from autopsy and location of death. From Kenward et al. (2000) with kind permission from John Wiley & Sons.*

we wondered whether increasing realisation of the protective legislation might have made people more careful about hiding evidence of illegally killed birds. Due to the radio-tags, we were able to find Buzzards that had been hidden after being deliberately killed (Plate 22; see colour section). We were working at a time when tracking technologies were not so well known, and so either those hiding the Buzzards did not notice the tag, or they thought that burying the bird or placing a large log on it that would stop the signal. Indeed, recent publicity over missing Hen Harriers in the UK has claimed that those with lost radios have been killed illegally, despite recoveries of most birds found dead being from natural causes (Stephen Murphy, pers. comm.). In fact, the tags for tracking harriers by satellite are different from the tags we used. To get a good life from tags whose transmissions must be strong enough to reach satellites requires solar power. Solar tags will soon stop if not exposed to sunlight, for example if a natural death resulted in the bird being on its back or in shade, as well as perhaps being purposefully hidden or smashed. However, this was not a problem with our non-solar tags and we also knew most of the keepers and landowners in our study area. As well as producing an annual newsletter on our more interesting and useful findings, such as how to protect Pheasant pens from Buzzards, we let it be known that we did not condone illegal killing of raptors but the scientific priority was to record deaths, which could be done if radios from killed birds were left by themselves or with birds on road verges.

Similar to the earlier work showing that there were a disproportionate number of juvenile Northern Goshawks trapped at game-release sites in Sweden (Kenward *et al.* 1999), all of the radio-tagged Common Buzzards that we discovered had been deliberately killed were first-year birds. This suggests that if Buzzards in our region got through their first year, they were more settled and so were less likely go to the few danger areas where Buzzards were still killed. On the other hand, young birds are likely to be attracted to Pheasant pens containing young Pheasant poults, which look like easy prey, especially if any territorial Buzzards had already been eliminated, leaving 'vacant possession'. A tendency for inexperienced birds to succumb disproportionately to human artefacts or to have their rings reported after illegal killing (perhaps without the true cause of death) could explain why mortality rates in the first year were overestimated from ring recoveries.

Robin Prytherch and Lyndon Roberts (2012) provide a worthwhile comparison with ring recoveries from the south of England, not so far from our study area, and include the period when we were recording causes of death. Excluding nine of 26 recoveries where causes of death were unknown, by far the biggest cause for the remainder was collision with cars (40 per cent); combined with trains, power lines, a fence and a water trough, human artefacts caused 73 per cent of deaths, very similar to the findings from our ring recovery analysis. Likewise, dying either from natural causes or deliberate killing were both between 10–20 per cent, again similar to the ring recovery analysis. So, given that both the ringing and tracking data had been collected after the inception of raptor protection that was started with the Protection of Birds Acts of 1954, and both studies were in southern England, this suggests that much of the difference we found was due to different tracking technologies rather than a change in illegal killing. That is not to say that illegal killing might not be higher in other places. A study by Krone *et al.* (2006) also showed a high incidence of collisions among 149 Common Buzzards found dead in the period 1993–2002, mimicking ring recoveries in that it is what people find without tracking technology. About 38 per cent (57) died of 'unspecific trauma' (where the cause of the trauma remained unclear) and one less (56) from collisions with road vehicles. The others were found electrocuted (six), entangled in barbed wire (three), and after collision with trains (one) or rotors of wind power plants (one). Only 10 per cent died of infections, although 94 per cent of the birds had internal parasites (mainly nematodes) and 8 per cent of deaths were unexplained.

As a summary of longevity, Common Buzzards can live for more than 20 years, fitting the trends found in a recent review of raptor survival rates (Newton *et al.* 2016) by having 1) a reasonably high survival compared with smaller species, 2) lower survival in the first year, 3) no obvious difference in survival between the sexes, and 4) an increased survival rate when older and breeding has started. Radio-tagging has shown that deaths caused by human activities are likely to be overrepresented in ringing returns and, as these deaths are most likely for young and inexperienced birds, this can lead to an overestimation of mortality, and hence an underestimation of the length of raptor lives and the number that are present in populations, which in turn has implications for their conservation (Chapter 10). We have shown that not only are Common Buzzards able to live a long time, but that a lot of them can live for a long time. Let's now look at the considerable variety of factors affecting survival that create these general patterns, starting with disease.

Disease

Some of the most thought-provoking studies of disease in Buzzards come from the research team studying the influence of morphs in Westphalia, Germany (Chapter 1). It is curious that the different morphs should persist. Even if imprinting on parents leads to biased mate-choice, as a short-term explanatory mechanism, if the extreme morphs were at a significant disadvantage they would still disappear over time through natural selection. However, it is clear that they do not. Why might that be?

Originally, the team proposed that the persistence of the morphs might be due to resistance to parasites (Chakarov *et al.* 2008). This is a reasonable hypothesis, because

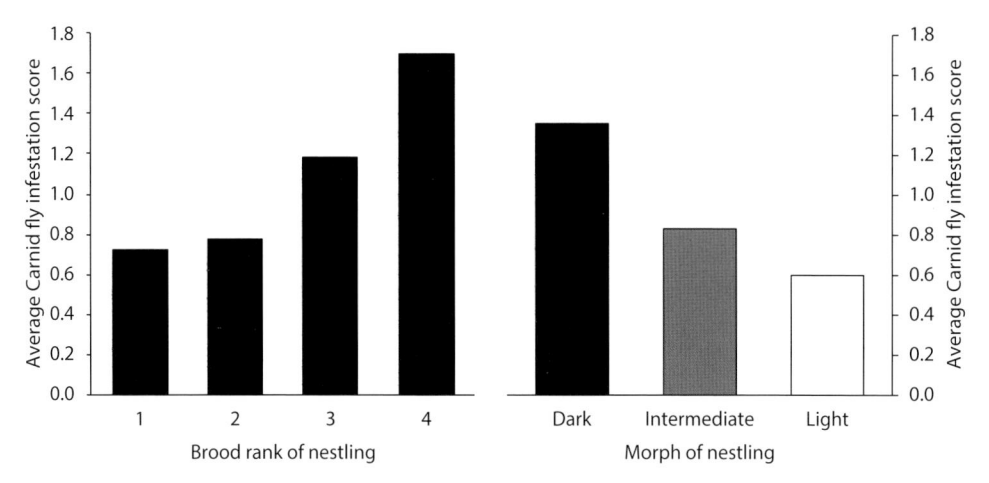

Figure 9.4. *The mean score for Carnid fly infestation in Buzzards according to the Buzzards' order of hatching in broods (left) and their colour morph (right). Data from Chakarov, N., Boerner, M. & Krüger, O. ('Fitness in common buzzards at the cross-point of opposite melanin-parasite interactions') in* Functional Ecology *(2008) with permission from John Wiley & Sons. Copyright Blackwell Publishing (on behalf of the British Ecological Society).*

parasites and other pathogens are thought to be primary drivers of evolution. The team therefore looked at the infestation rates of external blood-sucking Carnid flies (2mm dipterans in the family Carnidae found under the feathers of many birds) and an internal blood parasite (*Leucocytozoon toddi*) in nestlings. The strongest relationship with Carnid fly (family Carnidae) infestations was brood rank. The younger fledglings had bigger infestations, which would be consistent with younger birds being less able to compete for food and therefore being in poorer condition, so less able to get rid of parasites. However, there was also a very clear relationship between parasites and morph: dark morphs had the worst infestations, light morphs the lowest and the intermediates had an intermediate Carnid count (Figure 9.4). The relationship with the blood parasites was more complex, with prevalence depending on the year's rearing conditions, but the relationship was in the opposite direction. The intensity of the blood parasite infection was highest in the light morphs and lowest in the dark. Thus, there do appear to be advantages to the extreme morphs over the intermediate morph, depending on which parasite is most prevalent. Therefore, it is possible that the light and dark morphs each have their advantage during different periods, depending on the prevalence of the different parasites, and that's why the extreme morphs persist.

Evolution works by natural selection of the fittest heritable traits, which means that the morph would have to be genetically determined for any selection to occur. In recent years, it has become much easier to examine the genetic make-up of individuals. Thus, there is evidence from other species that individuals with greater genetic heterozygosity (e.g., mongrel pets) are fitter than those with less diverse genes (e.g., pure-breed pets). It was therefore logical for the team to investigate whether the colour morphs were genetically different. Unfortunately, despite the apparent selection for mates of the same colour morph and the indication of a genetic elements by Mendelian frequencies of morphs among offspring (Krüger *et al.* 2001), they were unable to differentiate the morphs genetically

(Boerner *et al.* 2013) with the techniques available to them. Therefore, the group also looked for any evidence that the more heterozygous an individual was, the greater resistance to disease it had. The only link that could be found was that in years of low vole abundance, those with greater heterozygosity had a slightly elevated resistance to infection with *Leucocytozoon buteonis*, another blood parasite. Nevertheless, this alone was unlikely to account for the differences in lifetime reproductive success. The team therefore concluded that there were unlikely to be significantly different susceptibilities to parasite load that were genetically determined. However, the Common Buzzard genome has yet to be sequenced to enable a complete analysis of its genetics, so this research may develop further.

We have seen that food can supply nutrients (such as the pigments melanin and the carotenoids, Chapter 2) for fighting disease, but the food may also be a source of disease. A study in the Czech Republic by Milena Svobodová *et al.* (2004) investigated the protozoan genus *Frenkelia*, which causes brain sarcosporidia and requires more than one host to complete its life-cycle. In raptors, it invades the intestinal epithelium and then multiplies before being distributed as infective sporocysts in the bird's faeces, which are picked up by intermediate small mammal hosts. In small mammals such as voles, the parasites first go to the liver to multiply before migrating to the brain, where they create cysts. The authors showed that Buzzard nestlings produced more sporocysts of the parasite *Frenkelia* in their faeces with each passing day in the nest. Moreover, by the time they left the nest, 100 per cent were infected with the parasite, suggesting that the food they were brought increased their infection. The prevalence within small mammals in the area was much lower than in Buzzards and it varied with season and habitat. In some areas, only 14 per cent of small mammals were infected. There is no evidence that Buzzards develop symptoms of an illness associated with these parasites, or that the latter cause an increased risk of death. However, given that all Buzzards were infected before they left the nest, it suggests that those small mammals infected with the parasite might be more vulnerable to Buzzard predation, perhaps an adaptation of the parasite to change the behaviour of the rodents in order to reach its next host. Vorísek *et al.* (1998), from the same Czech research group, showed that the voles Buzzards had caught were 2.4 times more likely to be infected with *Frenkelia* than those caught in snap-traps. In case those caught in snap traps were also not representative of the natural population, they also used a laboratory experiment to show that those mice that had been experimentally infected with a similar parasite, *Sarcocystis dispersa*, were more likely to be caught and eaten by Short-eared Owls, suggesting that the parasite made them more vulnerable to aerial predators. These are raptorial examples of the classic tendency for rodents infected with *Toxoplasma* to be most vulnerable to domestic cats.

Buzzard parasite loads can vary with region, too. Krone & Streich (2000) examined 83 Buzzards from three areas in Germany, looking for the trematode *Strigea falconispalumbi*, a parasite that potentially causes the disease trematodosis, which can lead to loss of body mass and reduced strength in flight. Out of all the study areas, only 8 per cent had died from diseases, as opposed to 84 per cent from trauma. In two areas, about one-third of the Buzzards contained the trematode, whereas only 3 per cent of Buzzards in the third region were infected. It's likely that the critical difference between the two areas was the amount of fresh water available, because the parasite has the first stage of its life-cycle in snails, then a stage in amphibians, before progressing on to the definitive host; in this case, it was likely that the Buzzards were eating infected frogs.

Another example of diseases in Buzzards having a link to water bodies was reported by Jean Hars *et al.* (2008). The extensive fishing lakes in the Dombes region of south-east France form an internationally important overwintering site for waterbirds, and bird surveys have been conducted there since 1987. In March 2006, there was an additional survey to collect well-preserved bird carcasses, due to the suspected presence of the infamous H5N1 avian influenza virus. This virus has recently done more than anything else to focus politicians around the world on birds, because of its potential to act like the plague among humans, but with winged avian vectors able to spread disease much faster than even the rats of the Middle Ages could manage. The survey was focused on waterbirds, and the main finding of the study was that the Mute Swan *Cygnus olor* was the ideal epidemiological sentinel for this disease, being extremely sensitive to the virus and convenient for virus detection. As a byproduct of the survey, non-waterbirds were collected if more than five were found dead at the same time. Forty Common Buzzards were found dead and collected, 16 were tested for the presence of H5N1 and one was found to be carrying the virus, which was only 6 per cent of the tested carcasses in an area where the disease was known to be prevalent. Probably, it is only those raptors preying on waterfowl that are likely to contract the disease. Even if the Buzzards were less likely to be killed by the disease than to carry it, it is unlikely that they would pass it back into a waterbird, because even faecal contamination when they bathe would be highly diluted. In March 2010, the H5N1 virus was isolated from a Buzzard in Bulgaria (Marinova-Petkova *et al.* 2012). Given that the nearest known outbreak was 250km away in Romania, and there were no poultry farms near where the carcass was found, it is possible this was a migratory bird, and possibly a Steppe Buzzard on its northerly spring migration. Among 385 dead Common Buzzards investigated for H5N1 only 3.1 per cent tested positive (and were killed by the virus) in Germany in 2006 (Van den Brand *et al.* 2015).

A further bird disease that is of wide human interest is avian tuberculosis (TB). Smit and colleagues examined more than 11,000 birds in the Netherlands in the period 1975–1985 (Smit *et al.* 1987). Only 82 birds (< 1 per cent) from all 25 species had avian TB, but it was noted that raptors had a higher incidence and 20 of 920 Common Buzzards (2.2 per cent) had the disease. Three Common Buzzards examined had related bacteria (*Mycobacterium fortuitum* and *Mycobacterium terrae*), thought to be incidental and something to do with the way in which they died rather than as possible infections from food. Their study also contrasted Common Buzzards with other species for the potential to pass disease on. They observed that infected Buzzards and Common Kestrels often had local wounds, and on Buzzards the wounds were usually around the leg area, which they attributed to the birds fighting over food in the nest, although fights with prey and over kills seem to us more likely. Thiede & Krone (2001) have also found a first instance of polygranulomatosis in a free-living Common Buzzard, which was in a terrible condition. This symptom is found in domestic poultry and is caused by *Escherichia coli* infections, which can lead to severe mortality in some flocks.

Many parasites are carried without causing birds any obvious problem until its condition is lost through lack of food or some other stress-inducing environmental factor. For example, a Scottish study found 20 per cent of 379 raptors from six species had intestinal parasites, with no difference between species. From this sample, Buzzards with less fat reserves, and therefore in poorer condition, were more likely to have parasites. It is possible that the parasite load could have been the cause, rather than an effect, of a poor condition (Barton &

Houston 2001). Bumblefoot (ulcerative pododermatitis) is a classic condition known by falconers to be a result of *Streptococcus aureus*, which is ever-present on the surface of the feet, but which only causes symptoms within a foot when a hawk has been stressed. It is far more common in captive than in wild hawks, but has been identified in free-living Buzzards (van Nie 1981).

Krone *et al.* (2001) analysed more than 1,000 raptors that had died after being presented to rehabilitation centres in Germany, usually because they had suffered a serious trauma. It is possible that some birds had accidents because of symptoms from the parasites, but most were considered likely to have died in collisions due to bad luck. Therefore, they probably reflected the ambient prevalence of blood parasites in raptors. On average, 11 per cent (less than the 20 per cent found by Barton & Houston 2001 in the UK) were found to have blood parasites. Generally, Buzzards were more frequently parasitised than all other hawk species. There were fluctuations during the year (Figure 9.5), with peaks of infection in spring (the 'spring relapse') and autumn. These peaks can be caused by seasonal hormonal changes that the parasites exploit to increase their chance to transfer to the next host (e.g., their own nestlings), and may also be affected by individuals getting stressed and more vulnerable at some times of year.

In our work, we observed a condition that we haven't found reported elsewhere. Some of the fledglings we radio-tagged, especially those from one particular nest, died soon after leaving the nest, and this happened several years in a row. When the carcasses were autopsied by Vic Simpson of the Veterinary Investigation Centre near Truro in Cornwall, we found that they had osteodystrophy, seemingly from birth (Simpson *et al.* 1997) and presumably due to a problem with their calcium metabolism. With that knowledge, it was easier to recognise the rickety legs that turned in and sideways due to the faulty bone development. Once we knew that young from this pair were susceptible, we noticed that they were making a 'yickering' alarm call when we handled them, which we can only assume was because they were suffering discomfort from their condition. Nestlings from these particular nests were

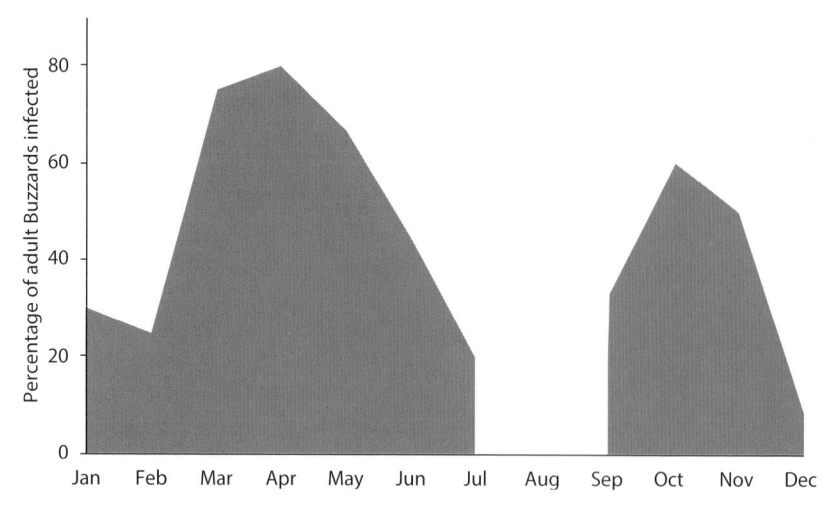

Figure 9.5. *The percentage of adult Buzzards with detectable parasite infections collected in different months of the year during the period 1993–2000, as recorded by Krone* et al. *(2001). Reproduced with kind permission from the Jagiellonian University.*

also noticeably more infested with external parasites, including Hippoboscid flies (family Hippoboscidae) and ticks (Ixodida), than those from other nests, perhaps because they could not groom their heads (where ticks tend to fester) so effectively with their talons. Here's a pathological description from one of the autopsies: '*Erysipelothrix rhusiopathiae* and *Ornithomyia lari* flies externally. Capillaria eggs & coccidial oocysts in intestines. Red nematode (*Cyathostoma lari*) found in the orbit of one bird. Splenic enlargement (possibly a Leucocytozoon infection)'. It is possible that some fledglings found by Jones (1985) may also have suffered from this condition, as his description closely matches what we observed in terms of the birds being malnourished and heavily infested with Hippoboscid flies (*Ornithomya avicularia*). His note even commented on the fact that Tony Hutson, who wrote an authoritative article on these parasites, had found them to be common parasites, quite often not having deleterious effects on the host. It may well be that the calcium deficiency led to a lack of resistance to the parasites.

Starvation

Weak Common Buzzards are sometimes found with a 'sharp keel', which indicates that not only have the fat deposits been used up, but that the breast muscles on the sternum have diminished such that they are concave and hugging the T-shaped bone. This is a bird on the verge of death from starvation, and it's not uncommon to find freshly dead Buzzards in this state. However, looking through all the literature, starvation is rarely given as a cause of death. There are often other associated factors, from trauma, to chronic or acute diseases. Of course, it's difficult to differentiate the cause and the effect. Did a disease cause the thinning first, or did starvation make the Buzzard more susceptible by compromising the immune system or creating problems among normally symbiotic bacteria in the gut? Likewise, as we know, Buzzards that are brought to our attention tend to have been associated with a human-made hazard, while those becoming weak through starvation and dying in the undergrowth are less likely to be found.

We therefore don't really know how many Common Buzzards starve to death as a result of poor foraging skills or lack of food alone. Starvation certainly affects nestlings, where the parents can only collect food within a certain area close enough to the nest. Adult Buzzards are able to maintain their mass on relatively little food, of types that would be inadequate for other species (Chapter 2). Dare (2015) even observed worms and a slug being taken to nests. However, although invertebrates may provide enough nourishment for an adult, they are not necessarily suitable for taking back to the nest in sufficient quantity to feed growing chicks. If food is scarce during this period, the parents simply may not be able to find food fast enough; younger members of the brood then starve, as all the food is eaten by the older siblings.

Despite their ability to fly to other areas in search of better food supplies, at least outside the breeding season, Common Buzzards do starve. One reason may be leaving departure until it's too late, when the weather prevents long-distance flight, with thermals absent and strong winds that they cannot easily fly with or against (Chapter 8). In the maritime climate of southern UK, the bad weather is not normally prolonged and it may usually be best for a Buzzard to save energy by fluffing out its feathers for insulation and finding a sheltered spot until the inclement weather passes. They can survive with little food for a few weeks and still make a full recovery. If it is frosty for several days, Buzzards can hunt birds and mammals that require unfrozen surfaces in which to find food. These species quite often have a more limited diet, and so without alternative nutrition they will become weak and vulnerable to predators such as Buzzards. Indeed, Common Buzzards are opportunistic, and readily eat the carrion of mammals and birds that die in the severe conditions; this can be a more efficient way of feeding than chasing lively prey. However, there will be times when the weather is unusually extreme, with an extended period too bad either for hunting or moving to another area – then Buzzards are in trouble. Dare (2015) reports particularly harsh conditions, such as the deep snow in south-west England in January and February 1963, when one-third of the adult Buzzard population died. When tracking, we noticed more frequent mortalities after two weeks of particularly harsh weather, such as the continuous frosts that keep worms deeper underground.

Natural enemies

As well as the parasites and bacteria that cause diseases, Common Buzzards are attacked by many other species; the latter may be defending territories, protecting their own young or have predatory intent. When very young, Buzzards may even be attacked by their own siblings while in the nest and when food is short (see Chapter 7). The danger doesn't stop when they leave the nest. As described in Chapter 5, Buzzards are highly territorial and will readily attack each other. We found freshly dead Buzzards that had likely been killed by raptors and one had Buzzard feathers in its talons, so we were pretty sure that its killer was the same species.

A meta-analysis of studies with accounts of raptors killing raptors found that Common Buzzards were reported only as the victims of diurnal raptors, not as the aggressors (Sergio & Hiraldo 2008). This is probably because their broad wings don't allow them to fly fast enough to pursue the smaller, more agile diurnal raptors. This has long been recognised, as Shakespeare's *The Taming of the Shrew* gives Petruchio words to mock Kate 'O slow-winged turtle [dove], shall a buzzard take thee?' However, in Chapter 2 we also described how Buzzards have been recorded attacking and killing nocturnal raptors.

There are direct observations of Common Buzzards being badly injured or killed by Northern Goshawks (Borowski 1978), White-tailed Eagles (Gradoz 1995) and Eurasian Eagle Owls (Mikkola 1976). Buzzards certainly don't fit the general trend from the meta-analysis that, on average, raptor victims are less than one-third of the mass of their killers (Sergio & Hiraldo 2008); Goshawks and Peregrines are similar in size. This maybe is where Common Buzzards' adaptation to feeding on invertebrates and fluctuating small mammal populations puts them at a disadvantage. They are not such fast fliers as many raptors, nor as powerful or manoeuvrable, and their small feet are no match for the big grasp of these other bird-hunting raptors. They are one rung down the hierarchy from an apex predator, as recent observations in Exeter have shown.

Since 1997, Nick Dixon and Edward Drewitt have regularly visited a Peregrine eyrie on St Michael's Church in Exeter, Devon, to assess the food brought to the nest. They had noticed that Peregrines drove passing Common Buzzards away, but it wasn't until 2012 that they began to find out quite how intolerant the Peregrine pair was (Dixon 2013). Just as the Peregrines' young were fledging from the nest, observers saw the Peregrine pair kill two

Figure 9.6. *A Buzzard successfully evading attack by a Peregrine.* © *Chris Wilson, with kind permission from British Birds.*

Buzzards in the same day, and their carcasses were found as proof. The pair of falcons worked as a team, as they can do when hunting other prey. The male Peregrine distracted the Buzzard from underneath, while the female launched attacks on its back, apparently hitting it so hard it broke the Buzzard's neck. One of these Buzzard victims was about 0.5km from the nest, quite some distance from the chicks the Peregrines were defending. At least two other Buzzards were killed that season, and many more attacks were observed, one of the later ones appearing to include the participation of a juvenile Peregrine.

Over three years, from 2012 to 2014, 98 Buzzards were seen to be attacked by the Exeter Peregrines; 77 cases involved both birds, the remainder launched by only one of the parents (Dixon & Gibbs 2015). Two-thirds had unknown outcomes, but at least 17 Buzzards were seen to be forced to the ground by the attack. Another nine Buzzards were found dead, stunned or injured, presumably from unseen attacks. Peregrines attack Buzzards all year round (Figure 9.6), but attacks are far more frequent during May and June, when less nest attendance is needed and the Peregrines spend more time soaring over their territory. The upsurge in attacks over the years has been attributed to a particularly aggressive female who arrived in 2009, and it seems her hatred of Buzzards has increased. In the first six months of 2015, at least 21 Buzzards were downed, nearly as many as had been grounded in the previous year. While this might be a particularly aggressive individual or pair, they are by no means unique in attacking Buzzards.

Peregrine attacks on Common Buzzards are not a new phenomenon. Derek Ratcliffe describes in his book *The Peregrine* (Ratcliffe 1980) how he had observed Common Buzzards being struck by cliff-nesting Peregrines, but he never saw the Buzzards injured and he thought that they were seldom killed. In fact, he thought that there was a collaborative association between nesting Peregrines, Northern Ravens and Buzzards. However, Peregrines have been nesting in cities again for many years now. There are many observers in cities, interest in raptors has grown and communication between those with a common interest has improved. The novelty is the collection of observations that show quite how frequent these attacks are, and what effect they can have. Thanks to online networking, it is now possible to hear of nesting Peregrines attacking Buzzards around no small number of ecclesiastical buildings in the UK – this is quite a spectacle, which we wouldn't have dreamed of seeing in a British city when we were growing up.

As with the Peregrines, Buzzards will attack other raptors that enter their territory, especially during the breeding season, and taking that risk can lead to injury and death. Lourenço *et al.* (2011) investigated how Common Buzzards, along with other diurnal raptors in Doñana National Park, responded to two species of owl. They presented breeding Buzzards with a Eurasian Eagle Owl, which is a known threat because remains of other raptors are found in its nest, and a Tawny Owl, which is too small to be a major threat, but would compete for food. While the number of Buzzard nests tested was small, the results were consistent with those from Booted Eagles, Red Kites and Black Kites. The Eagle Owl was aggressively attacked, whereas the Tawny Owl did not provoke such an overt response. These reactions make Eagle Owls good lures when trying to trap Buzzards in the suspended net of a Dho-Gazza (Zuberogoitia *et al.* 2008).

In areas where White-tailed Eagles are abundant, Common Buzzards nestlings are frequently brought to their nests. They are taken alive from the Buzzard nest and may not be killed immediately. Indeed, the young Buzzards may survive for a while, which suggests that they may even be fed by the adult eagles. Later, they are usually fed to the young eagles, but on rare occasions fully fledged Common Buzzards leave the eagles' nest (Chapter 7). A remarkable pair of White-tailed Eagles that specialise on nestlings of Common Buzzards in Lower Saxony, Germany, have been monitored by a camera trap, which revealed 17 Common Buzzard nestlings brought to the eagles' nest during the 2016 breeding season (Neumann & Schwarz 2017).

Noam Weiss and Reuven Yosef (2010), while watching the spring raptor migration through Eilat, Israel, observed a Steppe Eagle *Aquila nipalensis* stoop from a great height to attack a Steppe Buzzard that was in a small flock of conspecifics migrating north. The

Figure 9.7. *A Common Buzzard immobilised by a Four-lined Snake in southern Italy. Photo © Antonio Mazzei from Aloise* et al. *(2010) with kind permission from* Acta Herpetologica.

buzzard was brought to the ground, killed and eaten by the Steppe Eagle, which had to fend off another Steppe Eagle that wanted to steal the prey. Such attacks seem extremely rare in view of all the eyes that watch raptors migrating through the area every year.

There will also be occasions when Buzzards are killed while attacking their prey. Gaetano Aloise *et al.* (2010) reported finding a Common Buzzard partially entangled in a Four-lined Snake *Elaphe quatuorlineata* in southern Italy. Their dramatic picture (Figure 9.7) shows the snake wrapped around the Buzzard's neck, in such a way that the Buzzard could not get its feet onto the snake, and it is likely that if the fighting duo had not been disturbed, the Buzzard could have been asphyxiated, despite the observations that any attempt the snake made to bite the Buzzard tended to land on feathers.

Unnatural unintentional deaths

Humans appear to cause the deaths of many Common Buzzards, mainly by creating obstacles and driving vehicles that Buzzards collide with, rather than purposeful killings. Of course, Buzzards collide with natural features too, but we are far less likely to find them compared with where the death toll is unintentionally increased by human artefacts. These deaths are often associated with the power lines that traverse the countryside.

Power lines

It has been known for many years that power lines kill raptors. More than 60 years ago, an article by Benton (1954) established that power lines could be a significant cause of mortality for birds of prey in the USA. Much later, it was found that hundreds of raptors were being killed by certain stretches of power lines in Spain, and the problem was likely similar throughout Europe (Ferrer 2012). Hundreds of deaths of Saker Falcons *Falco cherrug* alone have been recorded on some sections of new power infrastructure being spread across Asia (Dixon *et al.* 2013), so that IUCN's World Congress in 2016 adopted a motion to take urgent steps to prevent and remediate the problems (IUCN 2016).

Common Buzzards are certainly susceptible to collision with power masts (Chiozzi & Marchetti 2000, Janss 2000, Krone *et al.* 2006). From our own study in Dorset, of 50 Buzzards found dead only one was electrocuted (Kenward *et al.* 2000), but we found other electrocutions from our release experiments that were not included in that particular analysis. Buzzards therefore do get electrocuted in the UK but, considering that mortality is low in the first place and that electrocution comprised only 2 per cent of the causes of death, it is unlikely that power lines have much of an effect on the local populations in Dorset. However, that is not necessarily the case elsewhere.

Guyonne Janss (2000) investigated which species were most at risk from power lines in central-west Spain. Common Buzzards and Northern Ravens were the species most often electrocuted, forming 60 per cent of the 471 electrocution victims from the study. However, these species were less susceptible to colliding with power lines than were species with higher wing loading, such as cranes, bustards and pigeons. An earlier study by Miguel Ferrer in 1987–88 found 44 electrocuted Common Buzzards, compared with only one that had died from collision. It is understandable that the Buzzards are drawn to sitting on the poles, because often they are the highest perch points in agricultural or pastoral fields. These 'artificial trees' may seem better than the real thing to raptors, because they don't have irritating branches in the way to restrict vision, and they are more stable for a large bird than the wires that swing between the poles. This can make power poles a magnet for species such as Buzzards that often hunt from perches.

In the UK and many other countries, high-tension pylons are safe for raptors, because the distance between live wires and potential earth points is too great for them to bridge; indeed, such pylons provide nest-sites for falcons. Cables to houses are also usually safe because the low-tension supply is within insulated cables. The problem occurs mainly with medium-tension cables, which are benign enough when suspended below supports, but less safe when on two insulators either side of a cross-beam, and most dangerous when there are three or more cables on top of a cross-beam (Plate 23; see colour section). The details were revealed elegantly in the 1970s by US falconer Morley Nelson, a film-maker with access to early high-speed cameras, which he used to film raptors as they settled on all sorts of power-poles (Nelson 1978), of course with the power turned off!

Much of the time, a Common Buzzard can perch on cross-beams with no ill-effects, presumably by folding its wings neatly so that no circuit is made between live wires and earth. The problem comes when birds stretch their wings, usually when taking off or landing, and then they may manage to reach between two charged lines on a wooden pole, or a charged line and a metal-strut pylon or, most often in our experience, the terminals of transformers. If unlucky enough to make the connection, the bird is instantly electrocuted. The radio-tagged birds that we recovered from under power poles had two burn marks, one on a wing and the other on the foot (Plate 23; see colour section), demonstrating how they had spanned two live points. These birds tended to smell like roast chicken, with a bit of burnt hair.

Although medium-tension lines of the insulators-on-cross-beam type are relatively common in the UK, and occurred within our study area, they caused no appreciable reduction in survival for our Buzzards. Perhaps this was partly because there were many trees in which to perch, and possibly also because the birds seemed to prefer perching on the more abundant single poles that support lower-tension power and telephone lines. Despite the fact that we have known the effect of power lines for so long, and there is motivation for electricity companies to avoid the problems where they can cause power outages, the knowledge transfer has been slow. There are many recommended modifications to make power poles safe from electrocution. Perch bars, or guards to take the birds away from live wires, can work for larger birds such as Buzzards, but smaller birds (e.g., Common Kestrels and Tawny Owls) still find their way into the danger zones, and so retrofitting insulation is the most effective way of protecting raptors (Janss & Ferrer 1999). This recommendation means that there is now a cost to the power companies of having to retrofit lines where the initial design was inadequate.

Wind turbines

More recently, it is not just power lines but the electricity generators themselves, in the form of wind turbines, which have become a hot topic for raptor biologists. Many biologists welcome means to generate power that are an alternative to burning fossil fuels, so in many ways wind turbines should be encouraged as a source of renewable energy. Ironically, the optimal places to install wind farms can coincide with areas used intensively by soaring birds. Those areas have considerable wind energy and therefore generate a good return on investment for both soaring birds trying to gain height, as well as conservationists trying to reduce carbon emissions and businesses trying to generate profits. There has been particular political heat at a few sites where a large number of turbines cover ridges that are used by migrating raptors, including Tarifa in southern Spain (c. 700 turbines) and Altamont Pass in California (c. 7,000 turbines). The reported collision rate (which will be less than the actual collision rate) is less than one bird per turbine per year, far fewer than at an offshore wind farm at Zeebrugge, Belgium, which had an estimated collision rate of 23 birds per turbine per year. Nevertheless, the fact that there are so many onshore turbines over such a wide area used by passing raptors means that the overall impact cannot be regarded as insignificant for rarer species. The potential of 7,000 raptors lost every year at Altamont Pass would have a serious impact on some species.

As another example, a wind farm at Smøla in Norway has reduced survival rates of young White-tailed Eagles locally. However, the remaining older birds seem to learn to avoid the turbines and some even nest within the wind farm boundaries, although other nesting pairs have gone. There will also be fewer non-breeders in the area due to reduced survival rates, but the area remains a good one in which to view eagles. For Buzzards that have healthy populations over an extensive area, it is unlikely that turbines would have any significant effect on a population, even if it is catastrophic for an individual. Where there have been the most severe problems (e.g., Altamont Pass and Tarifa), the turbines are also of an old design (faster turning rotors, nearer the ground) with a lattice work that provides perching points for raptors, and hence is especially attractive if there are abundant small mammals below (Percival 2005). When the thermals or orographic lift are good, then the birds soar high over the top, but when there is not so much wind, heavy birds struggle to gain height and are caught by the blades, the tips of which are often travelling at over 160km/h.

As well as outright death from collision, turbines may have an effect on density if some birds, including Common Buzzards, see the danger and avoid the wind farm. Whether Buzzards tend to avoid wind farms in response to the presence of the blades or because those that do try to use the area are hit and killed (Pearce-Higgins *et al.* 2009), there could be an ecological advantage for raptors in the surrounding area. This would occur if the lack of raptors in wind farms creates a refuge for small mammals and ground-nesting bird chicks, which then spill out to provide additional food in the surrounding area, like no-fishing zones in the ocean.

Luis Barrios and Alejandro Rodríguez (2004) studied the behaviour of different raptor species around wind turbines. Although Common Buzzards and Short-toed Snake Eagles often circled together with vultures on slope updraughts, they did not get too close to the turbine blades and rarely collided with them. These species have lower wing loadings than vultures, and apparently make a more efficient use of the ascending currents, gaining altitude quicker and farther from the turbines. Nevertheless, Buzzards do occasionally collide with the turbines. For example, among 149 Buzzard carcasses analysed in northern Germany, one was killed by a wind turbine (Krone *et al.* 2006).

Vehicle and artefact collisions

Inadvertent collisions with traffic also kill Common Buzzards. These often arise because cars or trains have hit other animals, especially Rabbits and released Ring-necked Pheasants, which then draw the scavenging Buzzards to feed on the road or rail track, in the path of oncoming vehicles. Road- and rail-side telegraph poles are popular with hunting Buzzards ('motorway sentinels', Hill 1990) since they provide high perches, often overlooking rough grass harbouring rodents. Some researchers have explored the possibility that even if Buzzards are not in the direct pathway of a vehicle, the turbulence that cars and trucks produce may push them into more dangerous places (Bosch 2004, Reicholf 2005).

While driving, we often see Buzzards sitting on poles. Unfortunately, it is also not uncommon to see them killed on the roads. It is less usual to observe Buzzards on railway tracks, because we can't scan the rails so easily, but radio-tracking did find them dead there, in fact more frequently than as road casualties. Of the 50 Buzzards from our area that were recovered dead, six were hit by trains and two by cars, providing 16 per cent of the causes of death (Kenward *et al.* 2000). Our electric trains, which approach more quickly and quietly than the old steam trains, are especially dangerous (Figure 9.8). One of the most unusual recoveries we had was of a bird removed from the front of an electric train that reached its final destination at Waterloo Station, London. It is difficult to see how Buzzards can benefit from the trains, compared with an exciting chance observation of a Peregrine that seemed to be using the temporary appearance of a train to block the escape route of some doves it had just flushed, so they had to turn back into its path (SW, pers. obs.).

Another six of the radio-tagged Buzzards we found had hit barbed wire fences or similar wires. One of these appeared 'alive' from its changing radio-tag signals as the bird swung about, so it was a while before we picked it up. There has been a more unusual report of a Buzzard getting caught in a kite string entangled between tree branches in the Netherlands

Figure 9.8. *A radio-tagged Buzzard killed on the railway in Dorset (© Sean Walls).*

(Mulder *et al.* 2001). Thankfully, it was found and released; we presume that this type of collision is very rare. We also found Buzzards caught in water, including livestock water troughs and the moats constructed around 'nodding-donkeys' used for extracting oil from below ground.

Considering the time that Common Buzzards spend soaring in summer, it seems likely that a few will be hit by aircraft. One such was recorded by Van Gasteren and colleagues (Van Gasteren *et al.* 2014), who were using GPS tags specifically to gain insights into the likelihood of bird-strikes around an airport. They tracked 12 Common Buzzards and found that breeding birds spent nearly all their time at the airport, and individuals could frequently cross the runway, even within five minutes of an aircraft passage. A particularly bold individual that spent too much time in close proximity to moving aircraft was eventually hit and killed. Nevertheless, their conclusion was that it was better to have established territorial pairs with experience of aircraft to keep other Buzzards away, rather than removing them and having inexperienced Buzzards move in.

Poison

Agriculture is probably the most widespread practice that results in Common Buzzards ingesting chemicals that can affect their health. Farmers use pesticides to increase crop yields by eliminating pests and diseases, thereby decreasing growing costs. It is ironic that providing more food or products that humans want also removes food for species that benefit an area's environmental health in other ways. Without healthy food chains based on adequate biodiversity, there may not be more fundamental ecosystem services such as healthy water and soil. The presence of raptors is an indicator of adequate prey species, which in turn have fed on a diversity of plants and insects in a stable food web.

Besides raptor numbers indicating that there is sufficient biodiversity, raptors are sometimes used as indicators for monitoring toxic chemicals (e.g., Newton 1979). Lowie Jager *et al.* (1996) made a systematic study of all dead Common Buzzards handed in to the Netherlands Veterinary Institute in 1992 to find out whether the carcasses contained heavy metals. About half of the Buzzards registered appreciable levels of poison. The authors found that although the levels of heavy metals (cadmium, copper, lead and manganese) were not fatal, they did reflect the environmental contamination, that is, those with higher heavy metal concentrations in the liver, kidneys and tibia were from areas known to have higher soil contamination. Consequently, Common Buzzards were considered a cost-effective biomonitor to signal gross changes of heavy metals that might have damaging effects on other species within the food web. López y López-Leitón *et al.* (2001) also used carcasses to look for the presence of aliphatic hydrocarbons in birds of prey from Galicia (north-west Spain). These contaminate the environment mainly as a result of oil exploitation and are usually more of an issue (from oil spills) for seabirds. Some are thought to be carcinogenic, and they have been shown to affect foraging ability and delivery of food to the nest in storm-petrels (family Hydrobatidae), leading to reduced breeding success (Fry 1995). Surprisingly, ten Buzzards (along with Barn Owls and Tawny Owls) showed reasonably high levels of these contaminants.

Carcasses are not necessarily needed for investigations of agricultural and other industrial pollutants. A study in Belgium showed that the concentrations of persistent organic pollutants (POPs), such as the commonly known polychlorinated biphenyls (PCBs), dichlorodiphenyltrichloroethane (DDT) and polybrominated diphenyl ethers (PBDEs), could be measured in tail feathers. In fact, only one feather is needed to extract sufficient material (Jaspers *et al.* 2006) and the levels found in the feather are proportional to the levels found in internal organs. The best correlation was between concentrations of PCB found in the bird's muscle and in the tail feathers, and the relationship was strongest if birds weren't starved. Feather concentrations correlated better with the bird's muscle than with its liver, probably because the liver replaces its tissues at a faster rate than muscle, which reminds us that when examining feathers, we are examining the levels during feather growth, not the rest of the year. This is particularly important for a species such as Common Buzzard, which is likely to eat young vertebrates during the breeding period when growing its feathers, but could eat a lot of invertebrates during the winter; different prey are likely to have different contamination levels. Likewise, there may be differences between Buzzards preying on grass-eating Rabbits and seed-eating small mammals, and those specialising on juvenile birds. Despite these minor caveats, the ability to sample both POPs and heavy metals (Jensen *et al.* 1972, Ellenberg & Dietrich 1981, Naccari *et al.* 2009) from feathers is great news, because there is no need for licences and it is much quicker to pick up Buzzard feathers around nests than attempting to catch them to take tissue samples. The short distance of prenuptial movements (Chapter 8) and the lack of migration in most areas make it very likely a Common Buzzard's feather growth was based on local foods, and this provides an opportunity to go back to museum specimens of breeding adults from known sources to check past contaminant levels.

The best-known example of pesticides affecting raptors concerned the use of DDT as an insecticide. This was highlighted in Rachel Carson's classic book, *Silent Spring* (Carson 1962). In the UK, it was found that the use of DDT resulted in Peregrine eggs being broken in the nest by the parents, because their shells had been thinned and weakened. The incidents were so common and widespread that they severely reduced Peregrine productivity, almost wiping out Peregrines from southern UK. A great account of the investigation that lead to this discovery, and then the subsequent rise of Peregrines after restrictions were placed on organochlorine insecticides, can be found in Derek Ratcliffe's *The Peregrine Falcon* (Ratcliffe 1980). The average egg shell thickness of Common Buzzards did not appear to have changed in England during the period before these organochlorines were banned, although there appeared to be an unusually high variation in thickness (Ratcliffe 1970). Pesticides were found in their eggs, but there were several factors in the Buzzard's favour. For a start, their mammal prey did not ingest the pesticides to the same extent as insectivorous birds, so the biomagnification of the concentrations was not as large in their primary food source (Tubbs 1974).

However, the much more toxic Dieldrin was used in sheep dips that could have led to the contamination of food that Common Buzzards ate when scavenging sheep carcasses. Indeed, freshly laid eggs were found to be contaminated with Dieldrin in the late 1960s, and it was suspected that the pesticide was responsible for addled eggs and breeding failure in Cumberland (Ratcliffe 1965). However, Dieldrin and other cyclodienes (e.g., Aldrin, Heptachlor) were also used as seed dressings, and it is this route that may have brought them to Buzzards elsewhere in Europe, if not so much in the UK, through eating animals that had fed on treated grain or perhaps through ingestion of contaminated soil by earthworms (see below for Carbofuran). There have been similar tales of raptors being endangered by agricultural chemicals, such as the Swainson's Hawks that were killed in large numbers when they migrated to winter in Argentina and ended up eating grubs that had been sprayed with pesticides banned in North America (Goldstein *et al.* 1999). The wider effects of pesticides are monitored these days and there is much more regulation on their use once effects are discovered. The UK restrictions on DDT have been in place for more than half a century; slowly, the chemicals have seeped out of the soil and Peregrine populations have recovered dramatically, to the joy of so many.

In other countries, both the use of pesticides and the diet of Common Buzzards can differ from the UK. For example, in north-east Spain there were still traces of the more persistent organochlorines found in Common Buzzard and Northern Goshawk eggs as late as 1999 (Mañosa *et al.* 2003). Interestingly, the levels in the two species were similarly low, perhaps because the Goshawks were primarily eating Rabbits. Those Goshawks that had more passerine remains at their nest had higher levels of contamination, which corroborates Ratcliffe's assessment that the Buzzard's diet helped them avoid the same fate as Peregrines. Buzzard eggs that contained dead embryos also had higher levels of pesticides and PCBs (mainly used industrially, e.g., in paints) compared with eggs that successfully hatched. However, the levels were still low enough to question whether the PCBs caused the embryos to die.

Carbofuran is a pesticide that has been used more recently on crops such as alfalfa, potatoes and corn, and has proved to be a hazard for many birds. Thankfully, it is now banned in Europe, but it is interesting to see how susceptible Buzzards are to its use. In the past, a granular form of Carbofuran was laid in furrows when seeding sugar beet fields. These granular formulations were known to be toxic to birds (Balcomb *et al.* 1984) and people had noted the sudden death of raptors in recently treated fields. In the 1980s, the Swiss Ornithological Institute found 90 dead or moribund Buzzards and kites, which prompted Dietrich *et al.* (1995) to look for the causal link. As the granules were not particularly attractive to a raptor used to eating animals and the application was in furrows, it seemed unlikely that the birds were simply eating the granules directly. However, the authors had noticed that Buzzards and kites often ate earthworms, especially on rainy days in spring. Earthworms can only ingest the Carbofuran once it has dissolved. Accordingly, the researchers both surveyed the worms in fields where a normal application of Carbofuran was conducted, and also set up experiments with worms exposed to different concentrations of Carbofuran in the soil. They found that the concentrations in the soil under a normal application were too low to kill earthworms, but the crops of dead Buzzards found on the field had much higher concentrations of Carbofuran. It turned out that if there was rain soon after the application, the earthworms could end up with a significant concentration of Carbofuran, and this became further concentrated if Buzzards ate several of those contaminated worms, enough to cause secondary poisoning. This is an instance where the Buzzard's ability to adapt to different foods may not have been an advantage. Although Carbofuran is now banned in the EU, and so the incidental poisoning will not be as great, it is interesting to see that Carbofuran was detected in 50 per cent of confirmed raptor poisoning cases in the UK in the early 2000s.

Rodents are pests that many people, not just farmers, try to eradicate because they eat food intended for humans, carry diseases and cause other damage. Although Common Buzzards seem not to have been greatly affected by DDT, because birds do not form a large part of their diet, unfortunately they can suffer secondary poisoning from anticoagulant

Figure 9.9. *Four Buzzards found dead (by tracking the radio-tag on the left-hand bird) in a location with poison left exposed for predators (© Robert Kenward).*

rodenticides when predating rats and mice. Doses of rodenticides are likely to slow or kill rats and mice, attracting scavenging Buzzards looking for an easy meal. Poisons are also sometimes used illegally in the open to kill pest-species, and these can impact local raptors (Figure 9.9).

Philippe Berny *et al.* (1997) used data collected over a four-year period in France (1991–94) to evaluate the detrimental effects of anticoagulants on non-target animals. They found that the liver concentrations of Bromadiolone were elevated in Buzzards and speculated that they would be susceptible to secondary poisoning. A similar study of carcasses (five diurnal raptors and six owls) collected in Denmark during 2000–2009, found that 84–100 per cent of individuals within each species had detectable quantities of anticoagulants, averaging 2.2 substances, mainly the rodenticides Difenacoum, Bromadiolone and Brodifacoum (Christensen *et al.* 2012). For 141 Common Buzzards, 94 per cent had detectable quantities of rodenticides. The level of exposure was so high that the authors of the paper suggested it could have local effects on reproduction, and therefore on population size. This situation is unlikely to change while municipalities and landowners are legally obliged to control rats. However, as Buzzards have spread rapidly back across southern England, a densely populated country with plenty of rodenticide use, it seems that current measures that discourage use of rodenticides in the open are probably enough to prevent much impact on Buzzards in the UK.

Common Buzzards are also susceptible to indirect poisoning when they scavenge animals that have been shot with lead bullets. Raptors sometimes target shot animals, if shooting is regular enough for them to form the association. At a local shoot, the guns firing on a shoot day can literally act like a Pavlov's dog stimulant, or 'dinner bell' for Buzzards, Red Kites and Northern Ravens, which gather above the shooting to collect any wounded or dead Ring-necked Pheasants that the hunting party doesn't pick up (Steve Double pers.

comm.). Lead has been banned from many products because of its poisonous properties, but lead-based ammunition is still widely used in shotgun pellets as well as in rifle bullets. Alternative ammunition is available, but sometimes the price, the fear of damage to shotguns or the shorter shooting range in shotgun ammunition prevent hunters from switching to lead-free ammunition. An analysis of ducks bought from game dealers in England in the 2013/14 winter revealed 68–70 per cent had been shot with lead, despite the introduction of the legislation restricting lead ammunition for shooting wildfowl in 1999 (Cromie *et al.* 2015).

Hernández-García *et al.* (2014) have shown that Buzzards are susceptible to this bio-accumulating heavy metal, especially when compared with Mallards *Anas platyrhynchos*, which seem to be more tolerant. Alessandra Battaglia *et al.* (2005), in northern Italy, found a Buzzard with acute lead toxicity thought to have been caused by the ingestion of more than ten lead shot. This would usually lead to the Buzzard dying within a few days. In contrast, lead shot in the muscle tissues (e.g., resulting from a Buzzard being shot but surviving) tends not to poison the bird. Mateo *et al.* (2003) looked for the presence of lead and arsenic in the bones of many raptor species in Spain and found that Booted Eagles and Common Buzzards had the lowest levels; only one of 107 Buzzards had >10mg/g of lead in its bone. This particular individual had been illegally removed from the nest, and it was suspected the bird could have been fed with shot-killed meat before being analysed. The use of lead for many human activities (fishing, transport, manufacturing, building etc.) is being phased out, or at least reduced, but the continuous use of lead-based ammunition in hunting poses a threat to birds of prey, especially scavengers such as Common Buzzards. A survey on raptor contaminant monitoring activities in Europe revealed the Common Buzzard as one of the most frequently studied species (Gómez-Ramírez *et al.* 2014), because it is so widespread and abundant, allowing robust comparisons between different countries.

Of course, poisoning can sometimes be deliberate, but that's a bigger subject with many more issues involved. We shall describe the history of illegal killing in the next chapter, together with a more thorough consideration of all the other factors that may affect Common Buzzard populations now and in the future.

Conclusions

1. We know from ring recoveries that Common Buzzards can live to be more than 30 years old. With much ringing in recent years and the latest record being set in 2016, older birds may yet be found.
2. Estimates for how many birds reach a particular age can depend on the method of study. Past estimates based only on ring recoveries appear to have underestimated survival, especially through the first year of life, compared with radio-tracking.
3. Radio-tracking also recorded a much higher proportion of deaths due to natural causes than among Buzzards reported dead from ring recoveries, which were most often found around artificial structures.
4. In unusually harsh conditions, appreciable numbers of Common Buzzards may starve to death; however, their capacity to absorb nutrients and conserve energy, combined with their ability to swap prey species and take advantage of animals worse affected by the conditions, may make them less vulnerable to starvation than some other raptors.

5. Common Buzzards are killed by other raptors, particularly Northern Goshawks, Eurasian Eagle Owls, White-tailed Eagles and Peregrines, and also by other Buzzards, both in the nest and in association with territorial defence.
6. The power industry's infrastructure causes some Buzzard deaths, mainly through electrocution when perching on poles and transformers, but also through collision with power lines and wind turbines.
7. Transport, including trains and road vehicles, may be the single human activity that inadvertently kills most Common Buzzards in southern England. Buzzards are attracted to road casualties for scavenging, and the poles along transport networks provide great perches for hunting, pulling Buzzards into harm's way.
8. In the past, agricultural pesticides such as organochlorines and Carbofuran have caused much secondary poisoning, and rodenticides remain a hazard for Buzzards, but the impact is reduced now that the risks are appreciated and mostly avoided.

CHAPTER 10
Common Buzzard populations

For us, large birds of prey are thrilling and beautiful, and it has been exciting to watch them at length during the course of our research, searching for their nests, getting close to the chicks as they grow and then being the first to find out where individuals go and how they spend their time. Nevertheless, sometimes the pleasure of working on Common Buzzards came in the office after some tedious number-crunching that often took countless attempts to make sure we had robust analyses. Only then could we be sure that what we were looking at was true, especially if it was against the perceived wisdom. Finding out that our tagged birds were, without doubt, surviving longer than expected gave us a problem. Why wasn't the number of nests increasing dramatically, and why weren't four-year old Buzzards starting to breed in our study area? What was happening to these birds, given that our tracking data suggested that most were not moving far from their hatch place? Over the years, we came to realise there were a lot of non-breeding Buzzards around, and eventually we designed an experiment to obtain a more realistic estimate of the number of birds on-site. We could combine walking transects for 'mark re-encounter' calculations with radio-tracking, to estimate the probability of the transect walker detecting a Buzzard while crossing its home range. This demonstrated that there was indeed a very large number of non-breeders, so many that it was difficult to reconcile with the traditional understanding of raptor populations, especially considering people's observations of exclusive territories around nests. Subsequent analyses showed us how much these 'territorial' birds could tolerate each other, and then the population structure could be modelled so we could understand the demography. There was no real 'Eureka!' moment, except maybe when we found that the estimated average size of an exclusive territory fitted almost perfectly with the population we had estimated in the study area, corroborating our other calculations. Piecing together the evidence and then realising that it provided a complete picture was very satisfying.

Estimating populations

Estimating populations is not as simple as counting individuals. For a start, there are three types of spatial distributions to consider. The most obvious is the spring and summer breeding distribution, within which researchers typically measure density as the number of nesting or territorial pairs. Secondly, we know from the previous chapter that some populations are migratory. The wintering distribution can therefore be quite separate, but nevertheless just as important as the breeding distribution. Thirdly, there is a resident distribution, where Buzzards can both breed and winter in the same area with no need to move. Then there are the obvious issues of counting an extremely widespread and numerous species, but one that spends a lot of time out of sight. As we'll see, these are not the only complications to finding out how many individuals occur in an area.

In this chapter we will explore how habitat and other environmental factors affect Common Buzzard presence and density. We need to bear in mind that those measures can be affected by how the studies have been conducted. If we try to compare densities across studies, we must understand that many will have chosen the study area according to requirements other than measuring presence and density. Past studies often focused on nest-sites, so the study site boundary was often created to encompass a large enough sample of nests. For small areas, this has sometimes been not much more land than the forested habitat where a number of Buzzard pairs nest, whereas those surveying a larger area such as a whole county or district, will inevitably also include habitat that is not suitable for nesting, and hence the overall density within such an area will be lower.

Using all the European studies that reported density, which is usually given for raptors as average numbers of active nests per 100km^2, Figure 10.1 shows how Common Buzzard projects appear to fit such a pattern of small study areas associating with high density. Indeed, there is a statistically significant trend ($p < 0.02$) for smaller areas to have higher densities. That result depended on four sites of less 100km^2 in which densities measured since 1990 substantially exceeded 50 nests per 100km^2. Short-term studies recorded 129 nests per 100km^2 in 71km^2 of Slovakia in 2007 (Šotnár & Topercer 2009), and an area of only 22km^2 in the Czech Republic achieved a remarkably high density of 230 pairs per 100km^2 in 1993 (Voříšek 2000). The last may not have been sustainable because by 1995 it had reduced to 144 per 100km^2. In the UK, there were 80 and 140 nests per 100km^2 in areas of 40km^2 and 75km^2, respectively (Sim *et al.* 2001, Prytherch 2013).

However, if one looks at densities of Buzzards in areas greater than 100km^2, there is no tendency to have lower densities in larger areas, nor is there evidence of such a trend in areas below 100km^2. What is clear is that Buzzards rarely exceed 30–40 pairs per 100km^2 over areas much above 100km^2, but that they can reach three or four times that density in smaller areas of very good habitat as large as 70–80km^2. In the long term, 75km^2 project in England, Prytherch (2013) has seen densities rise from 17 per 100km^2 in 1982 to 140 per 100km^2 (see long arrow in Figure 10.1). This is an average of only 0.7km^2 per pair, with some pairs apparently having less than 20ha for a defended territory. The increase in density does appear to have slowed substantially, possibly reaching the area's carrying capacity and there seems to be a decrease in productivity from the nests, suggesting a density-dependent effect on breeding.

These results give us a good idea of how densely Common Buzzards can nest when in ideal conditions – a mix of grassland, arable farming and woodland, in a temperate lowland

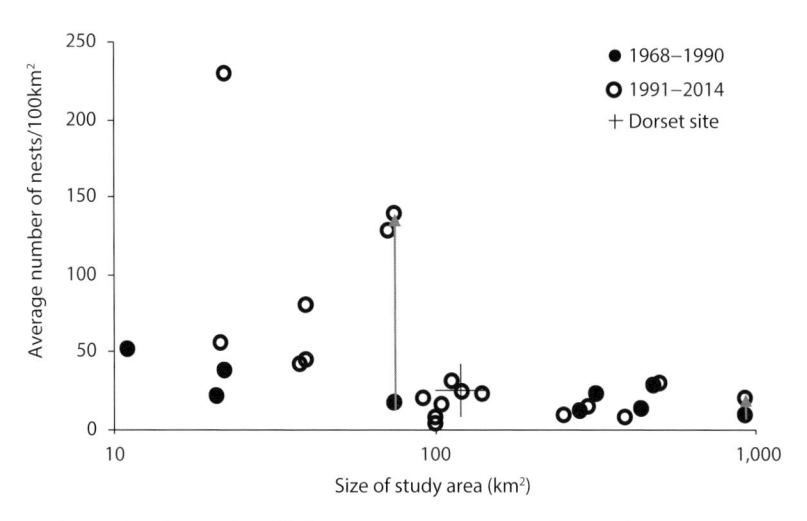

Figure 10.1. *Buzzard studies conducted before 1990, or covering more than 100km², averaged up to about 50 breeding pairs per 100km²; however, a number of later studies in smaller areas had appreciably greater densities. The two grey vertical arrows connect data from 1982 and 2012, and from 1990 and 2009, in two UK study areas of 75km² and 926km², respectively. The Dorset site (+) was close to median size and density. Data sources in Appendix 2.*

area where it rarely freezes for more than a few days at a time. To have seen a 700 per cent increase over 30 years near Bristol (Prytherch 2013), one can only conclude that Buzzard numbers were severely depressed before the study started. The land use has not changed dramatically, so Prytherch considers the Protection of Birds Acts (1954, 1967) and their incorporation into the 1981 Wildlife and Countryside Act as instrumental in allowing Buzzards back into the lowlands of central and eastern England. Others have suggested that the legalisation of Larsen traps, as an alternative to the illegal use of poisons, may also have been important for the recovery of Buzzards and for other scavenging birds not to be killed accidentally (Chapter 11).

With such a significant increase in Prytherch's study population near Bristol, it seems likely that other published studies have surveyed Buzzards below their real carrying capacity. Long-term studies in the last 30 years have tended to show that densities are increasing in many areas of western Europe, so those given in many past studies are not necessarily what they could have been. Furthermore, Buzzards can only be studied where they do occur, and in the past within the UK that was mainly in the lower hills of Wales and Scotland, which tend to experience a harsher winter and offer less food than the relatively more lush lowlands.

Where there is that ideal mosaic of trees and open farmland, it seems that Common Buzzards regularly nest 1–2km from each other. A typical nesting density, applied throughout the distribution, merely gives us one aspect of the breeding population. The total number of individuals will include not only the breeding population but also juveniles and unpaired individuals. Throughout much of the 20th century, people considered that only a small proportion of birds of breeding age did not breed. Indeed, during the period of population research in the 1960s and 1970s, spurred by fears about pesticides, the concept of breeding enough young as a 'replacement rate' to offset mortality rates (estimated by ringing) tacitly assumed that all mature birds were breeding.

Nevertheless, there has always been evidence for surplus raptors as well as breeders. For example, Newton's (1979) *Population Ecology of Raptors* summarised the evidence for a non-breeding surplus. There were plenty of observations that if a breeding raptor died, leaving one of the pair behind, a new bird would turn up and take over the role of the bird that was gone, often within the same breeding season and sometimes within a day. Likewise, if both birds died, new pairs could soon occupy the same nest-site. So, there was a sense that 'floaters' would be searching the area looking for their opportunity to settle and breed, keeping an eye on many territories for any opening that arose. It wasn't until our study in Dorset that it became clear what an enormous proportion of the population these non-breeders could form. Radio-tagging also showed them not to be floating around. There were so many potential partners nearby they didn't need to. To properly appreciate the proportion of non-breeders, we need to have a model of the population structure that estimates the relative numbers of males, females, breeders, juveniles and non-breeding adults.

Using 288 British ring recoveries (from 1969 to 1989), we collaborated with Stephen Freeman from the Institute of Mathematics at the University of Kent to estimate survival. A difference in our approach to previous models was that we considered not only survival to be age-specific (Chapter 9), but also that the proportion of birds breeding would vary with age and not be 100 per cent even for adults, and that their productivity too would be age-specific. When we assumed immigration and emigration to be equal (and thus to cancel each other out), we estimated from the ringing data that breeders during the 1990s formed only 33–38 per cent of all adults (Kenward *et al.* 2000) in our study area if the population was not growing. Indeed, there was no evidence that nest density was changing appreciably: the number of nests in which eggs were laid each year during 1990–97 (the eight years of intensive study) varied between 21–30 pairs, with an average of 25 and no tendency to increase. Therefore, from ringing data alone it seemed there were far more non-breeders than breeders, which was quite a surprise.

Moreover, data from radio-tracking had shown much better survival than had been reported from ring recoveries (Chapter 9) and we also now knew from radio-tracking that most Buzzards did not disperse far (Chapter 8), so appreciable immigration was unlikely. Thus, we could build an improved Age-specific survival and breeding (ASSAB) model from the radio-tagging (Kenward *et al.* 2000). Models for stable populations, in which population productivity matches mortality, can be seen graphically in Figure 10.2. With survival estimated using previous ringing data for Europe (Chapter 9, Bijlsma 1994), even with the relatively good Dorset productivity of 1.71 young fledging per clutch, the model was stable only with breeding rates for adults of 33 per cent, 79 per cent and 85 per cent in their second, third and older years, respectively (Figure 10.2, left). Such high breeding rates were achieved only by Buzzards at low population density as they recolonised eastern England (Figure 10.3). The model estimated 1.8 non-breeding Buzzards in spring for each pair nesting. However, recent ringing data from the UK, with Dorset breeding rates of 8 per cent and 14 per cent for Y2 and Y3 birds, estimated that only 43 per cent of older birds needed to breed for the population to be balanced (Figure 10.2, centre). With data from radio-tags in Dorset (Chapter 9), the model balanced if 29 per cent of birds more than three years old bred (Figure 10.2, right). Remarkably, this model indicated that only 16–21 per cent of Buzzards needed to breed each spring. If less than a quarter of the population bred, the non-breeding component was far from negligible!

Common Buzzard populations

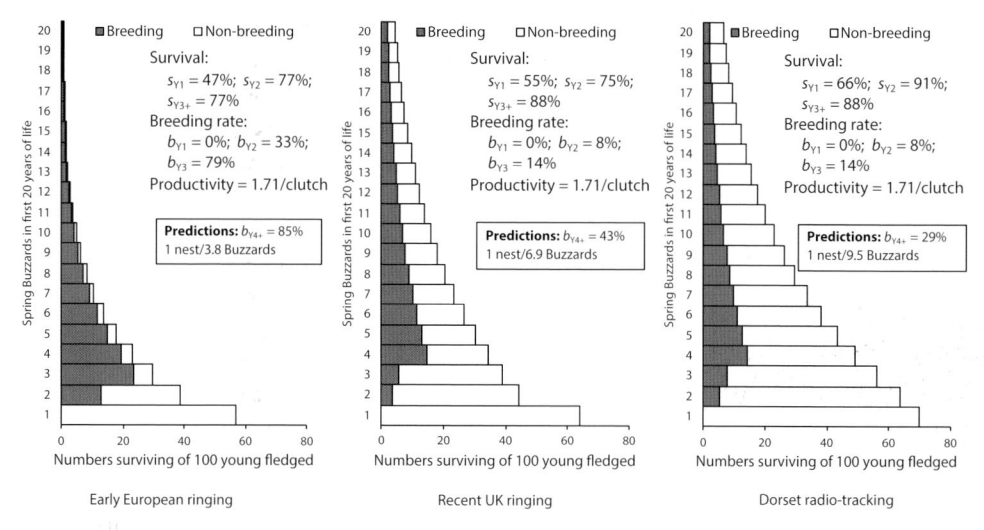

Figure 10.2. *Moving upwards from the bottom, the bars show the percentage of fledged Buzzards that remain alive each year, with those breeding shaded dark, using survival estimates for early (1911–1990) European ringing data (left), recent (1969–1989) UK ringing data (centre) and Dorset radio-tracking in the 1990s (right).*

Early European ringing: Survival: $s_{Y1} = 47\%$; $s_{Y2} = 77\%$; $s_{Y3+} = 77\%$. Breeding rate: $b_{Y1} = 0\%$; $b_{Y2} = 33\%$; $b_{Y3} = 79\%$. Productivity = 1.71/clutch. Predictions: $b_{Y4+} = 85\%$, 1 nest/3.8 Buzzards.

Recent UK ringing: Survival: $s_{Y1} = 55\%$; $s_{Y2} = 75\%$; $s_{Y3+} = 88\%$. Breeding rate: $b_{Y1} = 0\%$; $b_{Y2} = 8\%$; $b_{Y3} = 14\%$. Productivity = 1.71/clutch. Predictions: $b_{Y4+} = 43\%$, 1 nest/6.9 Buzzards.

Dorset radio-tracking: Survival: $s_{Y1} = 66\%$; $s_{Y2} = 91\%$; $s_{Y3+} = 88\%$. Breeding rate: $b_{Y1} = 0\%$; $b_{Y2} = 8\%$; $b_{Y3} = 14\%$. Productivity = 1.71/clutch. Predictions: $b_{Y4+} = 29\%$, 1 nest/9.5 Buzzards.

An earlier study of Northern Goshawks in Sweden estimated that 73 per cent of males were breeding, but only 41 per cent of females; due to the relatively low survival rate of the smaller males, not all females were able to find mates (Kenward *et al.* 1991, 1999). These estimates gained credibility when the proportions of male and female adults trapped away from nest areas in winter, and later radio-tracked to nests in spring, was very close to the model estimates of those expected to be breeding. Also, using a similar model, Grainger Hunt (1998) estimated that a stable population of Bald Eagles contained 45–51 per cent of non-breeding birds. Apparently, our Common Buzzard population contained an even higher proportion of non-breeders than among the Goshawks and eagles. Although some owl species can *occasionally* have even higher non-breeding rates in poor vole years, this *average* proportion of non-breeders was actually so far from expectations that we knew that some would find it difficult to believe without further evidence.

We therefore sought to test whether the population model was credible by getting an independent estimate of the population wintering within the study area (Kenward *et al.* 2000). As we have discussed, surveying by visual counting is fraught with problems, because Buzzards can easily go unnoticed when perched in woodland or other hidden spots. Even if you count at times when you expect Buzzards to be most visible, for example, in the middle of the day during the courtship period when they are displaying, you are only likely to be counting those with a nest to defend, not the non-breeders. However, we again used the radio-tags to our advantage. As part of an MSc student project, Maarit Paahkala walked east–west transects of the 6km across our study area, as close to the 1km grid lines as possible, counting all the Buzzards she saw. Maarit not only counted the Buzzards but estimated their distance from the transect line, so that she could use the generally respected *distance* technique (Buckland *et al.* 2001) to estimate density. She also noted whether the birds that she saw were carrying a radio-tag, usually detected by the antenna that extends from the tag on the Buzzard's back but backed-up by checking for radio signals with the receiver she carried. Meanwhile, we tracked radio-tagged Buzzards on the study area so we could plot home ranges that could estimate a probability of sighting them within a specified distance of transect lines; neither of us told Maarit where they were before she had finished her fieldwork.

These observations gave us the crucial information we needed to estimate the size of the Buzzard population in two different ways, not only during an autumn session, when Maarit walked 300km of transects across the whole study area, but also during the following winter when she covered 300km again, in half the area where most birds were being radio-tracked. Comparing how often Maarit saw radio-tagged Buzzards with how often she was expected to encounter them on transects (from the radio-tracking) gave her a detection rate that was only 20 per cent in autumn, when leaves were still on the trees, but which rose to 71 per cent in winter. From these estimates, her 49 sightings of Buzzards in autumn and 89 in winter gave 'corrected-transect' density estimates of 2.0 and 2.26 Buzzards per km^2. Of the total 138 sightings, 11 were of radio-tagged buzzards. Knowing the number of radio-tagged Buzzards available to be seen, effectively 23 in the whole 120 km^2 area and 16 in the 60km^2 higher-density area, gave re-encounter estimates of 1.95 and 2.15 Buzzards per km^2. The four remarkably similar density estimates were compared with the 25 nests in the whole area, and 16 in the partial area, to estimate that breeders made up 21 per cent of the autumn population and 25 per cent of the winter population, only slightly above the estimate from the ASSAB model. This greatly strengthened our confidence in the model. Further confidence is given by the fact that the model predicts that 6 per cent of our Buzzards should live to 20 years; we have already recorded 3 per cent at or beyond that age, with another six years in which recoveries from our 1990–95 cohorts can occur before they all would have reached 30.

Having estimated the proportion of non-breeders in our local population, and found it to be much larger than expected, what might be expected in other areas? All we can do is speculate, because studies in other areas only report the densities of nests. In areas with extensive illegal killing, the proportion of non-breeders might be much lower, perhaps limited to first-year birds not old enough to breed. Moreover, in areas being newly colonised there wouldn't be many non-breeders because there are more nest-sites available than breeders, and so there would not be any competition to hinder breeding.

So, how do we know what the real population is in other areas? Just counting the nests and multiplying by two for the number of individuals breeding, and then adding a bird or two per nest for the current year's young, is not likely to be accurate. We have shown the potential for a population four times the number of breeding birds (eight times the number of nests), but of course the multiple is likely to vary from place to place, depending on productivity, survival and emigration/immigration. Assessing a population in the way we have just described, with someone walking transects through an area with many radio-tagged Buzzards to provide robust population estimates, is costly and very time consuming. Organising systematic recording of visually tagged birds is a less onerous option, but still requires appreciable work to catch and tag the birds. This last method was repeated more recently and estimated that the number of Buzzards hunting on the Langholm estate, in the Scottish borders (Appendix 3), was nearly six times the number of nests within the project boundary (Francksen *et al.* 2016c).

An indicator that there is likely to be a large non-breeding population is the age at first breeding. This concept was introduced with the idea that unusually early breeding in an established population indicated excessive mortality. Early breeding could be estimated for Northern Goshawks with feathers of breeders collected at nests (Kenward & Marcström 1981, Grünhagen 1983), because first-year birds have identifiably different feathers. Now, with the use of modern tools for tracking, including inexpensive micro-transponders as well as radio-tags, individuals of any species tagged as nestlings can be observed entering the breeding population whatever their age, not just by identifying first-year breeders of those species that have obviously different juvenile plumage and can breed in their first year (Kenward *et al.* 2000). In our Dorset study, only 8 per cent of Common Buzzards with radio-tags were breeding in their second year, and only 22 per cent in their fourth year; unfortunately, the tags did not last longer than four years, so we don't know when older birds started breeding. For those from the same area that dispersed naturally to the east, where they were recolonising and therefore presumably not as abundant, breeding started at an earlier age. In this case, about 20 per cent were breeding in their second year, and more than 60 per cent by their fourth year (Figure 10.3). Furthermore, when we released 42 Buzzards in counties in the east of England, where there were very few breeding Buzzards, more than 30 per cent of 17 released birds still being tracked were breeding in the second year, 83 per cent of 12 in their third year, and although only three were still being tracked in their fourth year, all of them were breeding. It is unlikely that everyone will have the resources to radio-tag sufficient birds, and currently there is no reliable way to identify feathers of second-year buzzards, so the comparison of nest numbers with a thorough mark-re-encounter survey during one winter is probably the best way to estimate non-breeder prevalence in a resident population.

Although no definite records have been found for Common Buzzards breeding in their first year, there are a few other estimates for age at first breeding. In North Wales (Appendix 3), Davis & Davis (1992) found that seven wing-tagged males started breeding at 3.1 years on average, whereas four females started later, at an average 4.0 years old and with ages varying between two and five years. Dare (2015) noted that in Suffolk, where

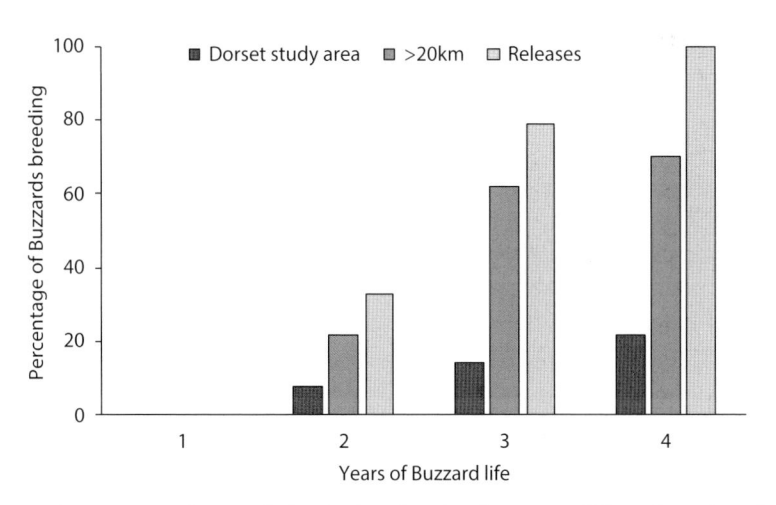

Figure 10.3. *The percentage of Buzzards that bred in their first four years of life, as shown by radio-tracking, was lowest in birds in the Dorset study area (high breeding density), greatest among birds released into areas with no known breeders, and intermediate with those from the Dorset study that settled after travelling more than 20km into areas being colonised (medium breeding density).*

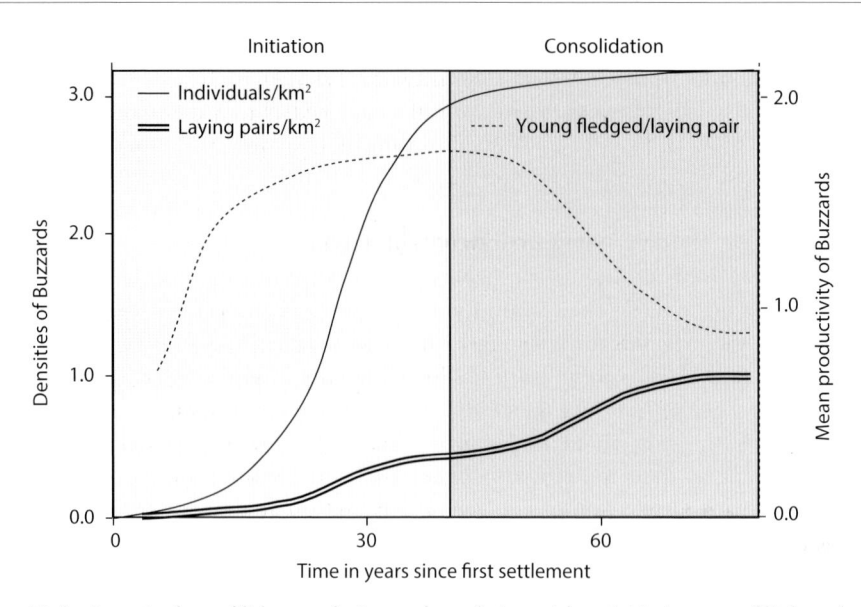

Figure 10.4. *Stages in the establishment of a Buzzard population, with an initiation stage of high productivity in which non-breeding individuals increase in number (as observed in our Dorset study) followed by a consolidation stage in which productivity declines and the proportion breeding may increase again.*

Buzzards were recolonising, new pairs often bred within 12 months of arriving on a territory. Unfortunately, we don't know the exact age of such birds; however, given our knowledge that Buzzards move furthest in their first autumn (Chapter 8), these new breeders could well be second-year birds, that is, arriving in their first year and breeding in their second.

An interesting question here is whether a Buzzard population at the high density of more than one pair per 1km^2, for example, Voříšek (2000) and Prytherch (2013), contains as many non-breeding adults as our study population in Dorset. Prytherch (2013) has shown that with a rising density, the percentage of successful nests and brood size reduces, which would reduce the number of non-breeders entering the population. So, perhaps when we studied our Dorset population it was when the proportion of non-breeders was especially high, but as the nest density rose, the proportion of non-breeders fell, as inputs to the population from breeding declined. Although two of our Dorset Buzzards tagged in 1990 bred in their second year, none of our tagged birds were breeding at that age by the end of the 1990s. However, the proportion of birds in their second year would then have been smaller if high density had reduced productivity, in which case non-breeders that stayed and delayed breeding for several years (rather than contest a territory) may have become skilled enough hunters or relaxed enough with their neighbours to manage to nest later on. Thus, a 1.5km strip of woodland that hosted a nest at one end each year in the 1990s and intermittent nesting at the other end, 15 years later has a regular nest in the middle, too, and a new nest 500m away to one side of the strip.

We suggest that there are two stages to the establishment of a new breeding population of such a long-lived species. The first could be considered an initiation ('start-up') stage, in which breeding starts at high productivity and builds a network of non-breeders in the interstices between the initial distribution of nests, followed by a consolidation ('catch-up') stage in which settled non-breeders gradually start breeding and productivity becomes strongly constrained by competition for food and nesting sites (Figure 10.4). A consolidated population might not be able to tolerate as many non-breeders as the initial population, but also would not have such a high proportion of non-breeders to accommodate due to the reduced productivity and emigration from a consolidated population.

Current Common Buzzard populations

BirdLife International has used estimated numbers of breeding pairs to propose that the world population of mature Common Buzzards (including the Steppe Buzzard subspecies) in 2015 was 2.1–3.7 million and increasing, breeding over an area of 33.9 million km^2. The maximum estimate is almost twice the minimum, reflecting a great deal of uncertainty. If the Steppe Buzzard is considered a discrete species in the future, and its population estimated separately, which seems unlikely (Chapter 1), then these numbers will have to be reconsidered. The species as it is considered today has a very healthy global population for a large, territorial predator. Compare this with the Rough-legged Buzzard, which was estimated to have 0.3–1.0 million mature individuals breeding over 41.6 million km^2, including the Americas as well as Eurasia. In Eurasia, the Rough-legged Buzzard does not extend so far south, so its breeding range is less than that of the Common Buzzard (Figure 1), and it breeds at a lower density. Considering other similar-sized raptors, the Common Buzzard is probably also appreciably more numerous than the more widespread Northern Goshawk (1.0–2.5 million) or Black Kite (1.0–2.5 million). Even looking across to the Americas, the ubiquitous Red-tailed Hawk is estimated to have a population of 2.3 million and the Broad-winged Hawk *Buteo platypterus*, which can be seen in kettles of tens of thousands when on migration, only has a population of 1.3 million. The Common Buzzard could be the most numerous hawk (Accipitriform raptor) in the world. That's very difficult to prove, because of the inaccurate nature of population estimates on many species. However, the only hawk that seems to come close is the Eurasian Sparrowhawk with a population estimated at 2.3–3.3 million mature individuals. Among Falconiform raptors, Common Kestrels are currently twice as numerous (4.0–6.5 million), although their populations are decreasing while Common Buzzard populations are increasing. The more similar-sized and globally distributed Peregrine is much less common, with only 0.1–0.5 million individuals.

With their huge population and increasing trend, Common Buzzard certainly fit easily into the category of 'Least Concern' for the Red List process of the International Union for Conservation of Nature. This is similar to other widespread *Buteo*s. Of 27 species recognised in the *Buteo* genus by the BirdLife International Taxonomic Checklist (BirdLife International 2015a), 21 are of Least Concern, two Near Threatened, three Vulnerable and one Critically Endangered. The 'Critically Endangered' Ridgeway's Hawk *Buteo ridgwayi* in the Dominican Republic, plus the 'Vulnerable' Socotra Buzzard (which may well actually be part of the *Buteo* superspecies, see Chapter 1) and Galapagos Hawk exist on small islands and so are always going to be at risk from locally severe events. The other Vulnerable species is the Rufous-tailed Hawk *Buteo ventralis* of Patagonia, but data on this species are very sparse so it is difficult to assess what restricts its population to less than the estimated 1,000 individuals. Most other *Buteo* species have large populations distributed over wide areas, making them robust to extreme weather events and more general environmental change.

Looking in more detail at European Common Buzzard populations, there is a mixed

The Common Buzzard

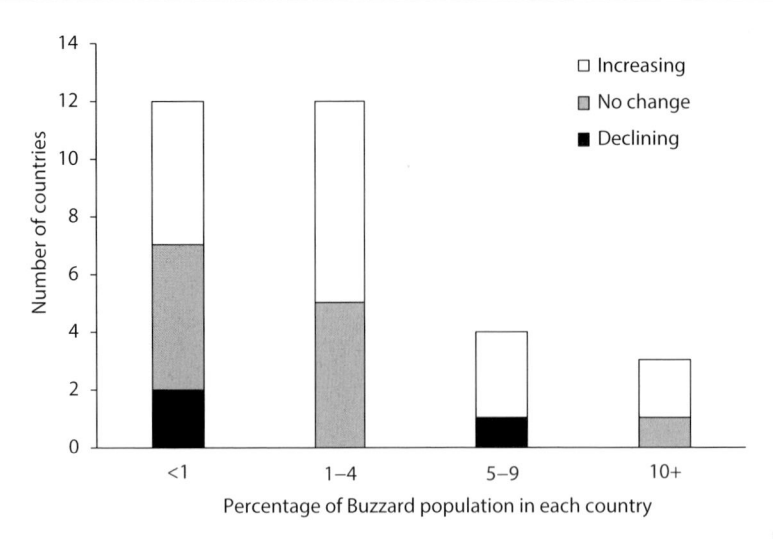

Figure 10.5. *The trend during 1980–2012 of estimated Common Buzzard populations shows greatest tendency to increase in those European countries that hold more than 1 per cent of the Buzzard population. Only Finland, Albania and Latvia (an unlikely member of the 5–9 per cent population category) reported a decline.*

story. Referring to the *European Red List of Birds Fact File for Common Buzzards* (BirdLife International 2015b), there are individual country estimates for the 27 EU members. An estimated 30 per cent of the breeding Buzzard population is in Russia, which is a huge area over which to get a population estimate. France (12 per cent) and Germany (10 per cent) are next in terms of the population size, which is reasonable given their size relative to other countries. The UK then follows with 6 per cent of the European Population.

In countries where there are reasonable data, long-term trends since 1980 are generally positive, and quite noticeably so in western Europe and Hungary (Figure 10.4). Six out of seven countries with sufficient data suggest an increase, and only two of 26 with medium quality data show a decrease. On the other hand, looking at short-term trend data for periods since 2000, there are slightly more countries with a decline than an increase. Of 11 countries where the data are deemed 'good', five are in decline, four are stable and only two are increasing. If 'medium' quality data are accepted, then of 23 countries 17 are stable/fluctuating, three are increasing and three are decreasing. The biggest increases are still in the west, noticeably the UK (with a 146 per cent increase since 2000, to potentially nearly 80,000 pairs) and the Netherlands (13–26 per cent increase to 12,000 pairs). The biggest short-term declines (20–49 per cent) have been in Germany (nevertheless holding between 80,000–135,000 pairs), Poland (more than 50,000 pairs), Latvia (34,000 pairs, an unusually large estimate for a small country and deemed 'medium' quality) and Finland (estimated 4,000 pairs in a relatively large country).

The sources are too varied, and too far from standardised, to really scrutinise comparison between countries. For example, in Finland there has been extensive monitoring for decades, which provides robust estimates, whereas in at least one other country a popular bird book is given as the reference, and books sometimes contain 'guestimates', which can quickly become outdated. Likewise, Sweden's Common Buzzard population is shown as increasing strongly over the medium term (1980–2012), but if data from autumn counts at Falsterbo in southern Sweden are used, then the long-term trend sees a decline from the 1940s to

1997 (Kjellén & Roos 2000). Some of the birds migrating through Falsterbo will have bred in Finland and Norway. Finland saw a 30–36 per cent decline during 1982–2012, but the population (approximately 4,000) is much smaller than in Sweden (17,000–45,000), whereas Norway's even smaller population (1,000–2,000) is considered to have been increasing over the long and short term, so the effect of Fennoscandian countries is likely small compared with the global population. Moreover, the decline in Scandinavian birds migrating through Falsterbo (Kjellén & Roos 2000) seems to reflect exceptionally high tallies in early years (especially in 1950, with 41,000 counted), whereas since 1980 the counts have been much more stable, averaging about 12,000 per year and with an increasing trend in the most recent decades (Figure 8.14 in Chapter 8). As discussed in Chapter 8, Nils Kjellén and Gunnar Roos also noted that the proportion of first-year birds soaring past was increasing. It is therefore possible that reduced counts at Falsterbo did not represent declining populations, but that an increasing number of more mature territory holders have been remaining in southern Scania to defend their territory year-round now they can survive the (milder) winters.

Looking at all the countries together, the long-term increases run parallel with three factors: (i) the increased enforcement of European laws together with a change in the public's attitude to killing raptors, (ii) global warming and (iii) widespread changes in land management. It is more difficult to say what has caused some of the more recent declines. Obviously, places to start looking are national changes in farming practices and raptor guilds, or perhaps even a reversal in attitudes now Buzzards have become common enough to start having a bigger impact on game conservation in some countries. Let's now explore in detail how some of these factors may affect the future of Common Buzzard populations.

Factors affecting populations

Within regions, Common Buzzard densities have fluctuated substantially. To analyse why that may be, we need to consider what promotes or limits their populations. Many of the factors affecting populations are themselves being changed by human activity and can be regarded as indirect human impacts. We will leave the direct human impacts until the next chapter where we can see them in the context of current conservation and potential future relations with humans. Here, we have included natural and indirect (accidental) human effects. We combine these with elements from previous chapters, especially concerning breeding (Chapter 7) and survival (Chapter 9) to summarise how various factors interact to affect not simply individual Common Buzzards, but their species' distribution and population density. Thus, we now consider how resources, competition, predation, pollution, disease and climate affect Buzzard numbers.

Resources

The most fundamental resource for survival and reproduction is food. We know that Common Buzzards are very adaptable, eating anything from worms and insects to birds as large as male Ring-necked Pheasants, mammals up to the size of a young Rabbit and carrion of even larger species. Those raptor populations that exploit fluctuating populations of small mammals are sometimes thought to have a nomadic strategy to cope with the changing resource (Newton 1998), or to maintain their distribution by taking alternative prey if it is available when the main prey is scarce. The latter is the case for the Common Buzzard, although breeding may suffer when food is scarce (Reif *et al.* 2001, 2004b, Selås 2001). Although this adaptability means that Buzzards can usually find suitable prey in a wide variety of landscapes, and thus have a wide distribution, we also know that changes in abundance of particular prey can potentially affect breeding performance and Buzzard density. For example, when myxomatosis suddenly suppressed the Rabbit population (Dare 1957, Moore 1957, Davis & Saunders 1965), the effect on Buzzard populations in Devon lasted into the 1970s; this is thought to have been due to the frequent reoccurrence of the disease keeping Rabbit numbers suppressed (Dare 2015). Despite such local effects, the unchanged distribution of Common Buzzards in south-west England was a testament to their ability to find food wherever they are.

Even where breeding success of Common Buzzards depends greatly on cyclic resources, such as voles, their density is unlikely to be greatly affected by the fluctuating numbers of prey. This is because Buzzards are a long-lived species, with the potential for many breeding seasons. It is breeding that really creates pressure on parent birds, which lose condition during the process, through laying eggs (females) and hunting hard to feed female and young (males). In low prey years, it may be a better option for an adult to abstain from breeding, and thereby not risk lowering its chance of survival at a time when poorly nourished fledglings also won't survive. If adult survival is relatively unaffected by the poor prey supply, then the population would be unlikely to decline by more than a breeding season's worth of young, in which case the total population in areas with many non-breeders would reduce by 12–16 per cent at most. In these circumstances, the core of the population will be maintained by occasional productivity boosts during those years in which prey are abundant during breeding. Thus, due to a varied diet combined with its digestive efficiency, fluctuating food is one of the softer influences on density within the Common Buzzard's broad distribution, despite temporarily strong local effects on breeding.

Nevertheless, nothing stands still, and our land use is likely to continue to change, so it is interesting to look at the more recent changes and what impact they might have. A recent study shows how a crop that has really expanded throughout Europe has benefits for Common Buzzard populations. In an area of Poland where Oilseed Rape was nearly one-fifth of the agricultural crop, a seven-year study found that Common Vole numbers appeared to be higher in rape fields and that the breeding success of Buzzard nests nearby was improved (Panek & Hušek 2014). Although the amount of Oilseed Rape did not affect whether a pair attempted to breed, it did improve the chances of success and increased the brood size of those that raised young (Figure 10.6).

Habitat changes can also affect the number and quality of nesting sites, an important resource for population growth and, ultimately, the maximum density of Buzzards in an area. Often, Common Buzzards are associated with forest edge because it provides nesting habitat and good perches for hunting (Chapter 6). Humans have made changes to the countryside, felling forests to give numerous open areas next to the shelter of woodlands. As with food, the Common Buzzard appears remarkably adaptable when it comes to where to nest; it is able to use cliffs, buildings or even the ground if no trees are available and ground predators are absent (Chapter 6). Provided there are some trees, Common Buzzards appear able to use arable landscapes successfully, at least for hunting. However, when land is

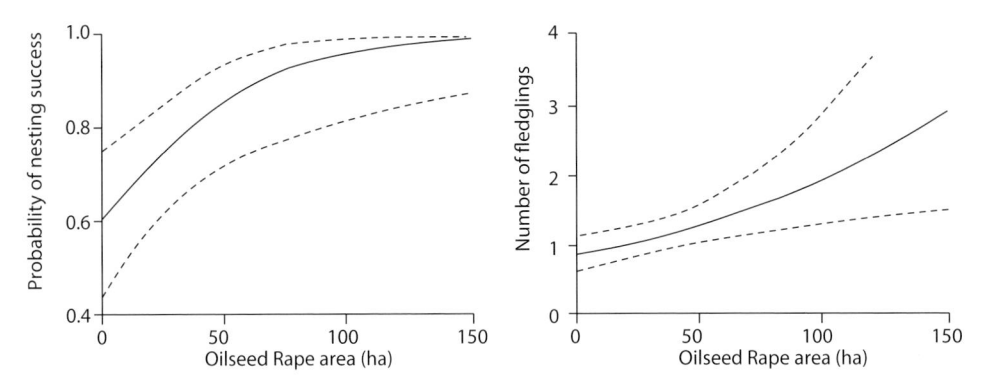

Figure 10.6. *In western Poland, the more Oilseed Rape there was within 1km of a Buzzard nest, the less likely the nest was to fail (left) and the greater the number of fledglings produced (right). Dotted lines show 95 per cent confidence intervals. Redrawn from Panek & Hušek (2014) with kind permission from* Bird Study.

extensively deforested for agriculture there may be too few nest-sites, perhaps with compounding effects of disturbance by humans and other species. It will therefore be interesting to see what densities Buzzards can achieve in eastern England, for example, as they continue to spread back into East Anglia, a region that has large tracts of land with comparatively few trees.

As well as adverse effects in areas where trees are eradicated, some researchers have been concerned by changes in sylviculture within woodland. Wojciech Bielański (2004) thought that tidying woodlands of malformed trees would reduce places where Buzzards could build their relatively large nests. Likewise, Sergio in Italy (Sergio *et al.* 2005) and Zuberogoitia *et al.* (2006) in Spain both provide evidence that timber management can have an effect on density. Sometimes, changes can be positive, as proposed by Taylor *et al.* (1988), who thought that afforestation was allowing Common Buzzards to recolonise Northern Ireland, although they also moderated these views with the opinion that too much afforestation can restrict open hunting areas.

The development of woodlands and fields for housing reduces the resources we associate with flourishing Buzzard populations. There are indications that the inevitable increase in urban development has an indirect negative effect on Common Buzzard populations (Palomino & Carrascal 2007). Recently in the UK we have seen a lot of pressure to abandon the greenbelts around towns and cities. However, even if our towns and cities are set to sprawl out and cover more rural areas, Buzzards are heading in the opposite direction and filtering into suburbia like other raptor species have done in the past. The most renowned raptor species in both Europe and North America, brought to the attention of the wider community by webcams at their nests, is the Peregrine. This falcon used to nest on tall building such as cathedrals, and it now nests in major cities on the plethora of tall buildings we have today. Less well known, but certainly not uncommon even in the 1980s, were Eurasian Sparrowhawks nesting in urban copses or churchyards (Newton 1986). More recently, even the Northern Goshawk, a species traditionally renowned by falconers for its 'highly strung' behaviour and habit of nesting deep in woodland to avoid humans, has built populations in major cities such as Hamburg (Rutz 2008). Likewise, Common Buzzards are now found on the edge of the largest conurbations on the south coast of England. Indeed, analyses of radio locations showed that although our Buzzards on the whole avoided suburban habitats, a minority found small patches of them to be a useful resource (Chapter 4).

It will be interesting to see how a species known for its robust tolerance and even exploitation of human activity will adapt to very urban landscapes. One problem may be a lack of easy food in the modern 'concrete jungle'. In the past, Red Kites were well known as city birds that fed on rubbish, and Black Kites have joined with crows to replace vultures in this role in some Asian cities. Thankfully for humans, the rubbish (especially food waste) is cleaned up and controlled to prevent disease spreading in the modern urban environment. However, that also means a lack of food to scavenge. In contrast to the countryside, worms are not as abundant in cities to act as a safety net during winter, when there are few inexperienced young bird and mammal prey. Nevertheless, we have observed Buzzards worming on sports fields, and resting on football goal-posts. Having recently walked around a local suburban park and seen a Common Buzzard boldly sitting in an oak tree above the people walking below, perhaps this species, too, can become a common suburban bird, as is the case with other *Buteo*s in North America, including the Red-shouldered Hawk (Bloom & McCrary 1996, Dykstra *et al.* 2000) and Red-tailed Hawk (Bosakowski & Smith 1997). On the other hand, Rough-legged Buzzards and Ferruginous Hawks are rather scarce in the urban environment (Berry *et al.* 1998). It may be that the Common Buzzard will remain uncommon in built-up areas, because it is only very rarely found nesting on artificial structures, compared with other raptors that have been successful in towns (Chace & Walsh 2006).

Beyond the socio-economic requirements that drive urbanisation, changes in land use broadly reflect changes in the main countryside industries. The importance of cultivation and energy industries is clear. We need to feed and warm ourselves. If the UK population continues to rise, we may need to intensify land use further. The predictive modelling of Buzzard populations (Chapter 4) showed the importance of grassland and rough ground for Buzzards. Agricultural intensification could involve more food crops or biofuel and less grass for dairy and meat production if cattle are moved indoors. Extensive afforestation to provide wood, which is considered environmentally friendly for providing building materials that lock up carbon, instead of releasing it through production of cement and firing bricks, may also increase. Thus, changes in the countryside could become less favourable for Common Buzzard foraging.

There could also be some positive impacts for the Common Buzzard from changes in land use. We have shown how current arable land has some attraction for Buzzards, and perhaps new cultivation techniques may enhance the availability of worms. To be prosperous, a crowded island such as Great Britain may require more land for recreation and to attract tourists. State subsidies for crop production have been declining, and are already less for agriculture across the European Union than the combined private spending on angling, hunting and gathering, and watching wildlife (Papathanasiou & Kenward 2014). Taken together with continuing state spending to ensure clean water and other 'ecosystem services' for humans, private recreational spending may further motivate landowners to maintain their diverse land-holdings in ways that benefit generalist species. These are complex socio-environmental issues. However, given the adaptability of the Common Buzzard, it does seem reasonable to assume that the indirect impacts of land-use changes are unlikely to be sufficient to make this resilient species rare again in the UK.

Competition

Availability of food and nest-sites depends not only on the abundance of these resources, but also on whether there is competition from other animals that are trying to use the same resources. We have reviewed the evidence of Common Buzzards competing among themselves in Chapter 7, but they also compete with other species. The main interspecific competition is likely to come from very similar species, especially the Rough-legged Buzzard, whose summer breeding populations predominantly occur north of the Common Buzzard. In an area where both species' distributions overlap, Sylvén (1978) investigated how the two coexist, which is by having exclusive territories (both intra- and inter-species). Unfortunately, we really don't have much historical data on Common Buzzard distributions in the north, merely observations that they now appear to have a northerly limit, beyond which only Rough-legs are found. Without Rough-legged Buzzards occupying those northerly areas, maybe the distribution of Common Buzzard would extend further north.

In terms of competition for food, it is difficult to see any particular prey species having much effect on Common Buzzard distribution, because of the ability they have to switch to alternative prey. Clearly, Buzzard productivity (and hence numbers) does respond to the fluctuations in vole and Rabbit numbers. When these animals are abundant, there may be enough for all the predators in a guild to benefit. On the other hand, the unintended benefit that Rabbits receive from game-keeping indicates that a complete guild of predators, including the Stoats *Mustela erminea* and Red Foxes that keepers legally control, can affect Rabbit abundance. The effects of food and intra-guild predation on vole population cycles are still far from clear.

Even if vertebrate prey are not abundant enough, Buzzards can survive through the winter on worms and other invertebrates, but is there also competition for these resources? With the present extent of stock-farming in Great Britain, it is hard to imagine that Eurasian Badgers or other worm-eaters are going to reduce the worms in farmland to such an extent that Buzzards can't survive the winter. However, there are other raptors with a very similar niche, such as Red Kites, which can also cause interference competition. Although there are areas where Common Buzzards and Red Kites coexist, such as the Chilterns and North Wales, it is not clear what effect a high density of kites may have on the density of Buzzards, nor how this may change when Northern Goshawks become abundant in such an area (Plate 24; see colour section).

We conclude that the presence of avian and mammalian competitors may prove in future to have noticeable impacts on Buzzard breeding and density. Changes in distribution due to competition seem less likely.

Predation

Although it can be difficult to see subtle shifts caused by competition, there is evidence that Common Buzzards are killed by other raptors, either for territorial advantage or as prey. Just as Buzzards prey upon relatively defenceless nestlings and fledglings (Chapter 2), so buzzard chicks are preyed upon by other raptors, corvids and martens (Chapter 9). Nayden Chakarov and Oliver Krüger (2010) found lower Buzzard nesting success within 1.5km of Eurasian Eagle Owl nests, and Eagle Owls are notorious for raiding nests or killing recently fledged raptors within their hunting range. Nevertheless, no study has shown that Buzzard distributions have been limited by predators. For sure, they choose nest-sites that are further from larger predatory birds, which limit their density but not their distribution. Most studies investigating the interactions are naturally conducted where both species exist, and some have observed the increase in one of the predators while the Buzzard's density has declined. None have shown the predatory species to have exterminated Buzzards to the extent it affects the latter's wider distribution. The same can be true for cases where mammalian predators may eat unguarded eggs; although they may have a local impact, they are unlikely to affect the wider distribution.

In fact, predators can have an unexpected positive effect on populations of other species. An example is 'mesopredator release', where the arrival of an apex predator affecting a lower predator then releases another from the guild to be more successful. Eagle Owls will eat any bird and are renowned for their capacity to eat other raptors. On a simple level, one might therefore expect that if Eagle Owls are present, then Common Buzzard density and/or reproductive success should decrease. Indeed, this is what we have discussed when Northern Goshawks cohabit with Buzzards. Chakarov & Krüger (2010) were able to combine many years of hard toil recording Buzzard and Goshawk populations with the serendipitous colonisation of one of their study areas by Eagle Owls. Goshawk density was reduced in the area near the Eagle Owl nests, but increased further away, as though birds had been displaced. What about the Buzzards? We might expect them to reduce in all areas due to the Eagle Owls in the centre and increased Goshawks in the surrounding areas. Instead, they increased density slightly around the Eagle Owls, suggesting that the suppression of the Goshawks took the pressure off the Buzzards, and didn't replace it with equal predation from Eagle Owls.

Other species (apart from humans) probably do kill Common Buzzards, either as the latter are trying to kill them (e.g., snakes, Figure 9.7 in Chapter 9) or because they are caught on the ground, perhaps while scavenging, by a fox or other mammalian predator. However, those incidences are relatively rare and unlikely to have any measurable effect on populations, even locally, just as more frequent attrition by avian predators is unlikely to affect Common Buzzard distributions.

Poisoning

Poisoning has the potential to cause widespread devastation and therefore dramatically alter a species' population and distribution. The combined effect of DDT, dieldrin and aldrin in terms of both direct mortality and reduced productivity was so severe that populations of bird-eating raptors crashed in the late 1950s and 1960s (Ratcliffe 1980, Newton 1986). However, in common with other raptors that ate few birds, UK Common Buzzard populations did not appear to have suffered much. Nevertheless, there were declines in the Netherlands, possibly caused by direct mortality from dieldrin, aldrin, organo-mercury compounds and carbofuran (Chapter 9) (Dietrich *et al.* 1995). Fortunately, no single chemical yet seems to have had a widespread effect on British Common Buzzards, which at that time had not yet recolonised the eastern areas where pesticides were used most profusely.

So far, so good for the Common Buzzard, but we must not be complacent. Imagine a pesticide not found to be toxic for vertebrates but that, when used on insect crop pests, leaves long-term residues in soil that affect the abundance of invertebrates, including earthworms. There could be an effect on all species that eat earthworms, including Common Buzzards, which might not be detected until large areas of land had been affected. Pesticides like neonicotinoids appear to have such effects on pollinators, though secondary effects on their predators have been harder to pin down. The less obvious side effects of new agricultural treatments on the lower levels of food chains and soil fertility need effective monitoring. We may hope that with each new environmental pollutant, our society becomes increasingly cautious and better at prediction-based avoidance of such issues, but will it?

Looking at other raptors, even in recent times we have seen dramatic crashes of *Gyps* vulture populations in the Indian subcontinent, with two species declining by more than 97 per cent between 1992–2007 due to the use of Diclofenac as a non-steroidal, anti-inflammatory drug (NSAID) for cattle. This is a drug like Ibroprufen and is also used for humans, so the early loss of vultures from the 'towers of the winds', where Pharsees (followers of the Zoroastrian religion) were laid to be consumed after death, should have raised the alarm. However, the replacement of vultures by scavenging crows and kites was at first blamed on persecution and unidentified disease, before Oaks *et al.* (2004) isolated Diclofenac as the real cause. It is ironic that the European Union, which is enthusiastic about legislation to prevent direct human impacts on raptors, recently approved the veterinary use of Diclofenac in southern Europe, including Spain, which holds the majority of Europe's vultures. While this could be a concern for other species, it is unlikely that Diclofenac will impact Buzzard populations greatly, because although Buzzards will scavenge on domestic animals, leaving dead livestock in the open is not as common as it once was.

Focusing blame on deliberate killing by humans when raptors decline is not uncommon, but it is dangerous for conservation because it often creates conflict between groups that could be fighting the real problem together. The blame-game in such a conflict also distracts the public from the genuine issue. This was the case for DDT, with falconers still being blamed for the decline in Peregrine populations in Germany 30 years after organochlorines had been found guilty in the UK. Falconers were the solution, not the problem, through huge efforts restoring the tree-nesting Peregrine population of east–central Europe, yet the trade in falcons is still considered a risk for the Peregrine in the Convention on International Trade in Endangered Species (CITES). Trade is also still blamed for declines in Saker Falcon populations, although research funded by falconers has now identified the dangers of massive killing of falcons and vultures on poorly designed power lines (Dixon *et al.* 2013) and is devising rectification. Treating people as problems rather than solutions is especially unhelpful when it is in error.

Disease

Although we saw in Chapter 9 that a number of diseases have been identified in Common Buzzards, there seems to be no epidemic that has yet affected any population over a wide area. Indeed, most diseases evolve to persist in their host population; they do not kill all their hosts, because that could lead in turn to the extinction of the pathogen. However, this is not a threat to underestimate in wildlife as a whole, because disease that is alien to an area or still adapting to a new host species can have a devastating effect. The introduction of diseases such as Eurasian measles to the Americas, and Ebola to cities in West Africa, are clear examples of both effects for human populations. However, transmission of disease is at its least effective in species that live at relatively low density and seldom come very close to each other due to territoriality, like Common Buzzards, although scavenging of carcasses infected during a massive bird 'flu' epidemic could potentially threaten Buzzards.

Transmission by insect vectors could also overcome the separation of individuals in a territorial species. In southerly areas where mosquitoes transmit diseases such as the zoonotic White Nile fever, which kills wild raptors, it is interesting to wonder whether the separation of some avian species might be mediated by disease. Are there wild species that can live in an area in winter, when mosquitoes are less active, but have to move to high latitudes without the disease in summer? It would be interesting to examine, by looking at immunity-conferring genes, whether any such effects might account for separation of buzzard species.

Climate

Climate change is very likely to continue to have an effect on the population dynamics of Common Buzzards. With their breeding distribution apparently limited both by cold in the north and aridity in the south, we would expect any change in climate to alter the Buzzard distribution, although it's not so easy to work out the details of such change, especially as counter-effects may occur. One might suppose that global warming would benefit Common Buzzards in the north of their range, because generally plants grow faster and produce more vegetation that in turn can produce more herbivorous and granivorous prey for Buzzards. However, in Finland, at the northern edge of the Buzzard's distribution, and hence somewhere expected to gain from the warming (Figure 10.7, left), there appears to be a long-term decline of the number of territorial pairs since the early 1990s. Aleksi Lehikoinen *et al.* (2009) investigated the relationship between declining brood size and the laying date, which

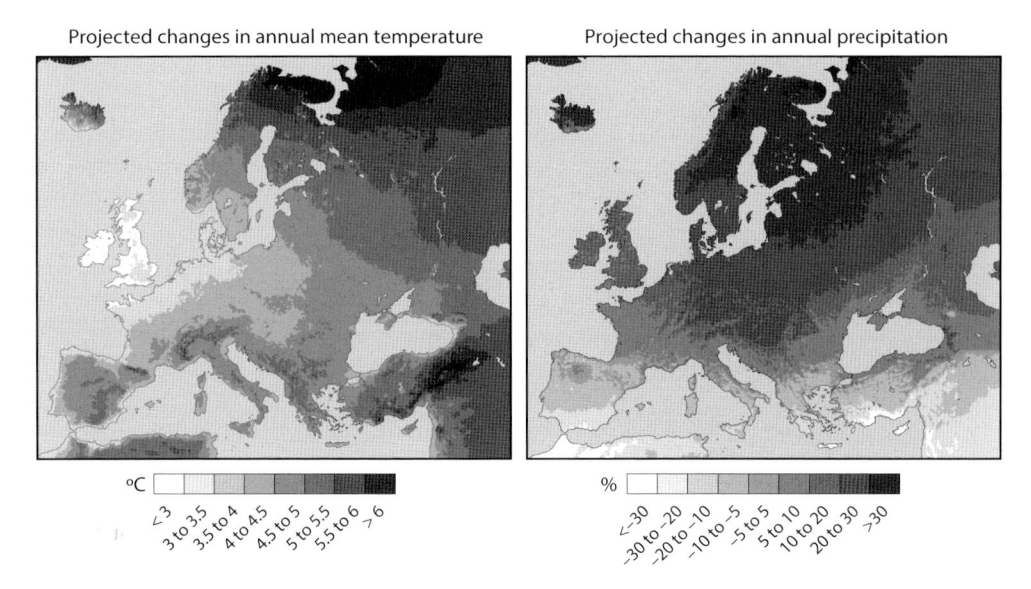

Projected changes in annual mean temperature   Projected changes in annual precipitation

°C <3 3 to 3.5 3.5 to 4 4 to 4.5 4.5 to 5 5 to 5.5 5.5 to 6 >6

% <-30 -30 to -20 -20 to -10 -10 to -5 -5 to 5 5 to 10 10 to 20 20 to 30 >30

Figure 10.7. *Effects of global warming are likely to be least in north-west Europe, and may benefit Common Buzzard distribution to the north, provided that substantially increased rainfall there and in eastern Europe does not impact productivity too severely. Data from Coordinated Regional Climate Downscaling Experiment CORDEX (partners in Appendix 2), with kind permission also from the European Environment Agency.*

was getting earlier. They found that hatching is earlier in warmer Aprils, and there is a trend for Aprils to be warmer over time. However, while April temperatures have increased, summer temperatures have not, so the chicks hatch in less favourable conditions. Combined with increased precipitation produced by global warming (Figure 10.7, right), conditions overall look as though they have been getting worse for Common Buzzards in Finland, and that may be a cause of the apparent decline in their breeding success there.

It has been suggested that an increase in Common Buzzard numbers in northern Europe depends on the decrease in the North Atlantic Oscillation (NAO) (Jonker *et al.* 2014), which affects the stability of Europe's weather. There is certainly evidence for the hypothesis, given that NAO affects weather, and weather affects survival and breeding success – key elements of population dynamics. It is true that the NAO has decreased in strength, and that Buzzard populations have increased, as have those of other raptors throughout Europe. However, one must be very careful about assuming that associations between two factors are causal. In fact, the correlation between NAO and survival varied within the regions studied by Jonker *et al.*, sometimes even having a correlation in the opposite direction at a more local scale. It is therefore likely that other factors contribute substantially to the change. Nevertheless, it is clear that weather does have an effect on survival and breeding success, and with global changes it could have dramatic effects on populations.

In coastal temperate areas with little snow cover, pressures from changing land use by humans may well be similar to those in the UK, and could compensate for climatic changes or exacerbate them. In harsher climes, an increase in snow cover would probably affect Buzzard migratory behaviour, as they find hunting in snow difficult (Sonerud 1986). These migratory populations may be more vulnerable to developments further south in the wintering areas, where we have seen increasing aridity and human pressures on land.

However, if climate change also increases the length and vigour of growing seasons in their breeding areas, negative impacts in wintering areas could be compensated by increased breeding productivity.

The migratory nature of some populations, combined with the long-distance dispersal of a proportion of sedentary populations, is adaptive when there are gradual changes in climate and associated land use. It means that the species is constantly exploring new sites by chance, and what may not have been suitable a decade earlier could become a haven, compared with areas that had been strongholds in the past but have now become less hospitable. This gives the Common Buzzard a lot of resilience. While its distribution may change and its population density may fluctuate, the species will not disappear quickly unless there is a sudden apocalyptic change. An analysis by Krüger & Radford (2008), looking at the vulnerability of certain raptor species found 'less habitat specialization, a larger clutch size and more plumage polymorphism were associated with lower extinction risk and larger population and range sizes'. The Common Buzzard therefore matches the characteristics of a 'survivor species'. The environment may change (with or without humans) but as a species, Common Buzzards are adaptable and are likely to sustain populations that breed in temperate areas.

Despite the difficulty of predicting how global warming may affect Buzzards at local scales, it is likely to encourage a northward shift in the breeding distribution. There is evidence from the Americas that the wintering distribution of the migratory Rough-legged Buzzard has moved north (Figure 10.8) as fast as 8km per year (Paprocki *et al.* 2014). Whether the Common Buzzard will push north to further overlap or displace the Rough-legged Buzzard from its breeding areas remains to be seen, but it would not be surprising if the generalist Common Buzzard outcompetes the specialist Rough-leg as northern latitudes warm. That seems to be the basis on which the Red Fox is spreading north, causing the northward movement of the southern edge of the Arctic Fox *Vulpes lagopus* distribution (Fuglei & Ims 2008). On the other hand, at the southern end of the Common Buzzard's

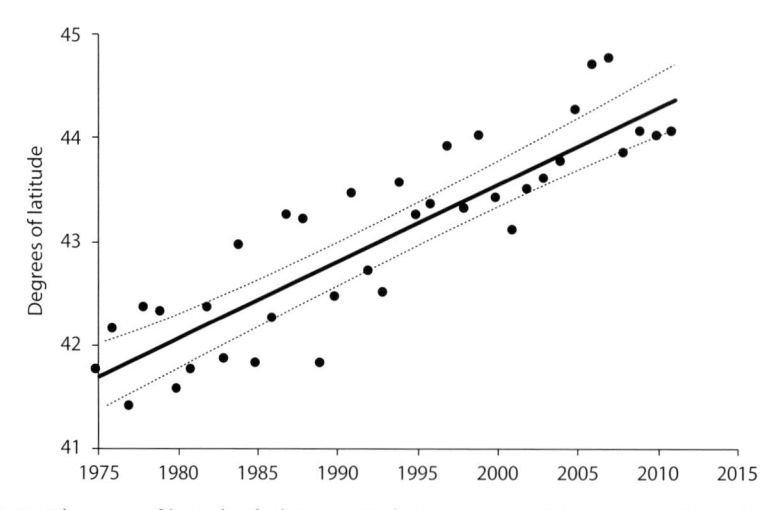

Figure 10.8. *The centre of latitude of Christmas Bird Count reports of the wintering Rough-legged Buzzard population distribution in North America has shifted northwards, as shown by trend-line (solid) and 95 per cent confidence intervals (dotted) from Paprocki* et al. *(2014), with permission from the Public Library of Sciences.*

distribution, the drying of the Mediterranean region could favour the Long-legged Buzzard over the Common Buzzard, leading to a potential decline of Common Buzzards in the south. Indeed, Long-legged Buzzards are already colonising Iberia, and modelling predicts a wider spread north of the Mediterranean (Chamorro *et al.* 2017). It will also be interesting to see, albeit probably not in our lifetimes, whether the Upland Buzzard could spread off the high steppe, or alternatively if it will become more restricted with global warming.

As well as climate change aiding the competition with the genetically very similar Rough-legged Buzzard (Chapter 1), it may also be creating more overlap with other similar species, and the outcome may not be merely competitive. Common Buzzards are known to hybridise both with Rough-legged Buzzards (Gjershaug *et al.* 2006) and Long-legged Buzzards (Dudas & Janos-Toth 1999, Corso 2009, Elorriaga & Muñoz 2010, 2013). The northward spread of Long-legged Buzzards into Common Buzzard distributions, and Common Buzzards into Rough-legged Buzzard distributions, might increase the area of overlap of similar species and actually suit hybrids better (Elorriaga & Muñoz 2013). If so, the Common Buzzard would not continue with its current genetic composition. However, being surrounded by other very similar species that seem better adapted to those conditions, it is likely that any reduction in the distribution of Common Buzzards will be filled by very similar species such as Long-legged and Rough-legged Buzzards.

Many of the factors affecting populations are in fact dependent on human activity and can be seen as unwitting indirect effects of our species. Humanity as a whole is gradually coming to terms with evidence that we are now affecting the climate (Solomon *et al.* 2007) and that this is influencing species distributions. In the next chapter we look at the more direct impact humans have had on the Common Buzzard in the past, and how our current activities and relationship with this species is developing.

Conclusions

1. Estimating population sizes is problematic, especially when it is difficult to count non-breeders. The use of radio-tracking studies has given us a better understanding of Common Buzzards, including better knowledge of densities, survival and social behaviour.
2. With an improved understanding of how non-breeders can share areas, there is scope for estimating population density with more precision than merely counting the number of breeding pairs.
3. Since the 1980s, Common Buzzard populations have been increasing in most European countries where there are sufficient data, although there have been more recent short-term declines in some countries.
4. Common Buzzards are of Least Concern according to the IUCN Red List process, and in fact may be the most numerous hawk species in the world.
5. Other raptor species are also recovering to re-create raptor guilds, which may in turn change Common Buzzard nesting density, either by reducing Buzzard nesting through predation, dominance or interference competition, or potentially increasing it locally through mesopredator release.
6. Common Buzzards are very adaptable at using the resources available. They can exploit many different prey species, are not too specific in their nesting habits and are

long-lived. They can therefore delay breeding until later years rather than endanger themselves. This makes them resilient to much environmental change.
7. Weather is the environmental component that has most effect on their productivity and global distribution. Ultimately, it is climate change that is likely to have the biggest influence on their distribution and that of other closely related *Buteo* species.

CHAPTER 11
Our relationship with the Common Buzzard

The Common Buzzard that caught our interest when we were young did not have the mystique of an elusive and rare raptor in a remote part of the world. It was not a bird that could only be relished on the wildlife programmes that have entertained and educated us. It was an impressive wild raptor that could be readily seen and heard in certain regions of the UK, and transformed a pleasant walk in the country into something more exciting.

The high visibility of the Common Buzzard, and its tolerance of a human-created landscape, made it widely available for us to enjoy. In contrast, human intolerance of raptors as a group had totally shaped the distribution of this conspicuous raptor in the past, as shown most clearly in the UK with its especially dense human population. When starting the study in Dorset in the early 1990s, brandishing Yagi antennae on the edge of roads to radio-track our tagged birds, we were often stopped by folk curious about our activity. Some were suspicious that we were looking for people watching television without a licence. When we explained that we were tracking Buzzards they often looked baffled. Apart from not being aware of the technology available to track birds, quite often they didn't believe there were Buzzards in those parts!

Thankfully, even that is now history. A new tolerance allows many others to enjoy these mini-eagles across the UK. Buzzards and other raptors have also expanded their distributions in many European countries where populations are monitored. They can once again provoke awe and wonder in many people. For some, they might just help to begin a love affair with nature, which we believe is crucial if humanity is to remain a comfortable, integral part of natural systems on Earth.

Throughout this book, we have established what Common Buzzards need in order to survive and breed, how they defend what they need and the challenges they face. Humans have had an enormous impact on Buzzards, just as they have with many species. In this chapter, we look at the way in which people have directly interacted with Buzzards, current attitudes towards them and how our relationship might continue into the future. As a widespread, successful generalist, we can use the Common Buzzard as an analogy for the future of predators in general in our increasingly anthropogenic landscapes. They are common enough to engage the wider community. They are also common enough to trigger conflict with some interests, and common enough to provide sufficient data for evidence-based solutions.

Direct human impacts on Common Buzzards

Historical human relations

Humans have previously had a major impact on raptor numbers around the world, because they were seen by many as competition for resources in ways that affected human livelihoods. The human species now competes so effectively for global resources that numbers of most animal and plant species are declining, and numbers of some raptors are well below their former glory. So what's the story with Common Buzzards?

In *The Buzzard*, Colin Tubbs (1974) gave a detailed history of past persecution of this species in the British Isles, and how this led to population declines from 1600 to the beginning of the First World War. Tubbs used the accounts of the church wardens of Tenterden, in Kent, as a way to look back at what was being killed in pursuit of 'vermin' in England. The reason these accounts are useful is because they record the bounties paid on the submission of parts of freshly killed animals. Vermin were anything that ate what people wanted to eat, and in the case of Buzzards they were takers of lambs and domestic fowl (Tubbs 1974). Although it is difficult to extrapolate out to the whole of England, Tubbs used other records around the country to show that similar patterns of bounty payment were occurring elsewhere, and therefore presumably the same efforts were being made for pest control. Although Buzzards occasionally appeared in the records before 1676, it was later that the species' prospects changed, with the start of 'an intensive campaign for the thinning-out of vermin'. Although it is probable that a higher proportion of people could differentiate bird species in the past than in today's predominantly urban-based lifestyle, the nomenclature was less consolidated. Therefore, there is likely confusion between buzzards, harriers, hawks and other large birds of prey in the records kept by church clerks. However, it is certainly clear that the number of large hawks recorded as killed increased dramatically during this attempt at pest eradication. Many were taken as nestlings rather than adults, presumably because that was an easier way of obtaining a bounty than trying to hit a flying buzzard with the inaccurate and low-powered projectile weapons that were available before the 19th century.

In the mid-1700s the numbers of killed raptors on which bounty was being paid declined quite dramatically, perhaps because the vermin control had precipitated the desired effect, and there were fewer raptors to kill. Persecution turned to House Sparrows *Passer domesticus*, whose populations had exploded with the modernisation of farming practices (and maybe also the reduction in predators). Although it is impossible to establish precisely the real numbers being killed, there is evidence that Buzzards, kites and Northern Ravens had become much rarer by then. Books were being published about how to kill birds of prey, with an increasing use of firearms to shoot 'flying vermin'. There were also two important changes to land use; farming started to enclose land and game preservation became more popular with country estates. In Scotland, the highland clearances in the 18th and early 19th centuries created large estates used initially for sheep farming, but which later developed shooting interests. Game shoots increased to their heyday from 1880 to 1914, with thousands of gamebirds being shot in a day on some estates. Of course, this intensified the competition with anything that might damage the population of gamebirds, and so there was effective attrition of all competition, whether mammalian carnivores, raptors or even poaching humans. The improvement in gun technology, combined with the Buzzard's large size and soaring behaviour, was a fatal combination, and the demise of the Buzzard over much of England was assured. Tubbs also points to increasing sheep grazing on uplands, where Buzzards were seen as lamb predators, and the fashion for taxidermy and naturalists' collections as further facilitators in the decimation of Buzzards from much of England.

A. G. More mapped bird distributions systematically for the British Isles in 1865. He showed Common Buzzards only being common in the west (Devon, Cornwall, Wales, north-west England, western Scotland and Northern Ireland), and virtually eliminated from southern and eastern England, except for a pocket around the New Forest, in Hampshire. Even in Devon, where they were not uncommon in the wilder places, they were rarely seen in cultivated areas (D'Urban & Mathew 1892), suggesting the influence of humans. Surprisingly for the period, some keepers were fond of Buzzards and prevented the common practice of nest disturbance (Hurrell 1929). The rather separate population in the New Forest persisted only thanks to the stewardship of the Honourable Gerald Lascelles, a falconer who enjoyed and wanted to preserve all wildlife, not just game. Appointed Deputy Surveyor of the New Forest in 1880, he encouraged the Forest's keepers to not harm raptors,

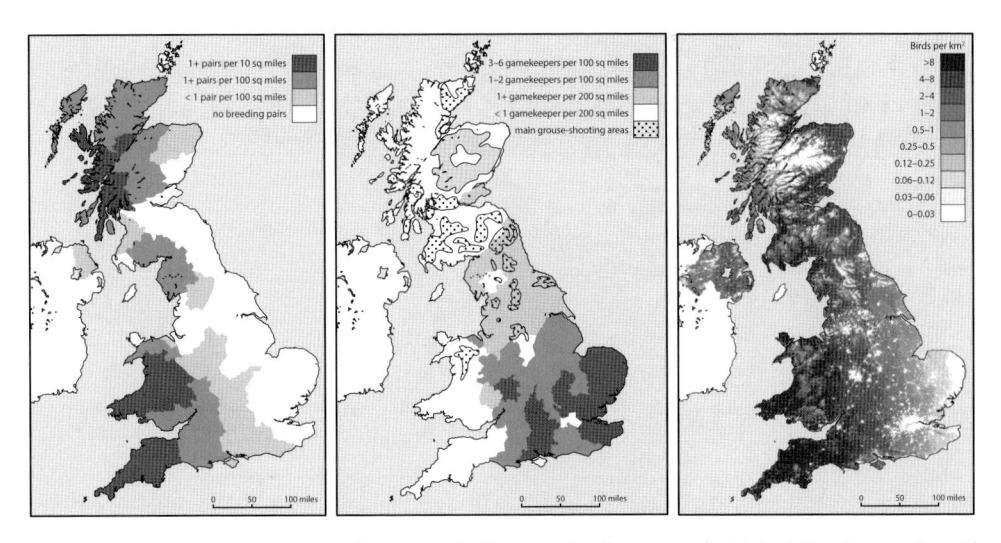

Figure 11.1. *Norman Moore pointed out that the Buzzard distribution in the United Kingdom in the mid-20th century (left) reflected the areas with the lowest density of gamekeepers (centre). The map on the right, reproduced with permission from the British Trust for Ornithology from the* Bird Atlas 2007–2011 *(Balmer et al. 2013, a joint project between BTO, BirdWatch Ireland and the Scottish Ornithologists' Club) shows how Buzzards are now spreading back to the east.*

an heretical attitude at the time (Tubbs 1974). A. G. More's distribution mapping was reconstructed by Norman Moore for his 1957 article which, combined with later distribution maps, was used to show a strong correlation between the density of gamekeepers and a lack of Buzzards (Figure 11.1). When gamekeepers were conscripted to fight World War I and World War II, Buzzards started to spread from the west into central England and east along the south coast, but they remained absent, or very rare, in areas where there was more than one gamekeeper per 100 square miles. Other factors may have affected Buzzard populations, but the case that persecution had severely affected the species' distribution was compelling.

In the early 20th century, the collection of data on the human killing of Buzzards remained disparate and not very methodical. The establishment of the British bird-ringing scheme in 1909 introduced a more consistent method that meant numbers ringed and numbers found dead between years and regions could be compared more robustly. After the founding of the BTO in 1932, ringing became even more systematic and useful. An analysis by Newton (1979) showed that 48 per cent of 33 ringed Common Buzzards that had been found before 1954 were reported as killed, whereas after the initial protective legislation in the 1954 Act only 14 per cent of 173 recovered Buzzards with rings were reported as deliberately killed. Ironically, in 1953 Moore considered Buzzards to be one of the commonest birds of prey in the UK, after which their numbers apparently dropped, coinciding with the spread of myxomatosis, which demolished the Rabbit population. As Rabbits are important as a food source for producing young, but not so necessary for survival, the response of Buzzards to the spread of myxomatosis may show that their population was still being limited by persecution, because without Rabbit-enhanced productivity they appeared unable to persist in areas where they now thrive.

Even into the 1970s, the deliberate killing of Common Buzzards was still extensive in areas of Scotland where the species had persisted during the eradication from much of England. Picozzi & Weir (1976) looked at causes of death of 52 Buzzards in Speyside, Scotland, between 1964–1972. They had been informed about particular patches of land where it seemed Buzzards were being poisoned, so they set about investigating those areas more thoroughly by using dogs to find hidden carcasses. They found 28 poisoned birds (54 per cent of the 52 found) and eight (15 per cent) that had been shot. This was a high percentage of deliberate killing, and it was by finding out what was happening on neighbouring estates that established the true extent of Buzzard extermination. There was confirmed poisoning on two estates of 600ha and 800ha, where two keepers admitted killing 75 and 84 Buzzards, most in early autumn. They were also told of two other estates where 24 and 40 individuals had been killed. So, in one year at least 223 Buzzards had been killed deliberately in relatively small areas of Scotland. In 1968–1971, they established that alphachlorolose and mevinphos were used as poisons on 12 of 15 estates within 30km of their study area. On four estates they found 28 dead Buzzards near baits. Before the poisoning started (1966–1969) there were six pairs each spring, fledging an average of 2.3 young, but subsequently there were only four pairs, and an average of 0.5 young fledged per year.

There was a very similar history of raptor persecution in other European countries (Newton 1979b), if not everywhere on the Continent, with bounties still being used to encourage the killing. Buzzards feature in bounty lists in Norway and the Netherlands in the 1800s, while between 1948 and 1968 Austria was officially killing between 3,500 and 7,500

Buzzards a year. Viking Olsson (1958) found that out of 473 recovered carcasses reported in Sweden, 42 per cent had been shot and another 20 per cent had died due to other human activity. Few people defended the Common Buzzard as a benign species or one that was useful to humans, but a handful of supporters did emerge from the mid-20th century. Dare (1957) pointed out the potential merits of Buzzards for finishing the job that myxomatosis had started, by finding the remaining few Rabbits to kill. In Poland, Truszkowski (1976) suggested that Buzzards were not only having little effect on gamebird numbers, but were also an important control for voles. He calculated that 56,157 voles were taken from a 30km^2 study site in six months alone, between October and the end of March 1970.

What has happened in the years since Colin Tubbs' historical account, and Ian Newton's assessment of raptor persecution in Europe at the end of the 1970s? Elliott & Avery (1991) made a widespread investigation into reports of Common Buzzard persecution in Great Britain between 1975 and 1989, and found that 238 were recorded to have been killed illegally. Given that Picozzi & Weir (1976) estimated that 223 Buzzards had been killed in only three years in a much smaller area of Scotland, this is a pretty low number. It is likely Elliott & Avery's recording may have substantially underestimated the total being killed as they had purposely omitted any unproven records to give a minimum estimate. Illegal killing may already have started to decrease south of the Scottish border, as indicated by Newton (1979), and the highest number of reports came from Scottish regions in Elliot and Avery's review. Without having a comparable recording of the number of Buzzards dying of other causes, one-off surveys of illegal killing only indicate that the crime persists, and generally do not reveal the magnitude of its effect for the populations concerned. Reports of killed Buzzards were more common on the eastern edge of the known Buzzard distribution than in areas central to the distribution, further west (Elliott & Avery 1991). There could have been various reasons for that, such as higher human density resulting in more finds, or because people were more likely to report dead Buzzards where they were an uncommon species, or due to less illegal killing in Wales. However, the rapid expansion of Buzzards eastwards since the survey, along with the historical evidence, supports their notion that recolonisation was still being hindered by human killing on the eastern front in the 1980s.

David Gibbons *et al.* (1995) revisited Norman Moore's relationship between density of Buzzards and gamekeepers for the uplands only, identifying suitable upland habitat using the ITE land classification scheme (Chapter 4). They found that, in keeping with Moore (1957), the presence of grouse moors was the most consistent association with the absence of Common Buzzards and Northern Ravens, along with sheep-stocking density and coniferous woodlands. They were careful to clarify that the relationship did not necessarily prove a causal link of persecution leading to absence; there could be alternative explanations, so they considered the data further. Theoretically removing half the grouse moors and then predicting Buzzard densities from the other variables showed that Buzzards would likely occur in much greater densities throughout Scotland and northern England, so there was reason to consider that numbers were still being suppressed by illegal killing. Interestingly, the Buzzard distribution did reach the east of Scotland, seemingly just after the time when Larsen traps were legalised.

Larsen traps catch large birds, and they are aimed at corvids, especially Hooded Crows in Scotland and Magpies in England. The Larsen trap was approved under the UK Wildlife and Countryside Act for general use in England, Scotland and Wales during 1990–1991;

keepers could then legitimately kill the corvids, which were designated as pests, and release other species unharmed. This reduced the desire of some for illegal and unselective techniques, such as pole traps or poisoning, which had been used to kill species accepted as pests, but which also resulted in raptors and other birds being killed. As well as legalising Larsen traps, training for gamekeepers was increased at agricultural colleges during this period, and a strong stance against illegal killing of raptors has been taken by the National Gamekeepers' Organisation and the Scottish Gamekeepers' Association since both were founded in the 1990s.

A decade later, Sim *et al.* (2000) investigated the changes in persecution levels, particularly on the Welsh–English border (Appendix 3). Following up on Elliott & Avery's 1991 assertion that Common Buzzards were being prevented from recolonising to the east, they surveyed for soaring Buzzards during the breeding season to get some idea of the density of nests. They found that there was a good correlation between the soaring count and the number of pairs breeding, so they could then compare the 1996 survey data with that collected in 1983. In Wales, soaring Buzzards had increased by 118 per cent, and on the border of the known distribution in England, they had increased by 348 per cent. Moreover, the distribution was moving east. Some of the increase could have reflected increasing Rabbit density as the impact of myxomatosis tailed off (because Rabbits had developed some immunity to the disease). However, the extent of the increase suggested to them that something else was involved. Although it was difficult to assess the scale of persecution, the percentage of the population reported killed by humans had decreased by 75 per cent when comparing the years 1990–95 with the period 1975–89. The number of reported poisonings dropped dramatically after 1991, they thought due to a combination of higher penalties for offenders, the campaign led by the Ministry of Agriculture, Fisheries and Food (MAFF) in 1990 against illegal poisoning, and the associated legalisation of the Larsen trap (DETR & JNCC 2000). Supporting the surmise that allowing legal management of appropriate species could benefit conservation, a survey of government conservation departments across Europe around the millennium showed that poisoning of raptors was most prevalent in countries where there were fewest legal methods of pest management (Kenward 2004).

Thus, there was evidence that a sea-change in both attitudes and permitted management techniques was closing the door on the persecution of raptors, in terms of the intention of eliminating them from large areas. The Convention on Biological Diversity (CBD) (United Nations 1992) had not only put a focus on maintaining species diversity, but also made the sustainable use of wildlife resources, such as abundant game species, respectable among conservationists. Hunting organisations made it very clear that persecution of raptors was abhorrent and that any residual destruction of raptors by ignorant or law-breaking management should instead be called illegal killing. While we know that occasional illegal killing persists to this day, the signs are that it had reduced sufficiently by the 1990s for Common Buzzards to build in numbers and start recolonising the east of England.

Indeed, from our study in the 1990s we found illegal killing to be responsible for a relatively small proportion of deaths (Chapter 9, Figure 9.3). At the start of our work, it accounted for only 24 per cent of deaths, primarily in the first year of life when mortality over the year was about 33 per cent. Thus less than 8 per cent of the first-year cohort was killed illegally initially; this probably represented approximately 1.5 per cent of the study area's total population. The occasional illegal killing (e.g., Plate 22; see colour section) was far from preventing our study population from increasing. On the other hand, Alan

Stewart's book *Wildlife Detective* (Stewart 2007) provided many examples of continued illegal killing of Buzzards in Scotland at the turn of the 21st century. He gave an interesting insight into why the tradition of killing might be continuing, by recounting how some farmworkers with 'an abundance of anthropomorphic sentiment' described the horrors of a Northern Lapwing or leveret being caught by a Buzzard and screaming like a baby. The increasing human empathy for wildlife is heartening, but it can work both for and against predatory species.

In the UK as a whole, there is very clear evidence that persecution was previously high enough to eliminate Buzzards from considerable areas, as was likely in more populated areas of western Europe. There is a rationale to extreme management of this type, which was used for large mammalian predators as well as their avian cousins, in that eliminating a species entirely is in the long run less work than continuous control. That may be an acceptable practice for weeds in gardens, but is certainly not acceptable for maintaining countryside biodiversity. It is better recognised today that the riches of nature are there for many different interests, now that more of us have leisure time to enjoy the countryside.

The UK has not been the only country to change its attitudes to wild raptors in recent decades. People and laws in other European countries have also become much more conservation oriented. Pan-European legislation dating from the Bern Convention in 1979 has had an effect, as has withdrawal of organochlorine pesticides, so that raptor populations have increased in total density (including non-breeders), and are even recolonising some areas from which they were absent, as we saw in the last chapter. For example, the Canary Island Buzzard subspecies *Buteo buteo insularum* on Fuerteventura was on the brink of extinction in the 1970s, when there were fewer than ten pairs. In 1988, the population there was calculated at 15–20 pairs, and by two decades later there were more than 85 pairs (Palacios 2004).

Current attitudes

In the last few centuries humans have had an enormous impact on Common Buzzard populations. How are we treating the species now, and how might we influence it in the future?

Although persecution of raptors has become totally unacceptable in most of western Europe, appreciable illegal killing continues in parts of the Common Buzzard's distribution. For example, Stauros Kalpakis *et al.* (2009) studied 2,829 buzzards (Common and Long-Legged) admitted to a wildlife hospital in Greece during 1996–2005. An amazing 75 per cent of the Common Buzzards and 70 per cent of the Long-legged Buzzards had been shot (about 84 per cent of the adult Common Buzzards, and around 70 per cent of the younger birds). Large migrating birds have been used indiscriminately for target practice in some countries, such as Malta and Georgia. Although responsible hunting organisations have cooperated to reduce such behaviour, and the Federation of Associations for Hunting and Conservation in the EU (FACE) has worked with BirdLife under a memorandum of understanding signed with the European Commission, controversial local cultural issues can be slow to change. This will continue to affect raptor distributions in some regions, and sometimes even on a multinational scale for migrants.

In the 21st century, Common Buzzards have spread back into almost all the areas of the UK from which they were once extirpated. If all suitable habitat reached densities of the 120 pairs per 100km² found in Somerset (Prytherch 2013), then the British population could be nearer 150,000 territories, compared with the 44,000–60,000 estimated at the turn of the century (Clements 2002). Of course, we know that territorial pairs are only part of the population; the real population contains many juveniles and non-breeders, so 150,000 pairs might represent more than half a million Buzzards in the UK alone after a successful breeding season. Although that seems unlikely, because Common Buzzards will not reach 120 pairs per 100km² in all areas, it shows the potential for the population to continue increasing for a long time to come. It takes time to recolonise because of the relatively short distance the majority of Buzzards disperse, but the philopatric tendency of most to move only short distances also helps the density of the population to increase locally, expanding breeding areas on a slower 'rolling front' rather than quickly flooding out into the low density areas, as noted with Red Kite recolonisation (Newton *et al.* 1994).

Understandably, although a dramatic increase in Common Buzzard densities will be regarded as a conservation success by many, it can also result in a backlash from game managers who remember the 'good old days' when they didn't have to put up with so many raptors causing them additional hard work and economic impact. Dare (2015) describes how Buzzards were initially accepted by landowners and farmers when they returned to Suffolk, but within five years they had started to raise concerns about predation risk with him. With money and livelihoods at risk, politicians get involved. BirdLife Hungary elected the Common Buzzard as bird of the year in 2012, to raise awareness of a campaign by hunters to control the species, on the basis of a belief that its population had increased dramatically. Tibor Csörgő *et al.* (2012) used the Common Bird Census to show that the population had remained remarkably stable in Hungary since 2000, and had only increased by 20 per cent since 1999, at least as far as breeding pairs were concerned, so the call for control was unfounded. Even in the UK, where the majority of people have enjoyed the return of the Buzzard to their local area, there are still a few with anachronistic views. More usually, hunters enjoy seeing a Buzzard or two, though some may also consider the balance has swung too far. They may have a point, not only from a hunting standpoint, but also from a conservation perspective.

Possible impacts from the current suite of predators on ground-nesting game in the lowlands have been minimised by hunters putting protective pens around Ring-necked Pheasant poults until they have grown too large for some predators, and roost out of reach of others. However, these pens act as magnets for some raptors, which then kill the Pheasants while they are still vulnerable as small poults. In 2012, the British Department for the Environment Food and Rural Affairs (DEFRA), proposed a study on the effect of catching Buzzards near Pheasant pens and disturbing nearby Buzzard nests to reduce breeding. A debate ensued through the press and social media. On one side, the debate was about the ethics of killing an impressive-looking native raptor, which many people were beginning to enjoy again, in order to safeguard non-native gamebirds for a smaller number of people to enjoy shooting. On the other side, there was concern about livelihoods in rural communities.

To put some perspective on that, Pheasants contribute most to the total avian biomass in Great Britain (30 per cent), due to the release of between 20 million and 35 million individuals a year (Dolton & Brooke 1999, Park *et al.* 2008, Lees *et al.* 2013). Common Buzzards are hardly a major cause of Pheasant deaths. Our own study found that, according to keepers' estimates, an average of 4.3 per cent of their Pheasants were killed by Buzzards,

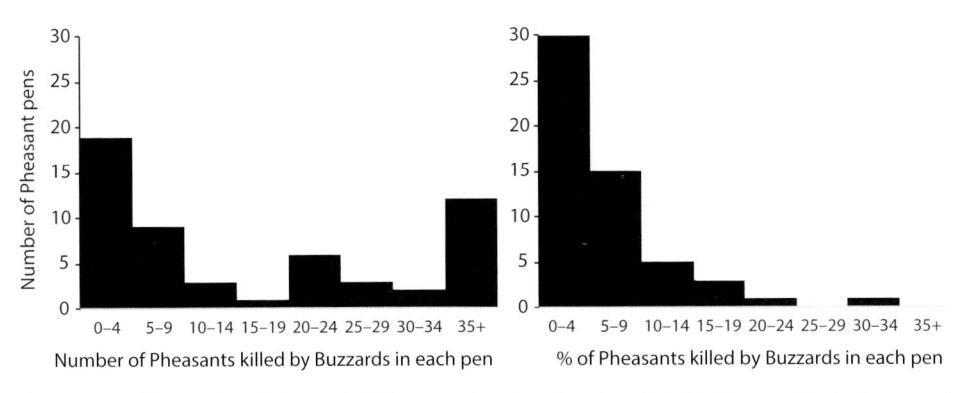

Figure 11.2. *The number of Ring-necked Pheasants thought to have been killed by Buzzards in the Dorset study area at 27 pens during 1994 and 28 pens in 1995 (left), and as a percentage of those released at the pens on each of the 55 occasions (right). Data from Kenward* et al. *(2001) with kind permission from John Wiley and Sons.*

although the estimate exceeded 10 per cent at 4 of 28 pens annually (Figure 11.2). In comparison, 3.2 per cent of the Pheasant poults were considered to have been killed by foxes, 1.4 per cent by other raptors and less than 1 per cent by other mammals (Kenward *et al.* 2001b). Among more than 100 young Buzzards that we tracked through the autumn, only about one-third had Pheasant pens in their home ranges and just 11 per cent were substantially associated with pens. Especially useful information was that heavy Buzzard predation was predicted by poor ground cover in pens, good perches in deciduous trees overhead and a relatively low density of released poults in the pens.

With so many Buzzards but relatively few likely to be taking Pheasants, widespread killing is neither a legal nor an efficient way of removing the few birds responsible (Arraut *et al.* 2015). The radio-tags actually enabled us to identify potential individuals, by assessing their attraction to the pens in a similar way to our assessment of their proximity to nests in Chapter 5. Among 136 home ranges of Buzzards that we tracked between August and October in 1990–95, 100 (74 per cent) had no pen in their home range, and another 25 (18 per cent) were recorded less than 20 per cent of the time within 200m of a pen (Figure 11.3). However, the remaining 11 (8 per cent) either spent more than 20 per cent of their time near pens, or showed much greater than random attraction to pens, with three spending more than 50 per cent of their time within 200m of the Pheasant pens; these three may have been subsisting on Pheasants. These birds were a small proportion of the local Buzzard population.

The example of a keeper convicted for the illegal killing of wildlife, who had kept a record of removing 100 Buzzards in six months (Lees *et al.* 2013) is therefore especially disheartening, not least for hunting organisations wishing to continue their sport and knowing that these unnecessary cases sway public opinion against them. We are way beyond a stage where complete eradication of Buzzards is acceptable, so killing some only results in others arriving. Steps can be taken to hinder Buzzard hunting at Pheasant pens, including increased ground cover and removal of nearby perches from which to launch attacks (Kenward *et al.* 2001) or covering pens. For the keeper whose livelihood is at risk due to an especially persistent Buzzard, and where no other alternatives are available, the ability of European law to accommodate selective removal is a pressure-release valve, even if it is not a panacea for all raptor predation issues.

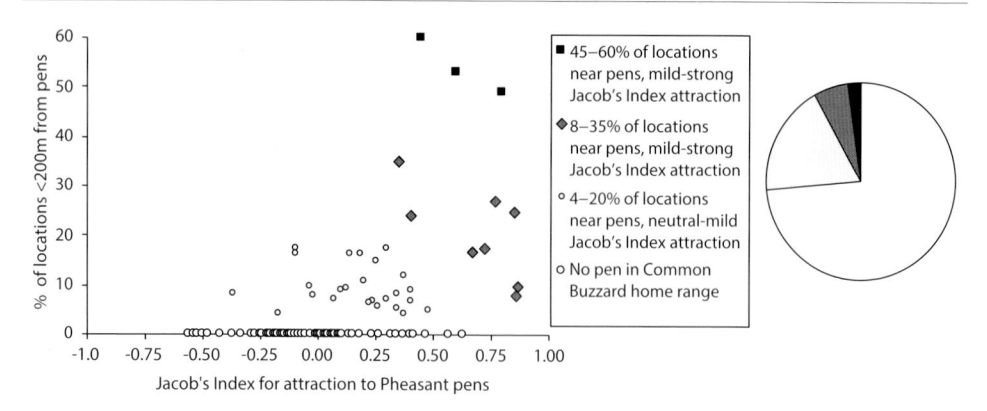

Figure 11.3. *The pie chart shows that only a small proportion of 136 young Buzzards associated appreciably with Pheasant pens in our area. Dark shading indicates a cluster of eight birds either spending 25–35 per cent of their time near pens or with strong values of Jacob's Index for attraction to pens, and a cluster of three spending 49–60 per cent of their time near pens and with similar high values of Jacob's Index. Data from Kenward et al. (2001) with kind permission from John Wiley & Sons.*

Moreover, other studies have found that Red Foxes are a much worse problem than most Buzzards. For example, in one study 23 per cent of 486 radio-tagged Pheasants were killed between release and the start of shooting (predominantly by foxes) and another 13 per cent predated during the shooting period, but only three (0.6 per cent) were thought to have been killed by raptors (Turner & Sage 2003). Overall, therefore, the losses from Buzzards are not great, and are probably mostly preventable by improving pen characteristics. It is also worth remembering that relatively high proportions of released Pheasants are killed on roads (Lees *et al.* 2013, Arraut *et al.* 2015).

Although a public uproar obliged withdrawal of the 2012 DEFRA proposal for experimental management of Buzzards at Pheasant pens, there was anger from the Countryside Alliance and other rural stakeholders that the idea was so disliked by many who were not affected and were considered not to have looked at any evidence. The following year, Natural England granted licences to remove four Common Buzzard nests and eggs from near a Pheasant shoot, and to trap and remove five birds from a poultry farm. The killing of hens stopped after the second bird was trapped, the female of the breeding pair (Graham Irving, pers. comm.).

Government departments for the environment are among those most aware of international thinking on conservation because their delegates attend regular conferences of parties to multilateral agreements, such as the Convention on Biological Diversity. Rather than just considering species, the CBD emphasises treatment of land as ecosystems on which populations of species depend. It mentions protecting species just once, while defining ecosystems and protected areas and mentioning these ten and five times respectively. An understanding that humans as well as other species are dependent on ecosystems and that we are important managers of them was reflected in CBD's Ecosystem Approach (1998) to conservation, and subsequently in the Millennium Ecosystem Assessment (2004), which defined 'supporting', 'regulating', 'productive' and 'cultural' services of ecosystems for humans. Biodiversity is important for all these ecosystem services. A government department that is managing human activities, and the land on which they occur, appreciates the need to consider all these services, including among them the recreational cultural services. These cultural services include both the aesthetic appeal of species for many people, and also the practical values for hunting, fishing and gathering, which include productive values too, for example, hunting for venison, fishing for wild salmon, gathering blackberries and chanterelles, and harvesting straw for thatch.

Therefore, a government department has to consider that there can be conservation benefits from Pheasant shoots, such as maintaining woodland for amenity and carbon sequestration, as well as predator control. A recent review of the scientific literature, investigating the effects of gamebird management on non-game species, showed that while there were some detrimental effects there could also be significant benefits, predominantly through predator control (Mustin *et al.* 2018). Common Buzzards are not the only species worth conserving, so we do need to consider the wider context. Controlling some of nature's predators can be helpful for fast-diminishing species, which may benefit from mammalian predator control and also if there are fewer Buzzards that take their nestlings (Chapter 2). Not all gamebirds are introduced, and some are in serious decline in the UK, such as the red-listed Grey Partridge and amber-listed Red Grouse. These smaller game species are easier for Common Buzzards to prey on, potentially affecting their natural populations. The question of a need to remove individual predators for conservation purposes has not been considered compelling for Common Buzzards yet, but could be reconsidered in the future for conservation purposes, as opposed to the prevention of serious damage to livestock.

Bringing the situation closer to home for many who live in towns and cities, it will be interesting to see how the urban dweller deals with Buzzards if they colonise suburbia to any great extent. These days, with little illegal killing outside of towns and cities, and none within, they may do well in such areas. Might some who now condemn killing raptors to protect gamebirds become averse to Buzzards in towns? Although some of us are elated and awe-struck to see Eurasian Sparrowhawks snatching birds attracted to bird tables, others find themselves passionately angry. A bird rescue centre once told us that they would not treat injured Sparrowhawks because they'd had to look after so many birds that had been injured by them and had seen the damage this raptor inflicts. It is natural for us to feel affronted when witnessing corvids taking passerines from nests we had enjoyed in our garden, or gulls stealing the bacon rind from garden feeding stations designed to attract songbirds. Is there a chance that some small dogs could be targeted by Buzzards that might not differentiate them from Rabbits? We have heard that Buzzards will dive on small dogs in the countryside; if they get used to humans, will that happen in gardens? We already see some conflict in the national news when Buzzards become bold enough to strike the scalps of people within their territories. While cases involving Buzzards are less frequent than the drama associated with gulls stealing chips from seaside trippers, there are cases of Buzzards attacking humans that have been badly reported, causing views to polarise. For instance, a BBC report of one incident (Maguire 2015) appeared to use the head of an eagle rather than a Common Buzzard to illustrate the piece. So, with the media ready to construct a story about Buzzards attacking humans, but not informed enough to recognise the difference between species, will a misinformed public know or care enough? How will we treat Buzzards in our very artificial environment? Could they become painted as villains, in the same way as urban gulls have been, leading once again to direct action against them, as in the past? Such questions will be answered as time goes by.

Attitudes are often swung by emotive reporting, including on social media, and news programmes are happy to have anecdotes to bring people with polarised views together for entertainment. It is rare that substantial evidence is considered in such debates, and to be fair there is often not enough clear-cut evidence, which might in any case make a story too complex even for a documentary. However, robust evidence is important if we are to manage the landscape for conservation and livelihoods, and that needs the backing of a majority that is informed, not swung by emotions. This brings us to an interesting experiment in Scotland that sheds light on how complex conservation can be, but also indicates some possible solutions if people are willing to consider them.

Raptors or grouse? Both or neither?

Scotland now has experimental evidence for what can happen on grouse moors in the absence of raptor persecution. Even though this may seem like a diversion from Common Buzzards, it is worth exploring this particular study because it highlights the issues we need to address for long-term conservation. Also, although the project had the higher-profile Hen Harrier for a flagship species initially, as the experiment continued the Buzzard's role has attracted increasing attention.

In 1992, the Joint Raptor Study was set up on the Langholm Estate in Scotland to investigate the impact of birds of prey on Red Grouse. It was led and documented by Steve Redpath and Simon Thirgood (Redpath & Thirgood 1997). Hen Harriers were the target raptor study species, being both rare (certainly compared with Common Buzzards) and more associated with taking grouse. The idea was to protect raptors effectively on Langholm and assess what happened to Red Grouse numbers on that estate compared with other moors nearby. The project needed to be run over a number of years because Red Grouse undergo population cycles of boom years with high populations due to successful breeding, followed by low years when breeding is very poor, sometimes due to weather but also to parasites when conditions favour these. It was also important to allow time for the raptor populations to stabilise. The study identified several factors that affected the breeding raptors that established themselves on the moors, such as the amount of grassland compared with heather on the estate, which in turn affected other raptor prey such as Meadow Pipits and voles. Nevertheless, taking these and other factors into account, it was clearly apparent that raptors could suppress a grouse population, to the extent that there was no Red Grouse boom year where raptors were abundant, whereas on nearby moors with few raptors then grouse did reach the expected highs. The study showed that the lack of grouse recovery was primarily due to Hen Harriers eating a high proportion of grouse chicks. In fact, once Hen Harrier numbers had built up from the one or two nests present previously to 20 nests, then the shoot became economically unviable and shooting stopped. With no immediate management solution agreed, gamekeeping jobs were lost, which is the concern that has driven gamekeepers to reduce raptor numbers in the first place.

Some may say, 'Never mind a few jobs, I don't shoot grouse and I'd rather watch raptors'. However, the story doesn't stop there. If grouse shooting is unviable, the main alternative land uses are forestry and sheep grazing. After grouse shooting ceased on Langholm, sheep and feral goats then increased in numbers, converting more heather to grassland. However, without predator control from gamekeepers there was also a considerable increase in the number of Red Foxes, major predators of breeding Red Grouse. This meant that not only was the habitat becoming poor for grouse, but also that predation by mammals increased for ground-nesting raptors, including Hen Harriers. Unlike a grouse, which incubates briefly on a well-hidden nest and then disperses with its brood, raptors have nests over which they obviously display, and then follow incubation by a long rearing period with smelly food remains and large chicks in the nest. What better signal to attract a fox with kits to feed? With enhanced fox predation, the breeding pairs of Hen Harriers dropped back to the original one or two pairs. This was bad news not only for those who enjoyed shooting and the wider economy that it supported, but also for raptorphiles.

After ten 'wilderness years', during which the estate was not managed for either grouse or raptors, the Langholm Moor Demonstration Project was started in 2007 to search for an effective means of resolving the controversy about raptors on grouse moors by restoring grouse moor management. The usual methods of heather burning (and mowing), and bracken and livestock control were practised, along with predator control of other 'pest' species such as Red Foxes, Stoats and Carrion Crows. To reduce predation by Hen Harriers, diversionary feeding by means of putting food (day-old chicks and laboratory rodents) near their nests was used to reduce the need to hunt grouse. This method had been tested and found to substantially reduce the number of grouse chicks fed to nestling harriers, although there is a cost in terms of both the food and labour to provide it (Ludwig *et al. n.d.*, Redpath *et al.* 2001).

However, it turned out that Hen Harriers were not the only raptor issue for grouse on Langholm Moor. Solving the Hen Harrier issue with diversionary feeding focused attention on the fact that, even with these mitigating measures, winter Red Grouse mortality was as high as if there had been shooting (Ludwig *et al.* 2017). This contributed to a dynamic whereby the grouse population could only recover from the low level it had reached if grouse breeding success considerably exceeded observed average productivity. Comparison of different survey techniques showed 75–80 per cent of the winter grouse deaths were associated with raptors – but which raptor species was predating the chicks? The number of Peregrines and Northern Goshawks being seen in standardised watches at Langholm had, if anything, tended to decrease, but numbers of Common Buzzards had increased. There were not many more nests, but Buzzards were by far the most abundant raptor seen on the moor. Re-sighting of marked Buzzards suggested that about 60 were hunting on the hills, which matched the 12 nests within the study area plus non-breeders (though a smaller proportion of the latter than in Dorset). Prey remains at Buzzard nests and in pellets from winter roosts indicated that most adult pairs and about half the non-breeding Buzzards were taking grouse (Figure 11.4), but only occasionally in the case of each bird (Francksen *et al.* 2016b). So, although Peregrines took appreciable numbers of grouse, it seemed that Buzzards were also responsible for the poor recovery of grouse stocks when foxes and corvids were removed.

Buzzard numbers were boosted at Langholm through the availability of nest-sites and roosts in plantations, tree-filled gullies and other natural woodland around the edge of the moorland, which was on relatively isolated high ground (Plate 20; see colour section). These Buzzards were most abundant on the hillsides in years with high vole populations, and in areas with most grass (Francksen *et al.* 2017). Unfortunately, following the removal of sheep, the heather was slow to recover from the previous heavy grazing, with grasses the first to benefit, which increased food for voles. Thus, ironically the steps taken to restore heather in the long term may in the medium term have made the moor maximally attractive for

Figure 11.4. *A plucking post with Red Grouse remains under a Buzzard nest at Langholm (left), and Richard Francksen collecting pellets at a roost site of a satellite-tagged non-breeder (right). Photos © Robert Kenward.*

Buzzards hunting voles, but reaching numbers high enough for each Buzzard's occasional grouse kill to impact the recovering grouse population. With another decade of fox control and heather restoration, perhaps grouse shooting could again have supported the moorland management. However, few landowners would tolerate such a slow economic return from restoring moorland for grouse shooting after heavy sheep grazing, unless raptors could also be managed. With the Langholm Estate now returned to an unkeepered condition, and a Hen Harrier nest already predated by a resurgent fox population, it seems likely that the recovered populations of Hen Harriers, Merlins *Falco columbarius* and Short-eared Owls, will suffer the same downward trajectory after the Langholm Moor Demonstration Project, as happened after the Joint Raptor Study. This time the problem will not be due to Hen Harriers. Indeed, diversionary feeding of small Hen Harrier populations on grouse moors could be a welcome solution for reducing the rarity of this species, assuming foxes and other mammalian predators are controlled there.

Perhaps cutting down the small areas of woodland used by Common Buzzards for nesting and roosting, as well as working on heather restoration in ways that are less favourable for the voles they predate, might in time help restore grouse shooting on Langholm Moor. However, the habitat management needed is reminiscent of a paradox that struck us when working on Grey Squirrels. Grey Squirrels can damage young Beech trees by stripping bark so severely that no beautiful Beech wood, or fine carbon-locked timber can be grown. Ideally, in order to retain our native Beech woods, the damage by Grey Squirrels should be managed, preferably without the considerable cost of killing squirrels each year during the decade or two in which young trees remained vulnerable. We identified that a solution was to grow the Beech as even-aged monocultures isolated from other woods.

A few woods grown already in that way had negligible squirrel damage. However, and here's the paradox, such isolated monocultures of Beech support little biodiversity compared to a mixture with oaks and other species (Kenward *et al.* 1992). So which is best? To have landscapes managed both for harvestable productivity and high biodiversity, but with management also of pest species, or to manage for harvestable productivity alone and without the need to manage problematic species, but without much other biodiversity? Removing woodland from the edges of high ground, in an attempt to reduce Buzzard presence, is also not environmentally desirable because it is liable to increase soil erosion and flooding downstream.

It isn't merely the Hen Harriers and other ground-nesting raptors that can benefit from predator control on grouse moors, but also the waders that breed and maintain populations most successfully in areas with few generalist predators (Fletcher *et al.* 2010, Franks *et al.* 2017). Where raptors are the major cause for making game shoots unsustainable, we have choices:

1. Lose the economic value of game rearing and shooting, so that we either lose areas favourable for ground-nesting species (the current situation at Langholm) or have to find money from elsewhere to pay for conserving them. Who is going to pay? Taxpayers, philanthropists, donations through bird charities? Can that funding be relied upon to be supplied consistently and forever over the large areas that are currently managed for grouse shooting?
2. Allow game rearing in some areas and encourage the conservation of raptors in others. This can set up a source-sink situation, which is not so easy to manage with such a mobile species; and how do you know which estate's bird you are controlling? Perhaps large-scale zoning may work, as for large carnivores (Linnell *et al.* 2005), but there will inevitably be people unhappy to be in the 'wrong zone'.
3. Create a quota system for breeding raptors, such that numbers are managed to conform to an area-based quota of nests, agreed on a case-by-case basis with all appropriate stakeholders (local for estates, national and regional for wider areas).

Although the quota system (Potts 1998, Elston *et al.* 2014) is intolerable for some, it may be the best compromise in terms of economic cost and affordability if people want to see more raptors in the north of England and in Scotland without losing other wildlife. We know that this idea will still be met with condemnation by some, because the Hawk and Owl Trust endured opprobrium in 2015 when supporting a scheme for brood management. However, the case of Red and Black Grouse is different from that of Ring-necked Pheasants because along with Hen Harriers and ground-nesting waders, the grouse are residents rather than introduced species that are farmed and put out in high densities. The grouse would have lower populations without habitat management, predator and disease control, and would be absent in forestry, but nevertheless we have a moral duty to conserve their populations, and hunting interests can probably do more in that respect than any other body at the moment.

It is important to note that the quota system suggested above cuts both ways. As well as permitting some management for the benefit of game and other species vulnerable to predation by raptors, it would impose on land managers the responsibility of accommodating a quota of raptors for the enjoyment of raptor enthusiasts and the maintenance of healthy ecosystems. Further agreements between all the conservation stakeholders could include the use of removal techniques that are selective and avoid shooting predators, which can put non-target species at risk.

For example, the primary losses due to Hen Harrier predation is of grouse chicks during the breeding season, whereas males and some female harriers migrate from the moors after breeding. A responsibility for diversionary feeding for quota pairs should accompany any removal of broods above quota, thereby not only reducing the need for adults to hunt grouse but also providing chicks to start or boost populations elsewhere. In the case of Buzzards, an accumulation of non-breeders hunting on Langholm Moor in winter could have been reduced through release elsewhere of those caught live in authorised cage traps, set with carrion for crows. It was extremely rare to catch adult Buzzards in cage or spring-net traps (Figure 11.5), so adults (and young Buzzards not hunting the moorland) would remain for those more keen to see raptors than to hunt grouse.

There are two very important research-based principles at work in these quota proposals. One is that they select the young sector of the predator population that can best sustain harvest. Yield modelling shows that removing young raptors on a local scale from a healthy raptor population will reduce the total population locally but not the breeding population (Kenward *et al.* 2007). The techniques proposed above cannot eliminate the adults and so,

Figure 11.5. *Setting a spring-net to catch a Buzzard for tagging. Even with superb camouflage of the trap's rectangular outline, adults are warier of traps than are younger birds (© Robert Kenward).*

because areas with intense game conservation are a minority of the landscape, the methods are unlikely to have an appreciable effect on the total breeding population, due to immigration of young birds from elsewhere that can replace adults when they die. The second principle is that, by not allowing shooting of the predators, not only is more labour required to manage the raptors, but it would also be necessary to care for the raptors prior to release and thus build an empathy with the birds, thereby making it psychologically harder to kill other raptors. It is also key that for a quota system to work, illegal killing must stop completely; otherwise the mortality rate estimates necessary for model-based regulation become unreliable.

The issue here is both morally and scientifically complex. If we are trying to increase the paltry Hen Harrier population for the least public cost, leaving grouse shooting to maintain the moorland ecosystem, and keeping in mind that Common Buzzards are widespread and abundant, do we selectively target Buzzard nests and leave the harrier nests? How do members of conservation bodies really feel about active management of raptors? Does the polarisation of opinions by narrow interests make pragmatic decisions so difficult that all we can do is resort to the law to stop all illegal killing (quite an economic investment in itself), make grouse shoots uneconomic and start to lose habitat that is crucial for some of our species of conservation concern? These tricky problems are taking a long time to resolve. Conservation translocation of Hen Harriers has been agreed in the UK's Hen Harrier Action Plan, and licenced, but currently it has not been started as we finish this book, so we wait with great interest to see what the outcome will be. What is crucial is that we continue to look objectively at the evidence. It is easier to be swayed by photos of dead raptors with evidence that they have been killed, than to spend time reading the drier scientific publications that assess such mortality as negligible at population levels, and perhaps even positive socio-ecologically (Kenward 2009, Webb 2014). It is important to remember that peer-reviewed publications, containing objective evidence that support different sides of the debate, help open the door to compromises.

Modern conservation

Conservation involves not only protection and evidence-based management, but also education to demonstrate the benefits of maintaining biodiversity and ecosystems. We live in an ever-changing world, with many new challenges as technology facilitates humans' ability to exploit, and sometimes ruin, our natural environment. Thankfully, conservation is evolving too, and new technologies can be used to enhance not only our biological knowledge, but also the impact of the conservation message.

Throughout this book we have shown how electronic tagging of Common Buzzards has allowed us to gather more realistic data about their survival, foraging and long-distance movements in the field; as those tracking devices get better, more detail will be added. In the laboratory, our ability to analyse species from a genetic basis is useful not only for recognising evolutionary origins, but also for conservation, primarily by helping us to understand more clearly what we are trying to conserve. So far, it has been difficult to separate the Common Buzzard genetically from other buzzard species of the Old World (Chapter 1). Moreover, there is increasing awareness of naturally occurring hybridisation between buzzard species. If we are content just to see an abundant large brown hawk that we call the Common

Buzzard, we will likely remain happy. But if we are concerned about a species or subspecies with a particular genetic make-up, we could have a hard job conserving it, especially as it has such variable plumage that it is difficult to identify purely on a visual basis. The Common Buzzard that we will live with in future may not be the Common Buzzard as we know it now; we cannot halt continuing evolution. That does not diminish the importance of the Common Buzzard, but recognising the status of this species in an evolutionary context makes it more interesting to study for insight into evolutionary processes.

Another key modern technology is the internet. 'Citizen science' has already begun to use the incredible power of the internet and mobile communications to harvest the observations of volunteers. The analysis by Neil Paprocki *et al.* (2017) used citizen science to reveal changes in migratory behaviour and distributions of a North American *Buteo* (Chapter 10). Christmas Bird Counts, which started in 1900, showed that Red-tailed Hawks were migrating shorter distances, or even not migrating from areas where they had in the past, and their distribution was moving northward. We look forward to similar analyses from the millions of records submitted over the internet by volunteers throughout Europe, which are now being collated in the Euro Bird Portal. BirdTrack is a software application that has been designed to facilitate the submission of data on the presence of species and hence is able to record distribution changes. Anyone can sign up and download the BirdTrack app to record birds and their positions on smartphones. The data are then uploaded to a central database, which the organisers of the system can access and which produces some very useful information for birders wanting to know when and where to see particular species. The benefit to a bird-lister is great – it is much easier to use and more long term than a notepad and pen, plus the birder is contributing to conservation on a bigger scale while enjoying birding. The collection of data through this app has also played a forensic role in a case involving Common Buzzards. In 2013, 10 Buzzards and one Eurasian Sparrowhawk were found poisoned in eastern England, and a gamekeeper was taken to court. In his defence, the gamekeeper claimed that the poisoned birds must have been planted by people wanting to incriminate him, because he couldn't possibly have poisoned that number of birds given such low populations in the area. Historically, there were hardly any Buzzards in the area, but the BTO gave expert evidence based on the Bird Atlas 2007–2011 (Balmer *et al.* 2013) and data from BirdTrack at the time of the findings. This showed there were enough Buzzards around for him to have killed that number, and that helped secure the conviction. Such convictions are educational, as unselective poisoning is totally unacceptable.

Nevertheless, conservation education should primarily be positive rather than punishment. For example, let us consider the Batumi Raptor Count, conducted at the migration bottleneck between the Black Sea coast and the mountainous Lesser Caucasus to the east, in Georgia, where Steppe Buzzards and other raptors are illegally shot on their migration. Sandor *et al.* (2017) have explored the social drivers for this killing by interviewing hunters throughout the region. Although the authors appreciate that the results may not be representative of hunters as a whole in the Batumi region, it is good to see that the causes are being investigated with a view to tackling the problem with understanding. Of 43 hunters interviewed, 53 per cent admitted shooting raptors. Of these, 89 per cent said their primary purpose was fun and only 7 per cent claimed that their first reason was for food, although 51 per cent of all 43 hunters said they ate the raptors they shot. This indicates that although raptors are eaten, it is not 'subsistence hunting', as had previously been thought, but more of a sport. Using this insight, an appeal is being made to more law-abiding hunters to persuade their fellow hunters and to educate the young about other ways to appreciate the raptor migration. Education of the children in the area is crucial (Figure 11.6), so that even if the old hunters are not being persuaded very quickly, there is a generation growing up who will enjoy seeing the raptors and benefiting from them in ways other than shooting. Look what effect a combination of education and legislation has had in many countries. There is hope.

Such hope extends to even more direct management. In Finland, at the northern extent of the Common Buzzard's distribution, some people have actively tried to increase Buzzard populations. Due to the concern that older forests were being felled, reducing the number of potential nest-sites, bird ringers have been erecting artificial nests since the late 1970s (Saurola 1978). Of 434 breeding attempts on 221 artificial platforms, 382 were successful, although nest success (raising at least one young) was slightly lower on platforms than at natural nests (Björklund *et al.* 2013). This suggests that these platforms can be used to boost populations. As a widespread species, Buzzards are also likely to benefit from measures to make poles on power lines safe perches for other raptors, and where perches are deliberately provided in open areas to enhance predation by perching raptors on mammalian pests in forestry or agriculture.

Beyond radio-tracking, the internet and the citizen science that it has enabled, the power of modern computers also allows us to simulate the real world and to model what may happen if we make changes. That gives us an opportunity to consider the consequences of changing land use or laws. As discussed in Chapter 4, an impressively true-to-life model indicates how Common Buzzards could colonise a landscape, based on very few rules (Arraut *et al.* 2015). With remote-sensing of land use we can now predict potential Buzzard

Figure 11.6. *Educating children about raptors at Batumi. Photo © Batumi Raptor Count.*

densities across existing landscapes where they are not present, or across fictitious landscapes based on possible future changes in land use. We can add to those habitat-based models with models of the population structures (Chapter 10), based on telemetry projects that deliver rapid assessments of survival and breeding (Kenward *et al.* 2007).

Buzzards do not stand alone; models for other species can be built too, and combined. Therefore, societies are now in a much better position to predict impacts and hence consider the possible consequences of their actions. This should mean a much better chance of having reasoned debate, with tolerance of other interests that are capable of contributing to conservation. The exchange of citizen science data for model-based decision support on land use was planned in a recent pan-European project designed by IUCN and part-funded by the European Commission (Kenward *et al.* 2013). Partners in the project set up a 'Naturalliance.eu' website in 23 languages to explain how a wide variety of activities benefitting from wild species – from farming and fishing to gamekeeping, and gathering fruit and fungi – could contribute to the restoration of habitats across Europe. From this beginning has sprung cooperation between international bird conservation groups, falconers (Figure 11.7) and game conservation interests for global-with-local networking through similar multilingual portals. There is now also a Sakerfalcon.org for the Saker Falcon, and a Naturalliance.org. Such conservation collaborations are much better than conflict.

It is fantastic to see the conservation message growing and pervading not just schools and visitor centres, but our wider community. However, education can also get drowned out by lots of other competing voices, promoting polarisation for organisational and personal ends. Sadly, the long human evolution as a social animal has resulted in tribal tendencies as well as empathy. So, while we have the tools, we need the political will and investment, and that comes with contact and appreciation of wildlife. Birds of prey are perfect species for stimulating that enjoyment.

Figure 11.7. *The conservation contributions of falconers are recognised by IUCN and the European Union's Ornis Committee (Kenward 2009). Photo © Sean Walls.*

The importance of the Common Buzzard

Perhaps everyone reading this book would support the conservation of other, usually less common, raptors and previously persecuted mammals such as martens and Grey Wolves *Canis lupus*. Yet our efforts to improve the conservation of other raptors that predate Common Buzzards may reduce Buzzard densities through intra-guild predation (Sergio & Hiraldo 2008). It has been exciting to see Northern Goshawk and Peregrine numbers increase in the UK in our lifetimes, and so far Common Buzzard numbers have increased even more dramatically. However, they may reduce again once other raptors also regain peak densities in their past distributions, or build new distributions enabled by our changing landscapes. The pair of Peregrines in Exeter are certainly culling Common Buzzards unfortunate enough to enter their airspace.

People who actively enjoy birds can appreciate large gatherings of wildlife at important resource areas or migration bottlenecks, but usually we also get enthusiastic about seeing different birds, and value avian species diversity, too. Even more importantly, the health of ecosystems locally and regionally is dependent on biodiversity. So, a mixture of raptor species is a worthwhile aspiration, even if the overall density of each species is lower. Indeed, we need biodiversity through all taxa, so some impacts on the density of a particular species, unless it is critically low, is not necessarily something to be concerned about, whereas eradication of any naturally occurring native species is now of great concern.

Common Buzzard currently represent a success story, due to changes in attitude that have allowed them back into old haunts from which they had vanished. They are not just the most common buzzard species, but probably the most common raptor of their kind. Common Buzzards are so visible, being large and aerial, that they connect many people in Europe with a real large raptor. How exciting! Conservation works by connecting people with wildlife, and Common Buzzards can provide that all important 'Wow!' factor on which we can build the conservation of other species and a care for the environment. Common can mean abundant, and that certainly suits the Common Buzzard at the moment. The English word 'common' originated from an old French *comun*, meaning belonging to all, and should remind us of our responsibilities to look after wildlife for everyone to enjoy. The Common Buzzard is currently a species that can inspire us all, and we hope it will continue to do so while humans live on Earth.

Conclusions

1. Historically, humans have had a great impact on Common Buzzard populations, intentionally eradicating them from large areas.
2. More recently across Europe, as illegal killing has become less acceptable, we have seen Common Buzzards returning quickly to areas where they haven't been seen for generations, and local population densities have hit new highs.
3. Common Buzzards could become resident in suburbia, which will likely delight some people, but may cause conflict with pet owners and bird feeders.
4. People who manage gamebirds are having to readjust to increasing raptor numbers; the re-establishing Common Buzzard can quickly be considered too common by those labouring to make a financially viable shoot, and this causes controversy.

5. Assessing the impact of Buzzards and other raptors on gamebirds should be based on sound evidence and considered in the context of other causes of death for gamebirds.
6. Likewise, solutions for possible conflicts need to be evidence based, because controlling raptors locally may not change the situation if it simply results in the arrival of others.
7. Predator control for gamebirds does have benefits for other wildlife; society therefore needs to consider the end result before imposing restrictions on hunting, while the hunting community needs to appreciate the impact that irresponsible illegal killing can have on public opinion. Understanding and compromise by all interests, with recognition of their different contributions to conservation, are needed for practical solutions.
8. Modern technologies are improving our understanding of changing bird distributions, their densities, what constitutes a species and how to solve socio-ecological problems, thereby enhancing capacities for conservation management.
9. Endangered species are used to drive public opinion with simple protection messages, but are often only accessible for most people as digital images. In contrast, many people in Europe can enjoy the conspicuous and abundant Common Buzzard in real life, providing an inspiration for conservation closer to home.

Appendix 1
Scientific names of species mentioned in the text

BIRDS
Atlantic Puffin *Fratercula arctica*
Augur Buzzard *Buteo augur*
Bald Eagle *Haliaeetus leucocephalus*
Barbary Falcon *Falco pelegrinoides*
Barn Owl *Tyto alba*
Black Grouse *Tetrao tetrix*
Black Kite *Milvus migrans*
Black Sparrowhawk *Accipiter melanoleucus*
Blackcap *Sylvia atricapilla*
Black-tailed Godwit *Limosa limosa*
Booted Eagle *Hieraaetus pennatus*
Broad-winged Hawk *Buteo platypterus*
Cape Verde Buzzard *Buteo bannermani*
Carrion Crow *Corvus corone*
Common Buzzard *Buteo buteo buteo*
Common Chaffinch *Fringilla coelebs*
Common Cuckoo *Cuculus canorus*
Common Kestrel *Falco tinnunculus*
Common Swift *Apus apus*
Crane Hawk *Geranospiza caerulescens*
Crested Caracara *Caracara cheriway*
Domestic Chicken (Red Junglefowl) *Gallus gallus domesticus*
Domestic Feral Dove *Columba livia domesticus*
Eastern Buzzard *Buteo japonicus*
Eurasian Eagle Owl *Bubo bubo*
Eurasian Scops Owl *Otus scops*
Eurasian Sparrowhawk *Accipiter nisus*
Eurasian Woodcock *Scolopax rusticola*
European Blackbird *Turdus merula*
European Honey-buzzard *Pernis apivorus*
European Starling *Sturnus vulgaris*
Ferruginous Hawk *Buteo regalis*
Forest Buzzard *Buteo trizonatus*
Galapagos Hawk *Buteo galapagoensis*
Golden Eagle *Aquila chrysaetos*
Great Black-backed Gull *Larus marinus*
Great Tit *Parus major*
Grey Heron *Ardea cinerea*
Grey Partridge *Perdix perdix*
Grey Phalarope *Phalaropus fulicarius*
Griffon Vulture *Gyps fulvus*
Gyrfalcon *Falco rusticolus*
Harpy Eagle *Harpia harpyja*
Hazel Grouse *Tetrastes bonasia*
Hen Harrier *Circus cyaneus*
Herring Gull *Larus argentatus*
Himalayan Buzzard *Buteo burmanicus*
Hooded Crow *Corvus cornix*
House Sparrow *Passer domesticus*
Indian Peafowl *Pavo cristatus*
Jackal Buzzard *Buteo rufofuscus*
Jay *Garrulus glandarius*
Lesser Kestrel *Falco naumanni*
Long-eared Owl *Asio otus*
Long-legged Buzzard *Buteo rufinus*
Madagascar Buzzard *Buteo brachypterus*
Magpie *Pica pica*
Mallard *Anas platyrhynchos*
Meadow Pipit *Anthus pratensis*
Merlin *Falco columbarius*
Montagu's Harrier *Circus pygargus*
Mountain Buzzard *Buteo oreophilus*
Mute Swan *Cygnus oler*
New Caledonian Crow *Corvus moneduloides*
Northern Goshawk *Accipiter gentilis*
Northern Lapwing *Vanellus vanellus*
Northern Raven *Corvus corax*
Osprey *Pandion haliaetus*
Ostrich *Struthio camelus*
Peregrine Falcon *Falco peregrinus*
Pied Avocet *Recurvirostra avosetta*
Pied Flycatcher *Ficedula hypoleuca*
Pied Wagtail *Motacilla alba yarrellii*
Red Grouse *Lagopus lagopus scotica*
Red Kite *Milvus milvus*
Red-breasted Goose *Branta ruficollis*
Red-legged Partridge *Alectoris rufa*
Red-tailed Hawk *Buteo jamaicensis*
Ridgway's Hawk *Buteo ridgwayi*
Ring-necked Pheasant *Phasianus colchicus*

Rook *Corvus frugilegus*
Rough-legged Buzzard *Buteo lagopus*
Rufous-tailed Hawk *Buteo ventralis*
Saker Falcon *Falco cherrug*
Sand Martin *Riparia riparia*
Short-eared Owl *Asio flammeus*
Short-toed Snake Eagle *Circaetus gallicus*
Snowy Owl *Bubo scandiacus*
Socotra Buzzard *Buteo socotraensis*
Song Thrush *Turdus philomelos*
Spanish Imperial Eagle *Aquila adalberti*
Steppe Buzzard *Buteo buteo vulpinus*
Steppe Eagle *Aquila nipalensis*
Swainson's Hawk *Buteo swainsoni*
Tawny Owl *Strix aluco*
Tufted Duck *Aythya fuligula*
Turkey Vulture *Cathartes aura*
Upland Buzzard *Buteo hemilasius*
Ural Owl *Strix uralensis*
Western Jackdaw *Corvus monedula*
White-tailed Eagle *Haliaeetus albicilla*
Wood Warbler *Phylloscopus sibilatrix*
Woodpigeon *Columba palumbus*

MAMMALS
Arctic Fox *Vulpes lagopus*
Eurasian Badger *Meles meles*
Bank Vole *Myodes glareolus*
Brown Hare *Lepus europaeus*
Common Vole *Microtus arvalis*
Coypu *Myocastor coypu*
Grey Squirrel *Sciurus carolinensis*
Lamb (Domestic Sheep) *Ovis aries*
Least Weasel *Mustela nivalis*
Lemmings *Lemmus lemmus, Myopus schisticolor*
Martens *Martes* spp.
Mole *Talpa europaea*
Mountain Hare *Lepus timidus*
Rabbit *Oryctolagus cuniculus*
Red Fox *Vulpes vulpes*
Red Squirrel *Sciurus vulgaris*
Short-tailed Vole *Microtus agrestis*
Stoat *Mustela erminea*
Water Vole *Arvicola amphibius*

West European Hedgehog *Erinaceus europaeus*
Wood Mouse *Apodemus sylvaticus*
Yellow-necked Mouse *Apodemus flavicollis*

REPTILES AND AMPHIBIANS
Adder *Vipera berus*
Common Frog *Rana temporaria*
Common Lizard *Lacerta vivipara*
Common Toad *Bufo bufo*
Four-lined Snake *Elaphe quatuorlineata*
Grass Snake *Natrix natrix*
Large Psammodromus *Psammodromus algirus*
Midwife Toad *Alytes obstetricans*
Ocellated Lizard *Timon lepidus*
Slow Worm *Anguis fragilis*

FISH
European Eel *Anguila anguila*

INVERTEBRATES
Carnid flies Family Carnidae
Carpenter ants *Camponotus* spp.
Dung beetles *Geotrupes* spp. and *Typhaeus typhaeus*
Field cricket *Gryllus campestris*
Horse-flies Family Tabanidae
Large White butterfly *Pieris brassicae*
Leatherjackets *Tipula* spp.

PLANTS
Aleppo Pine *Pinua halepensis*
Alfalfa *Medicago sativa*
Beech *Fagus sylvatica*
Corsican Pine *Pinus nigra*
Hawthorns *Crataegus* spp.
Ivy *Hedera helix*
Oilseed Rape *Brassica napus*
Poplars *Populus* spp.
Red Campion *Silene dioica*
Rowan *Sorbus aucuparia*
Scots Pine *Pinus sylvestris*
Silver Birch *Betula pendula*
Thrift *Armeria maritima*
Willows *Salix* spp.
Yorkshire Fog *Holcus lanatus*

Appendix 2
References for figures with many sources

CHAPTER 2

Figure 2.6. Common Buzzard prey frequency recorded by cameras or in pellets at nests in Europe:

Pinowski & Ryszkowski (1962), Truszkowski (1976), Sylvén (1978), Mañosa & Cordero (1992), Voříšek *et al.* (1997), Dare (1998), Skierczyński (2006), Selås *et al.* (2007), Tornberg & Reif (2007), Swan (2011), Jankowiak & Tryjanowski (2013), Francksen *et al.* (2016b).

Figure 2.6. Diet at nests in years or areas when small mammal counts were high or low:

Spidsø & Selås (1988), Dare (1998), Selås (2001), Zuberogoitia *et al.* (2006).

CHAPTER 4

Figures 4.10. Home range analysis for Dorset Buzzards was all done with software from Anatrack Ltd (www.anatrack.com) by Eduardo Arraut, Kathy Hodder, Robert Kenward and Sean Walls.

CHAPTER 5

Figure 5.4. Common Buzzard breeding territory sizes:

Tubbs (1974), Weir & Picozzi (1983), Halley *et al.* (1993), Hohmann (1995), Dare (1998), Prytherch (2013).

CHAPTER 7

Figure 7.3. The mean laying date of Common Buzzard populations across Europe:

Mebs (1964), Picozzi & Weir (1974), Maguire (1979), Jędrzejewski (1994), Austin & Houston (1997), Goszczyński (2001), Sergio (2006), Zuberogoitia (2006), Lehikoinen (2009), Rodríguez (2010), Rooney (2013).

Figure 7.12. Productivity of Common Buzzard pairs:

Davis & Saunders (1965), Tubbs (1967), Picozzi & Weir (1974), Newton *et al.* (1982), Jędrzejewski *et al.* (1994), Swann & Etheridge (1995), Cerasoli & Penteriani (1996), Austin & Houston (1997), Dare (1998), Goszczyński (2001), Sergio *et al.* (2002), Lõhmus (2003), Hakkarainen *et al.* (2004), Krüger (2004b), Zuberogoitia *et al.* (2006), Šotnár & Topercer (2009), Rodríguez *et al.* (2010), Prytherch (2013), Rooney (2013).

CHAPTER 10

Figure 10.2. Nest density in Common Buzzard studies:

Joensen (in Mendeley) (1968), Newton, Davis & Davis (1982), Weir & Picozzi (1983), Dare & Barry (1990), Kostrzewa & Kostrzewa (1990), Jędrzejewski (1994), Graham *et al.* (1995), Hohmann (1995), Cerasoli & Penteriani (1996), Penteriani & Faivre (1997), Sánchez-Zapata (1999), Kenward *et al.* (2000), Vorisek (2000), Goszczyński (2001), Selås (2001), Sim (2001), Sergio (2002), Krüger (2004), Zuberogoitia (2006), Turzański (2008), Driver & Dare (2009), Šotnár & Topercer (2009), Rodríguez (2010), Prytherch (2013), Panek & Hušek (2014).

Figure 10.7. We acknowledge the World Climate Research Programme's Working Group on Regional Climate, and the Working Group on Coupled Modelling, former coordinating body of CORDEX and responsible panel for CMIP5. We also acknowledge the Earth System Grid Federation infrastructure, an international effort led by the U.S. Department of Energy's Program for Climate Model Diagnosis and Intercomparison, the European Network for Earth System Modelling and other partners in the Global Organisation for Earth System Science Portals (GO-ESSP). We also thank the climate modelling groups for producing and making available their model output. Further information is available at https://www.eea.europa.eu/data-and-maps/figures/projected-change-in-annual-mean.

Appendix 3
Map of UK study sites and flightpaths

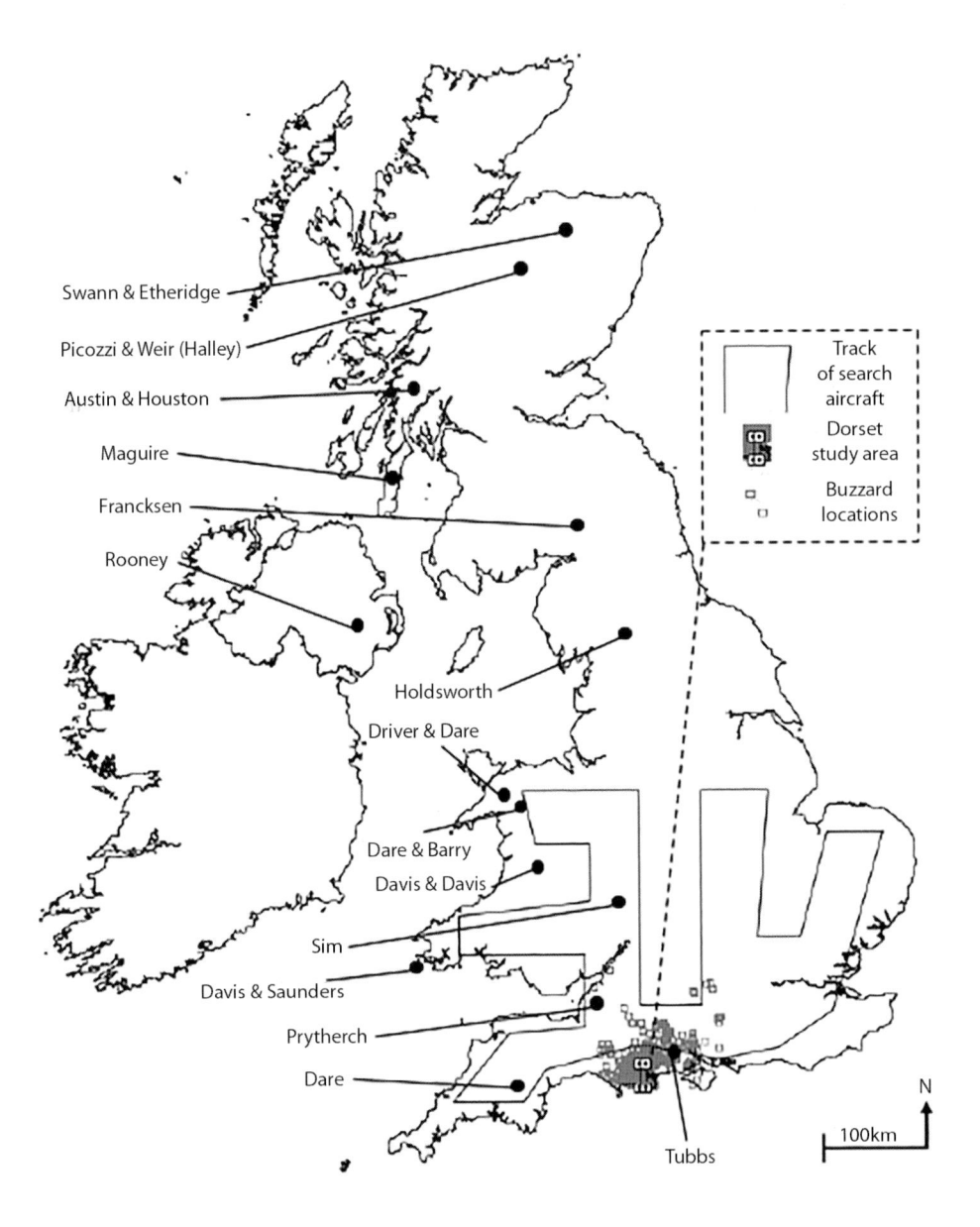

269

Appendix 4
Abbreviations

APH	Alternative Prey Hypothesis
ASSAB	Age-specific survival and breeding
BBC	British Broadcasting Company
BTO	British Trust for Ornithology
CBD	Convention on Biological Diversity
CEH	Centre for Ecology and Hydrology
CITES	Convention on International Trade in Endangered Species
CT	Computerised tomography
DEFRA	Department for Environment, Food and Rural Affairs
DDT	Dichlorodiphenyltrichloroethane
DETR	Department of the Environment, Transport and the Regions
DNA	Deoxyribonucleic acid
EU	European Union
GPS	Global positioning system
ITE	Institute of Terrestrial Ecology
IUCN	International Union for the Conservation of Nature
JNCC	Joint Nature Conservation Committee
LCMGB	Land Cover Map of Great Britain
LRS	Lifetime reproductive success
MAFF	Ministry of Agriculture, Fisheries and Food
MCP	Minimum convex polygon
MPH	Main Prey Hypothesis
NAO	North Atlantic Oscillation
NND	Nearest-neighbour distances
NSAID	Non-steroidal anti-inflammatory drug
PBDE	Polybrominated diphenyl
PCB	Polychlorinated biphenyl
PFH	Predator facilitation hypothesis
RADA	Resource-area-dependence analysis
RAF	Royal Air Force
RSPB	Royal Society for the Protection of Birds
SPH	Shared Prey Hypothesis
TB	Tuberculosis
UK	United Kingdom
US	United States of America
UV	Ultraviolet
VHE	Viral haemorrhagic enteritis
VHF	Very high frequency

References

Agostini, N., Amato, P., Provenza, A. & Panuccio, M. 2005. Do Common Buzzards *Buteo buteo* migrate across the Channel of Sicily during autumn? *Avocetta* 29: 19.

Alerstam, T. 1978. Analysis and a Theory of Visible Bird Migration. *Oikos* 30: 273–349.

Alerstam, T. 1990. *Bird Migration*. Cambridge Universtiy Press, Cambridge.

Aloise, G., Mazzei, A. & Brandmayr, P. 2010. An attempted predation on a Four-lined snake *Elaphe quatuorlineata* by a Common Buzzard Buteo buteo (Linnaeus, 1758). *Acta Herpetologica* 5: 103–106.

Amadon, D. 1982. A revision of the Sub-Buteonine hawks (Accipitridae, Aves). *American Museum Novitates* 2741: 1–20. New York, N.Y.: American Museum of Natural History.

Amar, A., Koeslag, A. & Curtis, O. 2013. Plumage polymorphism in a newly colonized black sparrowhawk population: Classification, temporal stability and inheritance patterns. *Journal of Zoology* 289: 60–67.

Andersson, S. & Wiklund, C. G. 1987. Sex role partitioning during offspring protection in the Rough-legged Buzzard *Buteo lagopus*. *Ibis* 129: 103–107.

Angelstam, P., Lindström, E. & Widén, P. 1984. Role of Predation in Short-term Population Fluctuations of Some Birds and Mammals in Fennoscandia. *Oecologia* 62: 199–208.

Aradis, A., Sarrocco, S. & Brunelli, M. 2012. *Analisi dello status e distribuzione dei rapaci diurni nel Lazio* (Quaderni). ARP Lazio.

Arraut, E. M., Macdonald, D. W. & Kenward, R. E. 2015. In the wake of buzzards: from modelling to conservation in a changing landscape. Pp. 203–221 *Wildlife Conservation on Farmland Vol. 2 Conflict in the Countryside*. OUP Oxford, Oxford.

Arroyo, B., Mougeot, F. & Bretagnolle, V. 2013. Characteristics and sexual functions of sky-dancing displays in a semi-colonial raptor, the Montagu's Harrier (*Circus pygargus*). *Journal of Raptor Research* 47: 185–196.

Austin, G. E. 1992. The distribution and breeding performance of the buzzard *Buteo buteo* in relation to habitat: an application using remote sensing and Geographical Information Systems. University of Glasgow.

Austin, G. E. & Houston, D. C. 1997. The breeding performance of the Buzzard *Buteo buteo* in Argyll, Scotland, and a comparison with other areas in Britain. *Bird Study* 44: 146–154.

Austin, G. E., Thomas, C. J., Houston, D. C. & Thompson, D. B. A. 1996. Predicting the spatial distribution of buzzard *Buteo buteo* nesting areas using a geographical information system and Remote Sensing. *The Journal of Applied Ecology* 33: 1541.

Balcomb, R., Bowen, C. A., Wright, D. & LAW, M. 1984. Effects on wildlife of AT-planting corn applications of granular carbofuran. *Journal of Wildlife Management* 48: 1353–1359.

Balmer, D. E., Gillings, S., Caffrey, B. J., Swann, R. L., Downie, I. S. & Fuller, R. J. 2013. *Bird Atlas 2007–11: the breeding and wintering birds of Britain and Ireland*. BTO Books, Thetford.

Barnard, C. C. 1978. Buzzard preying on Short-eared Owl. *British Birds* 74: 226.

Barrientos, R. 2006. A case of a polyandrous trio of Eurasian Buzzards (*Buteo buteo*) on Fuerteventura Island, Canary Islands. *Journal Of Raptor Research* 40: 43–46.

Barrios, L. & Rodríguez, A. 2004. Behavioural and environmental correlates of soaring-bird mortality and on-shore wind turbines. *Journal of Applied Ecology* 41: 72–81.

Barton, N. & Houston, D. 2001. The incidence of intestinal parasites in British birds of prey. *Journal of Raptor Research* 35: 71–73.

Barton, N. W. H. & Houston, D. C. 1993. A comparison of digestive efficiency in birds of prey. *Ibis* 135: 363–371. Blackwell Publishing.

Barton, N. W. H. & Houston, D. C. 1994. Morphological adaptation of the digestive tract in relation to feeding ecology of raptors. *Journal of Zoology* 232: 133–150.

Battaglia, A., Ghidini, S., Campanini, G. & Spaggiari, R. 2005. Heavy metal contamination in little owl (*Athene noctua*) and common buzzard (*Buteo buteo*) from northern Italy. *Ecotoxicology and Environmental Safety* 60: 61–66.

Beaman, M. & Galea, C. 1974. The visible migration of raptors over the Maltese islands. *Ibis* 116: 419–431.

Benton, A. H. 1954. Relationships of birds to power and communication lines. *Kingbird* 4: 65–66.

Berny, P. J., Buronfosse, T., Buronfosse, F., Lamarque, F. & Lorgue, G. 1997. Field evidence of secondary poisoning of foxes (*Vulpes vulpes*) and buzzards (*Buteo buteo*) by bromadiolone, a 4-year survey. *Chemosphere* 35: 1817–1829.

Berry, M. E., Bock, C. E. & Hair, S. L. 1998. Abundance of Diurnal Raptors on Open Space Grasslands in an Urbanized Landscape. *Condor* 100: 601–608.

Berthold, P. 1999. A comprehensive theory for the evolution, control and adaptability of avian migration. *Ostrich* 70: 1–11. Taylor & Francis Group, Abingdon.

Bielański, W. 2004. Impact of common silvicultural treatments on nest tree accessibility for Common Buzzard *Buteo buteo* and Goshawk *Accipiter gentilis*. *Ornis Fennica* 81: 180–185.

Bielański, W. 2006. Nesting preferences of common buzzard *Buteo buteo* and goshawk *Accipiter gentilis* in forest stands of different structure (Niepolomice Forest, Southern Poland). *Biologia* 61: 597–603. Versita.

Biljsma, R. G. 1999. Geslachtsdeterminatie van nestjonge Buizerds *Buteo buteo*. [Sex determination of nestling Common Buzzards Buteo buteo.] *Limosa* 72: 1–10.

BirdLife International. 2015a. The BirdLife checklist of the birds of the world: Version 8.

BirdLife International. 2015b. *Buteo buteo* (Eurasian Buzzard) European Red List of Birds Supplementary Material.

BirdLife International. 2017. Species factsheet: *Buteo buteo*.

Björklund, H., Valkama, J., Saurola, P. & Laaksonen, T. 2013. Evaluation of artificial nests as a conservation tool for three forest-dwelling raptors. *Animal Conservation* 16: 546–555.

Blezard, E. 1933. On the Buzzard. *Trans. Carlisle Nat. Hist. Soc.* 5: 31–66.

Bloom, P. H. & McCrary, M. D. 1996. The urban *Buteo*: Red-shouldered Hawks in southern California. Pp. 31–39 in Bird, D. M., Varland, D. E. & Negro, J. J. (eds). *Raptors in human landscapes: adaptation to built and cultivated environments.* Academic Press Limited, London, UK.

Bloomfield, A. 2013. Common Buzzards robbing Marsh Harriers of prey. *British Birds* 106: 411–412.

Boerner, M., Hoffman, J. I., Amos, W., Chakarov, N. & Krüger, O. 2013. No correlation between multi-locus heterozygosity and fitness in the common buzzard despite heterozygote advantage for plumage colour. *Journal of Evolutionary Biology* 26: 2233–2243.

Boerner, M. & Krüger, O. 2009. Aggression and fitness differences between plumage morphs in the common buzzard (*Buteo buteo*). *Behavioral Ecology* 20: 180–185.

Bohrer, G., Brandes, D., Mandel, J. T., Bildstein, K. L., Miller, T. A., Lanzone, M., Katzner, T., Maisonneuve, C. & Tremblay, J. A. 2011. Estimating updraft velocity components over large spatial scales: Contrasting migration strategies of golden eagles and turkey vultures. *Ecology Letters* 15: 96–103.

Bolton, M., Butcher, N., Sharpe, F., Stevens, D. & Fisher, G. 2007. Remote monitoring of nests using digital camera technology. *Journal of Field Ornithology* 78: 213–220.

Borowski, S. 1978. [Goshawk *Accipiter gentilis* (L., 1758) attacking and badly injuring the buzzard *Buteo buteo* (L., 1758)]. *Przeglad Zoologiczny* 22: 178–179.

Bosakowski, T. & Smith, D. 1997. Distribution and species richness of a forest raptor community in relation to urbanization. *Journal of Raptor Research* 31: 26–33.

Both, C., Tinbergen, J. M. M. & Noordwijk, A. J. Van. 1998. Offspring fitness and individual optimization of clutch size. *Proceedings of the Royal Society of London. Series B: Biological Sciences* 265: 2303–2307.

Van den Brand, J. M. A., Krone, O., Wolf, P. U., Van de Bildt, M. W. G., Van Amerongen, G., Osterhaus, A. D. M. E. & Kuiken, T. 2015. Host-specific exposure and fatal neurologic disease in wild raptors from highly pathogenic avian influenza virus H5N1 during the 2006 outbreak in Germany. *Veterinary Research* 46: 24.

Briggs, C. W., Collopy, M. W. & Woodbridge, B. 2011. Plumage polymorphism and fitness in Swainson's hawks. *Journal of Evolutionary Biology* 24: 2258–2268. Blackwell Publishing.

Briggs, K. B. 1983. Buzzard feeding on dung beetles. *British Birds* 76: 135–136.

Bright, G. 1955. Buzzard killing Tufted Duck. *British Birds* 48: 326.

Brown, D. 1975. A test of randomness on nest spacing. *Wildfowl* 26: 102–103.

Brown, L. & Amadon, D. 1968. *Eagles, Hawks and Falcons of the World*. McGraw-Hill, New York.

Brown, L. H. 1978. *British Birds of Prey*. Collins, London.

Bruderer, B. & Jenni, L. 1990. Migration Across the Alps. Pp. 60–77 *Bird Migration*. Springer, Berlin, Heidelberg, Germany.

Brüll, H. 1964. *Das Leben Deutscher Greifvögel*. Fischer, Stuttgart, Germany.

Buckland, S. T., Anderson, D. R., Burnham, K. P., Laake, J. L., Borchers, D. L. & Thomas, L. 2001. *Introduction to distance sampling: estimating abundance of biological populations*. New York.

Burnham, K. K., Burnham, W. A. & Newton, I. 2009. Gyrfalcon *Falco rusticolus* post-glacial colonization and extreme long term use of nest-sites in Greenland. *Ibis* 151: 514–522.

Burt, W. H. 1943. Territoriality and Home Range Concepts as Applied to Mammals. *Journal of Mammalogy* 24: 346–352.

Busch, W. 1997. A Common Buzzard (*Buteo buteo*) steals prey from a Eurasian Hobby (*Falco subbuteo*). *Ornithologische Schnellmitteilungen für Baden-Württemberg* 51/52: 41.

Bustamante, J. & Seoane, J. 2004. Predicting the distribution of four species of raptors (Aves: Accipitridae) in southern

Spain: Statistical models work better than existing maps. *Journal of Biogeography* 31: 295–306.

Bylicka, M., Wikar, D., Ciach, M. & Bylicka, M. 2007. Changes in density and behaviour of the Common Buzzard (*Buteo buteo*) during the non-breeding season. *Acta Zoologica Lituanica* 17: 286–291.

Carson, R. 1962. *Silent spring*. Houghton Mifflin, Boston.

Carter, I. & Grice, P. 2000. Studies of re-established Red Kites in England. *British Birds* 93: 304–322.

Castillo-Gómez, C. & Moreno-Rueda, G. 2011. A Record of a Common Buzzard (*Buteo buteo*) Nesting in an Abandoned Building. *Journal of Raptor Research* 45: 275–277.

Cerasoli, M. & Penteriani, V. 1996. Nest-site and aerial meeting point selection by common buzzards (*Buteo buteo*) in central Italy. *Journal Raptor Research* 30: 130–135.

Chace, J. F. & Walsh, J. J. 2006. Urban effects on native avifauna: A review. *Landscape and Urban Planning* 74: 46–69.

Chakarov, N., Boerner, M. & Krüger, O. 2008. Fitness in common buzzards at the cross-point of opposite melanin-parasite interactions. *Functional Ecology* 22: 1062–1069.

Chakarov, N., Jonker, R. M., Boerner, M., Hoffman, J. I. & Krüger, O. 2013. Variation at phenological candidate genes correlates with timing of dispersal and plumage morph in a sedentary bird of prey. *Molecular Ecology* 22: 5430–5440.

Chakarov, N. & Krüger, O. 2010. Mesopredator release by an emergent superpredator: A natural experiment of predation in a three level guild. *PLoS ONE* 5.

Chakarov, N., Pauli, M., Mueller, A. K., Potiek, A., Grünkorn, T., Dijkstra, C. & Krüger, O. 2015. Territory quality and plumage morph predict offspring sex ratio variation in a raptor. *PLoS ONE* 10: 1–14.

Chamorro, D., Olivero, J., Real, R. & Muñoz, A.-R. 2017. Environmental factors determining the establishment of the African Long-legged Buzzard *Buteo rufinus cirtensis* in Western Europe. *Ibis* 159: 331–342.

Chiozzi, G. & Marchetti, G. 2000. High mortality of Common Buzzards, *Buteo buteo*, following electrocution on a 15 kV power line. *Rivista Italiana di Ornitologia* 70: 172–173.

Christensen, T. K., Lassen, P. & Elmeros, M. 2012. High exposure rates of anticoagulant rodenticides in predatory bird species in intensively managed landscapes in Denmark. *Archives of Environmental Contamination and Toxicology* 63: 437–444.

Clements, R. 2002. The Common Buzzard in Britain: A new population estimate. *British Birds* 95: 377–383.

Corso, A. 2009. Sucessful mixed breeding of Atlas Long-legged Buzzard and Common Buzzard on Pantelleria, Italy, in 2008. *Dutch Birding* 31: 224–226.

Corso, A. & Gildi, R. 1998. Hybrids of Black Kite and Common Buzzard in Italy in 1996. *Dutch Birding* 20: 226–233.

Craighead, J. J. & Craighead, F. C. 1956. *Hawks, Owls and Wildlife*. Stackpole Co., Harrisburg, Pennsylvania.

Cramp, S. & Simmons, K. E. L. 1980. *Handbook of the Birds of Europe the Middle East and North Africa. The Birds of the Western Palearctic. Hawks to Bustards. Volume II.* University Press, Oxford.

Csermely, D. & Gaibani, G. 1998. Is foot squeezing pressure by two raptor species sufficient to subdue their prey? *The Condor* 100: 757–763.

Csörgő, T., Zornánszky, R., Szép, T. & Fehérvári, P. 2012. Should the Common

Buzzard be hunted? *Ornis Hungarica* 20: 1–12.

D'Urban, W. S. M. & Mathew, M. A. 1892. *The Birds of Devon*. R.H. Porter, London.

Daan, S., Dijkstra, C., Drent, R. & Meijer, T. 1989. Food Supply and the Annual Timing of Avian Reproduction. Pp. 392–407 in Ouellet, H. (ed.). *19th International Ornithological Congress, Ottawa, Canada*.

Dalrymple, T. 2014. Saline Lagoon News, Newport Wetlands. *The Dipper, Gwent Ornithological Society* 131.

Danchin, E., Heg, D. & Doligez, B. 2001. Public information and breeding habitat selection. Pp. 243–258 in Clobert, J., Danchin, E., Dhondt, A. A. & Nichols, J. D. (eds). *Dispersal*.

Dare, P. J. 1957. The post-myxomatosis diet of the Buzzard. *Devon Birds* 10: 2–6.

Dare, P. J. 1961. Ecological observations on a breeding population of the Common Buzzard, Buteo buteo, with particular reference to the diet and feeding habits. University of Exeter.

Dare, P. J. 1998. A buzzard population on Dartmoor, 1955-1993. *Devon Birds* 51: 4–31.

Dare, P. J. 2015. *The Life of Buzzards*. Whittles Publishing, Caithness.

Davies, N. B. 1976. Food, flocking and territorial behaviour of the pied wagtail (*Motacilla alba yarrellii*) in winter. *Journal of Animal Ecology* 45: 235–253.

Davis, D. R. & Saunders, T. A. . 1965. Buzzards on Skomer Island, 1954–1964. *Nature in Wales* 9: 116–124.

Davis, P. E. & Davis, J. E. 1992. Dispersal and age of first breeding of Buzzards in central Wales. *British Birds* 85: 578–587.

Davis, P. R. K. & Seel, D. C. 1976. Cuckoo taken by Buzzard. *British Birds* 76: 314.

Dawkins, R. 1976. *The Selfish Gene*. P. Oxford University Press, New York.

Demandt, C. 1934. Balzfliige des Mausebussards. Beitr. Fortpflbiol. *Vogel* 10: 144–145.

Demongin, L. 2016. *Identification guide to birds in the hand*. P. (Hervé Lelièvre and George Candelin, eds). Beauregard-Vendon.

Dickson, R. C. 1997. Buzzard robbing juvenile peregrine of prey. *Scottish Birds* 19: 124–125.

Dickson, R. C. 1998. Common Buzzards cartwheeling during a food pass. *Scottish Birds* 19: 166.

Dietrich, D. R., Schmid, P., Zweifel, U., Schlatter, C., Jenni-Eiermann, S., Bachmann, H., Bühler, U. & Zbinden, N. 1995. Mortality of birds of prey following field application of granular carbofuran: A case study. *Archives of Environmental Contamination and Toxicology* 29: 140–145.

Dittrich, W. 1985. Gefiedervariationen beim Mäusebussard (*Buteo buteo*) in Nordbayern. *Journal of Ornithology* 126: 93–97. Springer-Verlag, Stuttgart.

Dixon, A., Maming, R., Gunga, A., Purev-Ochir, G. & Batbayar, N. 2013. The problem of raptor electrocution in Asia: case studies from Mongolia and China. *Bird Conservation International* 23: 520–529. Cambridge University Press, Cambridge.

Dixon, N. 2013. Peregrine aggression towards Buzzards at St. Michael's Church, Exeter. *Devon Birds* 66: 30–31.

Dixon, N. & Gibbs, A. 2015. Territorial aggression shown by urban Peregrine Falcons towards Common Buzzards. *Devon Birds* 68: 13–20.

Dolton, C. S. & Brooke, M. D. L. 1999. Changes in the biomass of birds breeding in Great Britain, 1968–88. *Bird Study* 46: 274–278.

Dravecky, M. 2003. [An interesting observation of picking up fish from the water surface by Common Buzzard (*Buteo buteo*)]. *Buteo* 13: 105–106.

Driver, J. & Dare, P. J. 2009. Population increase of Buzzards in Snowdonia, 1977–2007. *Welsh Birds* 6: 38–48.

Dudas, M.-T. & Janos-Toth, I. 1999. Natural hybridization in buzzards. *Termézet* 6: 8–10.

Dwyer, J. F. 2014. Correlation of cere color with intra- and interspecific agonistic interactions of crested caracaras. *Journal of Raptor Research* 48: 240–247. The Raptor Research Foundation, Inc. 5400 Bosque Blvd., Suite 680, Waco TX 76710, US.

Dykstra, C. R., Hays, J. L., Daniel, F. B. & Simon, M. M. 2000. Nest Site Selection and Productivity of Suburban Red-Shouldered Hawks in Southern Ohio. *The Condor* 102: 401–408.

Einoder, L. & Richardson, A. 2006. An ecomorphological study of the raptorial digital tendon locking mechanism. *Ibis* 148: 515–525.

Elkins, N. 1983. *Weather and Bird Behaviour*. A & C Black, London.

Ellenberg, H. & Dietrich, J. 1981. The goshawk as a bioindicator. Pp. 69–88 in Kenward, R. E. & Lindsay, I. M. (eds). *Understanding the Goshawk*. International Association of Falconry and Conservation of Birds of Prey, Oxford.

Elliott, G. D. & Avery, M. I. 1991. A review of reports of Buzzard persecution 1975–1989. *Bird Study* 38: 52–56.

Elorriaga, J. & Muñoz, A.-R. 2013. Hybridisation between the Common Buzzard *Buteo buteo buteo* and the North African race of Long-legged Buzzard *Buteo rufinus cirtensis* in the Strait of Gibraltar: prelude or preclude to colonisation? *Ostrich: Journal of African Ornithology* 84: 41–45.

Elorriaga, J. & Muñoz, A. R. 2010. First breeding record of North African Long-legged Buzzard *Buteo rufinus cirtensis* in continental Europe. *British Birds* 103: 399–401.

Elston, D. A., Spezia, L., Baines, D. & Redpath, S. M. 2014. Working with stakeholders to reduce conflict – modelling the impact of varying hen harrier *Circus cyaneus* densities on red grouse *Lagopus lagopus* populations. *Journal of Applied Ecology* 51: 1236–1245.

Evans, H. AP. 1960. *Falconry for you*. Compton Printing Ltd, London and Aylesbury.

Evans, P. R. & Lathbury, G. W. 1973. Raptor migration across the Straits of Gibraltar. *Ibis* 115: 572–585.

Fathers, J. 2006. Polygyny in the Eurasian Sparrowhawk. *British Birds* 99: 265–266.

Fenyosi, L. & Stix, J. 1998. Megjegyzesek a 'Retisas (*Haliaeetus albicilla*) altat nevelt egereszolyv (*Buteo buteo*) fiokak' cimu irashoz. *Tuzok* 3: 64.

Fernández, C. & Azkona, P. 1993. Influencia del exito reproductor en la reutilizacion de los nidos por el aguila real (*Aquila chrysaetos* L.). *Aedeola* 40: 27–31.

Ferrer, M. 1993. Wind-Influenced juvenile dispersal of Spanish Imperial Eagles. *Ornis Scandinavica* 24: 330–333.

Fletcher, K., Aebischer, N. J., Baines, D., Foster, R. & Hoodless, A. N. 2010. Changes in breeding success and abundance of ground-nesting moorland birds in relation to the experimental deployment of legal predator control. *Journal of Applied Ecology* 47: 263–272.

Francksen, R. M. 2016. Exploring the impact of common buzzard *Buteo buteo* predation on red grouse *Lagopus lagopus*. University of Newcastle.

Francksen, R. M., Whittingham, M. J. & Baines, D. 2016a. Poster: Winter diet and home range of common buzzards *Buteo buteo*.

Francksen, R. M., Whittingham, M. J. & Baines, D. 2016b. Assessing prey provisioned to Common Buzzard *Buteo buteo* chicks: a comparison of methods. *Bird Study* 63: 303–310.

Francksen, R. M., Whittingham, M. J., Ludwig, S. C. & Baines, D. 2016c. Winter diet of Common Buzzards *Buteo buteo* on a Scottish grouse moor. *Bird Study* 63: 525–532. Taylor & Francis, Abingdon.

Francksen, R. M., Whittingham, M. J., Ludwig, S. C., Roos, S. & Baines, D. 2017. Numerical and functional responses of Common Buzzards Buteo buteo to prey abundance on a Scottish grouse moor. *Ibis* 159: 541–553.

Franks, S. E., Douglas, D. J. T., Gillings, S. & Pearce-Higgins, J. W. 2017. Environmental correlates of breeding abundance and population change of Eurasian Curlew *Numenius arquata* in Britain. *Bird Study* August: 1–17. Taylor & Francis, Abingdon.

Fry, D. M. 1995. Reproductive effects in birds exposed to pesticides and industrial chemicals. *Environmental health perspectives*: 165–71. National Institute of Environmental Health Science.

Fryer, G. 1986. Notes on the Breeding Biology of the Buzzard. *Ibis* 79: 18–28.

Fuglei, E. & Ims, R. A. 2008. Global warming and effects on the Arctic fox. *Science progress* 91: 175–191.

Fuller, R. M., Groom, G. B. & Jones, A. R. 1994. Land cover map of Great Britain. An automated classification of Landsat Thematic Mapper data. *Photogrammetric Engineering and Remote Sensing* 60.

García-Rodríguez, T., Ferrer, M., Carrillo, J. C. & Castroviejo, J. 1987. Metabolic responses of *Buteo buteo* to long-term fasting and refeeding. *Comparative Biochemistry and Physiology Part A: Physiology* 87: 381–386.

Van Gasteren, H., Both, I., Shamoun-Baranes, J., Laloë, J. O. & Bouten, W. 2014. GPS-logger onderzoek aan Buizerds helpt vogelaanvaringen op militaire vliegvelden te voorkomen. *Limosa* 87: 107–116.

Génsbøl, B. & Bertel, B. 2008. *Birds of prey*. Collins, London.

Gibbons, D., Gates, S., Green, R. E., Fuller, R. J. & Fuller, R. M. 1995. Buzzards *Buteo buteo* and ravens *Corvus corax* in the uplands of Britain: limits to distribution and abundance. *Ibis* 137: 75–84.

Gill, F. & Donsker, D. 2017. Birds of the world (v. 7.2): recommended English names. P. (F. Gill and D. Donsker, eds).

Gjershaug, J. O., Forset, O. A., Woldvik, K. & Espmark, Y. 2006. Hybridisation between Common Buzzard *Buteo buteo* and Rough-legged Buzzard *B. lagopus* in Norway. *Bulletin of the British Ornithologists' Club* 126: 73–80.

Glutz von Blotzheim, U., Bauer, K. & Bezzel, E. 1971. *Handbuch der Vögel Mitteleuropas. Volume 4: Falconiformes.* Akademische Verlagsgesellschaft, Frankfurt am Main, Germany.

Goldstein, M. I., Lacher, T. E., Woodbridge, B., Bechard, M. J., Canavelli, S. B., Zaccagnini, M. E., Cobb, G. P., Scollon, E. J., Tribolet, R. & Hopper, M. J. 1999. Monocrotophos-Induced Mass Mortality of Swainson's Hawks in Argentina, 1995–96. *Ecotoxicology* 8: 201–214. Kluwer Academic Publishers, Dordrecht.

Gómez-Ramírez, P., Shore, R. F., Van den Brink, N. W., Van Hattum, B., Bustnes, J. O., Duke, G., Fritsch, C., García-Fernández, A. J., Helander, B. O., Jaspers, V., Krone, O., Martínez-López, E., Mateo, R., Movalli, P. & Sonne, C. 2014. An overview of existing raptor contaminant monitoring activities in Europe. *Environment International* 67: 12–21.

Gorney, E. & Yom-Tov, Y. 1994. Fat, hydration, and moult of Steppe buzzards

Buteo buteo vulpinus on spring migration. *Ibis* 136: 185–192.

Goszczyński, J. 1997. Density and productivity of Common Buzzard buteo buteo and Goshawk accipiter gentilis populations in Rogów, central Poland. *Acta Ornithologica* 32: 149–155.

Goszczyński, J. 2001. The breeding performance of the Common Buzzard buteo buteo and Goshawk accipiter gentilis in central Poland. *Acta Ornithologica* 36: 105–110.

Goszczyński, J., Gryz, J. & Krauze, D. 2005. Fluctuations of a Common Buzzard *Buteo buteo* population in Central Poland. *Acta Ornithologica* 40: 75–78.

Goszczyński, J. & Pilatowski, T. 1986. Diet of common buzzards (*Buteo buteo*) and goshawks (*Accipiter gentilis*) in the nesting period. *Ekologia Polska* 34: 655–667.

Gradoz, P. 1995. White-tailed Eagle *Haliaeetus albicilla* catching a Common Buzzard *Buteo buteo* in flight. *Ornithos* 2: 180–181.

Graham, I. M., Redpath, S. M. & Thirgood, S. J. 1995. The diet and breeding density of Common Buzzards *Buteo buteo* in relation to indices of prey abundance. *Bird Study* 42: 165–173.

Greenwood, P. J. 1980. Mating systems, philopatry and dispersal in birds and mammals. *Animal Behaviour* 28: 1140–1162.

Greenwood, P. J., Harvey, P. H. & Perrins, C. M. 1978. Inbreeding and dispersal in the great tit. *Nature* 271: 52–54.

Griffith, S. C., Owens, I. P. F. & Thuman, K. A. 2002. Extra pair paternity in birds: a review of interspecific variation and adaptive function. *Molecular ecology* 11: 2195–212.

Griffiths, C. S., Barrowclough, G. F., Groth, J. G. & Mertz, L. A. 2007. Phylogeny, diversity, and classification of the Accipitridae based on DNA sequences of the RAG-1 exon. *Journal of Avian Biology* 38: 587–602.

Grünhagen, H. 1983. Regionale Unterschiede im Alter brütender Habichtweibchen (*Accipiter gentilis*). *Vogelwelt* 104: 208–214.

Hakkarainen, H., Mykrä, S., Kurki, S., Tornberg, R. & Jungell, S. 2004. Competitive interactions among raptors in boreal forests. *Oecologia* 141: 420–424.

Halley, D. J. 1993. Population changes and territorial distribution of Common Buzzards *Buteo buteo* in the Central Highlands, Scotland. *Bird Study* 40: 24–30.

Hampden Smith, M. 1986. Buzzards following and alighting on working plough. *British Birds* 79: 429.

Hardey, J., Crick, H., Wernham, C., Riley, H., Etheridge, B. & Thompson, D. 2009. *Raptors: a field guide for surveys and monitoring*. The Stationery Office, London.

Harel, R., Horvitz, N. & Nathan, R. 2016. Adult vultures outperform juveniles in challenging thermal soaring conditions. *Scientific Reports* 6: 27865. Nature Publishing Group, London.

Haring, E., Riesing, M. J., Pinsker, W. & Gamauf, A. 1999. Evolution of a pseudo-control region in the mitochondrial genome of Palearctic buzzards (genus *Buteo*). *Journal of Zoological Systematics and Evolutionary Research* 37: 185–194.

Härms, M. 1927. *Eesti linnustik*. Loodus, Tartu, Estonia.

Harris, S. & Yalden, D. W. (eds). 2008. *Mammals of the British Isles. Handbook 4th Edition*. The Mammal Society, London.

Hars, J., Ruette, S., Benmergui, M., Fouque, C., Fournier, J.-Y., Legouge, A., Cherbonnel, M., Daniel, B., Dupuy, C. & Jestin, V. 2008. The epidemiology of the highly pathogenic H5N1 avian influenza in Mute Swan (*Cygnus olor*) and other Anatidae in the Dombes region (France), 2006. *Journal of wildlife diseases* 44: 811–823.

Haukioja, E. & Haukioja, M. 1970. Mortality rates of Finnish and Swedish goshawks (*Accipiter gentilis*). *Finnish Game Research* 31: 13–20.

Hayman, R. W. 1970. Persistent ground-feeding by buzzards. *British Birds*: 132–133.

Hernández-García, A., Romero, D., Gómez-Ramírez, P., María-Mojica, P., Martínez-López, E. & García-Fernández, A. J. 2014. In vitro evaluation of cell death induced by cadmium, lead and their binary mixtures on erythrocytes of Common Buzzard (*Buteo buteo*). *Toxicology in Vitro* 28: 300–306.

Heroldová, M., Bryja, J., Zejda, J. & Tkadlec, E. 2007. Structure and diversity of small mammal communities in agriculture landscape. *Agriculture, Ecosystems and Environment* 120: 206–210.

Hewson, R. 1981. Scavenging of mammal carcasses by birds in West Scotland. *Journal of Zoology* 194: 525–537.

Heywood, A. 1986. Buzzards talon-grappling and tumbling to ground. *British Birds* 79: 429.

Hill, D. A. 1990. Motorway sentinels: Common Buzzards (*Buteo buteo* L.) on the Slavonski Brod Zagreb Highway in January 1989. *Troglodytes* 3: 14–15.

Hill, I. F. 1998. Post-nestling Mortality and Dispersal in Blackbirds and Song Thrushes. University of Oxford, Oxford.

Hodder, K. H. 2001. The common buzzard in lowland UK: relationships between food availability, habitat use and demography. University of Southampton, Southampton.

Hohmann, U. 1994. Status specific habitat use in the Common Buzzard. Pp. 359–365 in Meyburg, B. U. & Chancellor, R. D. (eds). *Raptor Conservation Today. World Working Group on Birds of Prey.* Pica Press, Crowborough.

Hohmann, U. 1995. Untersuchungen zur Raumnutzung und zur Brutbiologie des Mäusebussards (Buteo buteo) im Westen Schleswig-Holsteins. *Corax* 16: 94–104.

Holdsworth, M. 1971. Breeding biology of Buzzards at Sedbergh during 1937-67. *British Birds* 64: 412–420.

Honkavaara, J., Koivula, M., Korpimäki, E., Siitari, H. & Viitala, J. 2002. Ultraviolet vision and foraging in terrestrial vertebrates. *Oikos* 98: 505–511.

Horvarth, Z. 2009. White-tailed Sea Eagle (*Haliaeetus albicilla*) population in Hungary between 1987–2007. *Denisia* 27: 85–95.

Hubert, C. 1993. Nest-site habitat selected by Common Buzzard (*Buteo buteo*) in Southwestern France. *Journal of Raptor Research* 27: 102–105.

Hubert, C. H. & Carlier, P. 1992. Etude comparative des relations mâle–femelle chez la buse variable Buteo buteo et chez le faucon pèlerin *Falco peregrinus* au moment des éclosions. [Comparative study of male–female relationships in the common buzzard *Buteo buteo* and in the peregrine falco. *Cahiers d'éthologie* 12: 491–496.

Hume, R. A., Chater, M., Simms, C. & Simms, C. 1975. Short Notes. *Bird Study* 22: 260–261. Taylor & Francis Group, Abingdon.

Hunt, W. G. 1998. Raptor floaters at Moffat's equilibrium. *Oikos* 82: 191–197.

Hurrell, H. G. 1929. Census of Buzzards. *Western Morning News* 28 August: 6. Plymouth.

IUCN. 2016. Preventing electrocution and collision impacts of power infrastructure on birds. Recommendation 6.98. Gland, Switzerland.

Jager, L. P., Rijnierse, F. V. J., Esselink, H. & Baars, A. J. 1996. Biomonitoring with the Buzzard (*Buteo buteo*) in the

Netherlands: Heavy metals and sources of variation. *Journal of Ornithology* 137: 295–318.

James, P. C. & Oliphant, L. W. 1986. Extra birds and helpers at the nests of Richardson's Merlin. *The Condor* 88: 533–534.

Jankowiak, Ł. & Tryjanowski, P. 2013. Cooccurrence and food niche overlap of two common predators (red fox *Vulpes vulpes* and common buzzard *Buteo buteo*) in an agricultural landscape. *Turkish Journal of Zoology* 37: 157–162.

Janss, G. F. E. 2000. Avian mortality from power lines: A morphologic approach of a species-specific mortality. *Biological Conservation* 95: 353–359.

Janss, G. F. E. & Ferrer, M. 1999. Mitigation of raptor electrocution on steel power poles. *Wildlife Society Bulletin* 27: 263–273.

Jaspers, V. L. B., Voorspoels, S., Covaci, A. & Eens, M. 2006. Can predatory bird feathers be used as a non-destructive biomonitoring tool of organic pollutants? *Biology letters* 2: 283–285.

Jędrzejewski, W. & Jędrzejewska, B. 1993. Predation on rodents in Bialowieza primeval forest, Poland. *Ecography* 16: 47–64.

Jędrzejewski, W., Jędrzejewska, B. & Keller, M. 1988. Nest site selection by the buzzard Buteo buteo L. in the extensive forests of eastern Poland. *Biological Conservation* 43: 145–158.

Jędrzejewski, W., Szymura, A. & Jędrzejewska, B. 1994. Reproduction and food of the Buzzard *Buteo buteo* in relation to the abundance of rodents and birds in Białowieża National Park, Poland. *Ethology Ecology & Evolution* 6: 179–190. Taylor & Francis Group, Abingdon.

Jenkins, D. 1984. Common Buzzard eating dead salmon. *Scottish Birds* 13: 88.

Jensen, S., Johnels, A. G., Olsson, M. & Westmark, T. 1972. The avifauna of Sweden as bioindicators of contamination with mercury and chlorinated hydrocarbons. *Proceedings of the International Congress of Ornithology* 15:455–465.

Jeserich, E. 1970. Vergleichende Studien zur Bioakustik bei Sperber, Habicht und Maüserbussard. University of Tübingen, Germany.

Jiménez-Franco, M. V., Martínez, J. E. & Calvo, J. F. 2014a. Lifespan analyses of forest raptor nests: Patterns of creation, persistence and reuse. *PLoS ONE* 9: e93628.

Jiménez-Franco, M. V., Martínez, J. E. & Calvo, J. F. 2014b. Patterns of nest reuse in forest raptors and their effects on reproductive output. *Journal of Zoology* 292: 64–70.

Joenson, A. 1968. An investigation on the breeding population of the Buzzard (*Buteo buteo*) on the island Als in 1962 and 1963. *Dansk Ornith. Faren. Tidsskr.* 62: 17–31.

Jones, C. G. 1985. Heavy hippoboscid infestations on buzzards. *British Birds* 78: 592.

Jonker, R. M., Chakarov, N. & Krüger, O. 2014. Climate change and habitat heterogeneity drive a population increase in Common Buzzards *Buteo buteo* through effects on survival. *Ibis* 156: 97–106.

Joubert, B. 1989. Quelques données sur la reproduction d'une petite population de Buses variables (*Buteo buteo*) en Haute-Loire. *Le Grand Duc, N.* 35: 33–38.

Kalpakis, S., Mazaris, A. D., Mamakis, Y. & Poulopoulos, Y. 2009. A retrospective study of mortality and morbidity factors for Common Buzzards *Buteo buteo* and Long-legged Buzzards *Buteo rufinus* in Greece: 1996–2005. *Bird Conservation International* 19: 1–7.

Kappers, E. F., Chakarov, N., Krüger, O., Mueller, A. K., Valcu, M., Kempenaers, B. & Both, C. 2017. Classification and Temporal Stability of Plumage Variation in Common Buzzards Classification and temporal stability of plumage variation in Common Buzzards. *Ardea* 105.

Kasprzykowski, Z. & Cieśluk, P. 2011. Rough-legged Buzzard *Buteo lagopus* wintering in central eastern Poland: Population structure by age and sex, and the effect of weather conditions. *Ornis Fennica* 88: 98–103.

Kasprzykowski, Z. & Rzępała, M. 2002. Liczebność i preferencje siedliskowe ptaków szponiastych Falconiformes zimujących w środkowo-wschodniej Polsce [Numbers and habitat preferences of raptors Falconiformes wintering in central-eastern Poland].

Kenward, R. E. 1982. Goshawk hunting behaviour, and range size as a function of food and habitat availability. *Journal of Animal Ecology* 51: 69–80.

Kenward, R. E. 2001. *A Manual for Wildlife Radio Tagging*. Academic Press, London.

Kenward, R. E. 2004. Management tools for raptors. Pp. 329–339 in Chancellor, R. D. & Meyburg, B.-U. (eds). *Raptors Worldwide*. World Working Group on Birds of Prey and Owls, Berlin.

Kenward, R. E. 2006. *The Goshawk*. T & AD Poyser, London.

Kenward, R. E. 2009. Conservation values from falconry. Pp. 181–196 in Adams, W., Dixon, B. & Hutton, J. (eds). *Recreational Hunting, Conservation and Rural Livelihoods: Science and Practice*. Zoological Society of London & IUCN SSC/Sustainable Use Specialist Group & Blackwell Publishing.

Kenward, R. E., Clarke, R. T., Hodder, K. H. & Walls, S. S. 2001a. Density and linkage estimators of home range: Nearest-neighbor clustering defines multinuclear cores. *Ecology* 82: 1905–1920.

Kenward, R. E., Ewald, J. A. & Sharp, R. J. A. 2013. Pan-European analysis of environmental assessment processes. Pp. 120–133 *Transactional Environmental Support System Design: Global Solutions*. IGI-Global, Hershey, Pennsylvania.

Kenward, R. E., Hall, D. G., Walls, S. S. & Hodder, K. H. 2001b. Factors affecting predation by buzzards Buteo buteo on released pheasants Phasianus colchicus. *Journal of Applied Ecology* 38: 813–822.

Kenward, R. E. & Marcström, V. 1981. Goshawk predation on game and poultry: some problems and solutions. Pp. 152–162 in Kenward, R. E. & Lindsay, I. M. (eds). *Understanding the Goshawk*. International Association of Falconry and Conservation of Birds of Prey, Oxford.

Kenward, R. E. & Marcström, V. 1988. How differential competence could sustain suppressive predation on birds. Pp. 733–742 *Proceedings of the XIX International Ornithological Congress*.

Kenward, R. E., Marcström, V. & Karlbom, M. 1991. The goshawk (*Accipiter gentilis*) as predator and renewable resource. *Gibier Faune Sauvage* 8: 367–378.

Kenward, R. E., Marcström, V. & Karlbom, M. 1993a. Post-nestling behaviour in goshawks, Accipiter gentilis: I. The causes of dispersal. *Animal Behaviour* 46: 365–370.

Kenward, R. E., Marcström, V. & Karlbom, M. 1993b. Post-nestling behaviour in goshawks, *Accipiter gentilis*: II. Sex differences in sociality and nest-switching. *Animal Behaviour* 46: 371–378.

Kenward, R. E., Marcström, V. & Karlbom, M. 1999. Demographic estimates from radio-tagging: Models of age-specific survival and breeding in the goshawk. *Journal of Animal Ecology* 68: 1020–1033. Blackwell Science, Oxford.

Kenward, R. E., Parish, T. & Robertson, P. A. 1992. Are tree species mixtures too good for grey squirrels? Pp. 243–253 in Cannell, M. G. R., Malcolm, D. C. & Robertson, P. A. (eds). *The Ecology of Mixed-Species Stands of Trees: Special Publication Number 11 of the British Ecological Society*. Blackwell Science, Oxford.

Kenward, R. E., Pfeffer, R. H., Al-Bowardi, M. A., Fox, N. C., Riddle, K. E., Bragin, E. A., Levin, A., Walls, S. S. & Hodder, K. H. 2001c. Setting harness sizes and other marking techniques for a falcon with strong sexual dimorphism. *Journal of Field Ornithology* 72: 244–257.

Kenward, R. E., Walls, S. S. & Hodder, K. H. 2001d. Life path analysis: Scaling indicates priming effects of social and habitat factors on dispersal distances. *Journal of Animal Ecology* 70: 1–13.

Kenward, R. E., Walls, S. S., Hodder, K. H., Pahkala, M., Freeman, S. N. & Simpson, V. R. 2000. The prevalence of non-breeders in raptor populations: evidence from rings, radio-tags and transect surveys. *Oikos* 91: 271–279.

Kenward, R., Katzner, T., Wink, M., Marcström, V., Walls, S., Karlbom, M., Pfeffer, R., Bragin, E., Hodder, K. & Levin, A. 2007. Rapid sustainability modeling for raptors by radiotagging and DNA-fingerprinting. *Journal of Wildlife Management* 71: 238–245.

Kerlinger, P. 1989. *Flight Strategies of Migrating Hawks*. The University of Chicago Press, Chicago.

King, B. 1986. Food-seeking Buzzard following combine-harvester. *British Birds* 79: 429.

Kinley, R. I. & Thexton, D. B. 1985. Opportunistic food acquisition by Buzzard while mobbed by Peregrines. *British Birds* 78: 193.

Kirkwood, J. K. 1979. The partition of food energy for existence in the kestrel (Falco tinnunculus) and the barn owl (*Tyto alba*). *Comparative Biochemistry and Physiology Part A: Physiology* 63: 495–498.

Kjellén, N. 1994. Differences in Age and Sex-Ratio among Migrating and Wintering Raptors in Southern Sweden. *Auk* 111: 274–284.

Kjellén, N. & Roos, G. 2000. Population trends in Swedish raptors demonstrated by migration counts at Falsterbo, Sweden 1942–97. *Bird Study* 47: 195–211.

Koivula, M. & Viitala, J. 1999. Rough-Legged Buzzards Use Vole Scent Marks to Assess Hunting Areas. *Journal of Avian Biology* 30: 329–332.

Komdeur, J. & Pen, I. 2002. Adaptive sex allocation in birds: the complexities of linking theory and practice. *Philosophical transactions of the Royal Society of London. Series B, Biological sciences* 357: 373–380.

Kostrzewa, A. 1987. Territorialität, Konkurrenz und Horstnutzung dreier baumbrütender Greifvogelarten (Accipitres). *Journal für Ornithologie* 128: 495–496.

Kostrzewa, A. 1991. Interspecific interference competition in three European raptor species. *Ethology Ecology & Evolution* 3: 127–143.

Kostrzewa, A. & Kostrzewa, R. 1990. The relationship of spring and summer weather with density and breeding performance of the Buzzard *Buteo buteo*, Goshawk *Accipiter gentilis* and Kestrel *Falco tinnunculus*. *Ibis* 132: 550–559.

Krone, O., Priemer, J., Streich, J., Sömmer, P., Langgemach, T. & Lessow, O. 2001. Haemosporida of birds of prey and owls from Germany. *Acta Protozoologica* 40: 281–289.

Krone, O., Sömmer, P., Lessow, O. & Haas, D. 2006. Todesursachen, Krankheiten und Parasiten bei Mäusebussarden Buteo buteo in Deutschland, [Causes of death, diseases and parasites in Common

Buzzards Buteo buteo from Germany. *Populationsökologie Greifvogel und Eulenarten* 5: 439–448.

Krone, O. & Streich, W. J. 2000. Strigea falconispalumbi in Eurasian Buzzards from Germany. *Journal of Wildlife Diseases* 36: 559–561. Wildlife Disease Association.

Kruckenhauser, L., Haring, E., Pinsker, W., Riesing, M. J., Winkler, H., Wink, M. & Gamauf, A. 2004. Genetic vs. morphological differentiation of Old World buzzards (genus *Buteo*, Accipitridae). *Zoologica Scripta* 33: 197–211.

Krüger, O. 2002a. Interactions between common buzzard *Buteo buteo* and goshawk *Accipiter gentilis*: trade-offs revealed by a field experiment. *Oikos* 96: 441–452.

Krüger, O. 2002b. Dissecting common buzzard lifespan and lifetime reproductive success: The relative importance of food, competition, weather, habitat and individual attributes. *Oecologia* 133: 474–482.

Krüger, O. 2004a. The importance of competition, food, habitat, weather and phenotype for the reproduction of Buzzard *Buteo buteo*: Capsule Variation in reproduction between territories was strongly influenced by intra- and interspecific competition, phenotype, levels of r. *Bird Study* 51: 125–132.

Krüger, O. 2004b. The importance of competition, food, habitat, weather and phenotype for the reproduction of Buzzard *Buteo buteo*. *Bird Study* 51: 125–132.

Krüger, O. & Lindström, J. 2001. Lifetime reproductive success in common buzzard, Buteo buteo: from individual variation to population demography. *Oikos* 93: 260–273.

Krüger, O., Lindström, J. & Amos, W. 2001. Maladaptive mate choice maintained by heterozygote advantage. *Evolution* 55: 1207–1214.

Krüger, O. & Radford, A. N. 2008. Doomed to die? Predicting extinction risk in the true hawks Accipitridae. *Animal Conservation* 11: 83–91.

Lautenschlager, S., Bright, J. A. & Rayfield, E. J. 2013. Digital dissection – using contrast-enhanced computed tomography scanning to elucidate hard- and soft-tissue anatomy in the Common Buzzard Buteo buteo. *Journal of Anatomy* 224: 412–431.

Lees, A. C., Newton, I. & Balmford, A. 2013. Pheasants, buzzards, and trophic cascades. *Conservation Letters* 6: 141–144.

Lehikoinen, A., Byholm, P., Ranta, E., Saurola, P., Valkama, J., Korpimäki, E., Pietiäinen, H. & Henttonen, H. 2009. Reproduction of the common buzzard at its northern range margin under climatic change. *Oikos* 118: 829–836.

Lensink, R. 1997. Range Expansion of Raptors in Britain and the Netherlands Since the 1960s: Testing an Individual-Based Diffusion Model. *The Journal of Animal Ecology* 66: 811–826.

Leshem, Y. & Yom-Tov, Y. 1996. The magnitude and timing of migration by soaring raptors, pelicans and storks over Israel. *Ibis* 138: 188–203.

Likhopeck, E. A. 1970. Trophic links of the common buzzard in forests habitats (in Russian). *Biol Nauki* 8: 21–24.

Lind, O., Mitkus, M., Olsson, P. & Kelber, A. 2013. Ultraviolet sensitivity and colour vision in raptor foraging. *The Journal of Experimental Biology* 216: 1819–1826.

Linnell, J. D. C., Nilsen, E. B., Lande, U. S., Herfindal, I., Odden, J., Skogen, K., Andersen, R. & Breitenmoser, U. 2005. Zoning as a means of mitigating conflicts with large carnivores: principles and reality. *People and Wildlive, Conflict or Coexistence* 1: 165–172.

Literak, I. & Mraz, J. 2011. Adoptions of Young Common Buzzards in White-tailed Sea Eagle Nests. *The Wilson Journal of Ornithology* 123: 174–176.

Lõhmus, A. 2005. Are timber harvesting and conservation of nest sites of forest-dwelling raptors always mutually exclusive? *Animal Conservation* 8: 443–450.

Lõhmus, A. 2003. Are certain habitats better every year? A review and a case study on birds of prey. *Ecography* 26: 545–552.

López-Darias, M. 2007. First Documented Case of Double-brooding in the Eurasian Buzzard (*Buteo buteo*). *Journal of Raptor Research* 41: 340–341.

López y López-Leitón, T. J., Alvarez Piñeiro, M. E., Lage Yusty, M. A. & Simal Lozano, J. 2001. Aliphatic hydrocarbons in birds of prey from Galicia (NW Spain). *Ecotoxicology and Environmental Safety* 50: 44–47.

Lourenço, P. M. 2009. Rice field use by raptors in two Portuguese wetlands. *Airo* 19: 13–18.

Lourenço, R., Penteriani, V., Del mar Delgado, M., Marchi-Bartolozzi, M. & Rabaça, J. E. 2011a. Kill before being killed: An experimental approach supports the predator-removal hypothesis as a determinant of intraguild predation in top predators. *Behavioral Ecology and Sociobiology* 65: 1709–1714.

Lourenço, R., Santos, S. M., Rabaça, J. E. & Penteriani, V. 2011b. Superpredation patterns in four large European raptors. *Population Ecology* 53: 175–185.

Ludwig, S. C., McCluskie, A., KeanE, P., Barlow, C., Francksen, R. M., Bubb, D., Roos, S., Aebischer, N. J. & Baines, D. (n.d.). Diversionary feeding and nestling diet of Hen Harriers Circus cyaneus. *Bird Study*.

Ludwig, S. C., Roos, S., Bubb, D. & Baines, D. 2017. Long-term trends in abundance and breeding success of red grouse and hen harriers in relation to changing management of a Scottish grouse moor. *Wildlife Biology* 1: wlb.00246.

Madge, G. 1992. Buzzard with live common eel. *British Birds* 854: 187–188.

Maguire, E. J. 1979. Notes on the breeding buzzards in Kintyre. *Western Naturalist* 8: 313.

Maguire, J. 2015. Buzzard attack was 'like being hit by a rock' - BBC News.

Mallon, J. M., Bildstein, K. L. & Katzner, T. E. 2016. In-flight turbulence benefits soaring birds. *The Auk* 133: 79–85.

Mallord, J. W., Orsman, C. J., Cristinacce, A., Butcher, N., Stowe, T. J. & Charman, E. C. 2012. Mortality of Wood Warbler *Phylloscopus sibilatrix* nests in Welsh Oakwoods: predation rates and the identification of nest predators using miniature nest cameras. *Bird Study* 59: 286–295.

Malmiga, G., Nilsson, C., Bäckman, J. & Alerstam, T. 2014. Interspecific comparison of the flight performance between sparrowhawks and common buzzards migrating at the falsterbo peninsula: A radar study. *Current Zoology* 60: 670–679.

Mañosa, S. & Cordero, P. J. 1992. Seasonal and sexual variation in the diet of the common buzzard in northeastern spain. *The Journal of Raptor Research* 26: 235–238.

Mañosa, S., Mateo, R., Freixa, C. & Guitart, R. 2003. Persistent organochlorine contaminants in eggs of northern goshawk and Eurasian buzzard from northeastern Spain: Temporal trends related to changes in the diet. *Environmental Pollution* 122: 351–359.

Marcström, V., Kenward, R. E. & Engren, E. 1988. The impact of predation on boreal tetraonids during vole cycles: an experimental study. *Journal of Animal Ecology* 57: 859–872.

Marinova-Petkova, A., Georgiev, G., Seiler, P., Darnell, D., Franks, J., Krauss, S., Webby, R. J. & Webster, R. G. 2012. Spread of influenza virus A (H5N1) clade 2.3.2.1 to Bulgaria in Common Buzzards. *Emerging Infectious Diseases* 18: 1596–1602.

Del Marmol, P. 1997. Contribution to the study of the Buzzard (*Buteo buteo*) in Belgium: mortality and movements. *Aves* 34: 143–155.

Marsh, C. 1989. Talon grappling and cartwheeling by buzzards. *Devon Birds* 42: 48–49.

Martín, B., Onrubia, A. & Ferrer, M. 2014. Effects of climate change on the migratory behavior of the common buzzard *Buteo buteo*. *Climate Research* 60: 187–197.

Martin, G. R. 2015. What is binocular vision for? A birds' eye view. *Journal of vision* 9: 1–19.

Martínez-Padilla, J., Mougeot, F., García, J. T., Arroyo, B. & Bortolotti, G. R. 2013. Feather Corticosterone Levels and Carotenoid-Based Coloration in Common Buzzard (*Buteo buteo*) Nestlings. *Journal of Raptor Research* 47: 161–173.

Martínez, J. E., Jiménez-Franco, M. V., Zuberogoitia, I., León-Ortega, M. & Calvo, J. F. 2013. Assessing the short-term effects of an extreme storm on Mediterranean forest raptors. *Acta Oecologica* 48: 47–53.

Mateo, R., Taggart, M. & Meharg, A. A. 2003. Lead and arsenic in bones of birds of prey from Spain. *Environmental Pollution* 126: 107–114.

Maxwell, J. 2010. Report from the freezer. *Scottish Birds* 30: 39.

Mayfield, H. F. 1961. Nesting success calculated from exposure. *The Wilson Bulletin* 73: 255–261.

Mayfield, H. F. 1975. Suggestions for Calculating Nest Success. *The Wilson Bulletin* 87: 456–466.

Mead, C. J. 1973. Movements of British Raptors. *Bird Study* 20: 259–286. Taylor & Francis Group, Abingdon.

Mebs, T. 1964. Zur Biologie und Populationsdynamik des Mäusebussards (*Buteo buteo*) (Unter besonderer Berücksichtigung der Abhängigkeit vom Massenwechsel der FeldmausMicrotus arvalis). *Journal für Ornithologie* 105: 247–306. Springer-Verlag, Berlin, Heidelberg.

Melde, M. 1983. *Der Mäusebussard*. 2.Aufl. Neue Brehm- Bücherei, Wittenberg.

Mikkola, H. 1976. Owls killing and killed by other owls and raptors in Europe. *British Birds* 69: 144–154.

Mitchell, J. 1984. Common Buzzards feeding on fish carrion at Loch Lomond. *Scottish Birds* 13: 118.

Møller, A. P., Biard, C., Blount, J. D., Houston, D. C., Ninni, P., Saino, N. & Surai, P. F. 2000. Carotenoid-dependent Signals: Indicators of Foraging efficiency, Immunocompetence or Detoxification Ability? *Avian and Poultry Biology Reviews* 11: 137–159.

Moore, N. W. 1957. The past and present status of the buzzard in the British Isles. *British Birds* 50: 173–197.

Morris, F. O. 1862. *A History of British Birds, v. I*. Groombridge and Sons, London.

Mougeot, F. & Arroyo, B. E. 2006. Ultraviolet reflectance by the cere of raptors. *Biology letters* 2: 173–176.

Mulder, J., Mulder, J., Mulder, R. & Mulder, A. 2001. Buizerd *Buteo buteo* gestrikt in vliegertouw. *De Takkeling* 9: 156–158.

Mustin, K., Arroyo, B., Beja, P., Newey, S., Irivine, R. J., Kestler, J. & Redpath, S. M. 2018. Consequences of game bird management for non-game species in Europe. *Journal of Applied Ecology*: 1–11.

Naccari, C., Cristani, M., Cimino, F., Arcoraci, T. & Trombetta, D. 2009. Common buzzards (*Buteo buteo*) bio-indicators of heavy metals pollution in Sicily (Italy).

Nagy, K. A., Girard, I. A. & Brown, T. K. 1999. Energetics of free-ranging mammals, reptiles, and birds. *Annual Review of Nutrition* 19: 247–277.

Negro, J. J. & Grande, J. M. 2001. Territorial signalling: a new hypothesis to explain frequent copulation in raptorial birds. *Animal Behaviour* 62: 803–809.

Nelson, M. N. 1978. Preventing electrocution deaths and the use of nesting platforms over power poles. Pp. 42–46 in Geer, T. A. (ed.). *Bird of Prey Management Techniques.* British Falconer's Club, Oxford.

Neumann, J. & Schwarz, J. 2017. Seeadlerpaar mit einer besonderen Vorliebe für junge Mäusebussarde – Teil 3. Großvogelschutz im Wald. Jahresbericht 2017.

Newton, I. 1968. The temperatures, weights, and body composition of molting Bullfinches. *Condor* 70: 323–332.

Newton, I. 1979a. *Population Ecology of Raptors.* T & AD Poyser, Calton.

Newton, I. 1979b. Effects of Human Persecution on European Raptors. *Journal of Raptor Research* 13: 65–78.

Newton, I. 1985. Lifetime Reproductive Output of Female Sparrowhawks. *The Journal of Animal Ecology* 54: 241–253.

Newton, I. 1986. *The Sparrowhawk.* T. & A.D. Poyser, Calton.

Newton, I. 1998. *Population limitation in birds.* Academic Press, London.

Newton, I. 2001. Causes and Consequences of Breeding Dispersal in the Sparrowhawk *Accipiter nisus. Ardea* 89: 143–154.

Newton, I. 2010. *Bird migration.* Collins, London.

Newton, I., Davis, P. E. & Davis, J. E. 1982. Ravens and Buzzards in Relation to Sheep-Farming and Forestry in Wales. *The Journal of Applied Ecology* 19: 681–706.

Newton, I., Davis, P. E. & Moss, D. 1994. Philopatry and Population Growth of Red Kites, *Milvus milvus*, in Wales. *Proceedings of the Royal Society of London B: Biological Sciences* 257: 317–323.

Newton, I., McGrady, M. J. & Oli, M. K. 2016. A review of survival estimates for raptors and owls. *Ibis* 158: 227–248.

Van Nie, G. J. 1981. Avian tuberculosis in a free-living buzzard with bumblefoot (author's transl.). *Tijdschrift voor diergeneeskunde* 106: 1033–1036.

Nielsen, B. P. 1977. Danske Musvågers Buteo buteo trækforhold og spredning. *Dansk Orn. Foren. Tidsskr.* 71: 1–9.

Nielsen, B. P. & Christensen, S. 1969. On the autumn migration of Spotted Eagles and Buzzards in the Middle East. *Ibis* 111: 620–621.

Nikolov, S. C., Spasov, S. & Kambourova, N. 2006. Density, number and habitat use of Common Buzzard (*Buteo buteo*) wintering in the lowlands of Bulgaria. *Buteo* 15: 39–47.

Nore, T. 1979. Rapaces diurnes communs en Limousin pendant la période de nidification (Buse, Bondrée, Milan noir, Busards Saint-Martin et cendré). *Alauda* 47: 183–194.

Nore, T., Malafosse, J. P., Nore, G. & Buffard, E. 1992. La dispersion des jeunes de première année dans une population sédentaire de buse variable (*Buteo buteo*). *Rev. Ecol. (Terre Vie)* 47: 259–285.

Norrdahl, K. & Korpimaki, E. 2000. Do predators limit the abundance of alternative prey? Experiments with vole-eating avian and mammalian predators. *Oikos* 91: 528–540. Munksgaard International Publishers, Copenhagen.

Norris, K. J. 1990. Female choice and the quality of parental care in the great tit *Parus major*. *Behavioral Ecology and Sociobiology* 27: 275–281. Springer-Verlag, Berlin, Heidelberg.

Oaks, J. L., Gilbert, M., Virani, M. Z., Watson, R. T., Meteyer, C. U., Rideout, B. A., Shivaprasad, H. L., Ahmed, S., Chaudhry, M. J. I., Arshad, M., Mahmood, S., Ali, A. & Khan, A. A. 2004. Diclofenac residues as the cause of vulture population decline in Pakistan. *Nature* 427: 630–633. Nature Publishing Group, London.

Olech, B. & Pruszynski, M. 2000. Food caching or surplus killing in the Common Buzzard *Buteo buteo*? *Acta Ornithologica* 35: 215–216.

Olsson, O. 1958. Dispersal, migration, longevity and death causes of *Strix aluco*, *Buteo buteo*, *Ardea cinerea* and *Larus argentatus*. *Acta Vertebratica* 1: 91–189.

Palacios, C.-J. 2004. Current status and distribution of birds of prey in the Canary Islands. *Bird Conservation International* 14: 203–213.

Palko, S. 1997. Retisas (*Haliaeetus albicilla*) altal nevelt egereszolyv (*Buteo buteo*) fiokak. *Tuzok* 2: 109–111.

Palomino, D. & Carrascal, L. M. 2007. Habitat associations of a raptor community in a mosaic landscape of Central Spain under urban development. *Landscape and Urban Planning* 83: 268–274.

Panek, M. & Hušek, J. 2014. The effect of oilseed rape occurrence on main prey abundance and breeding success of the Common Buzzard *Buteo buteo*. *Bird Study* 61: 457–464.

Panting, P. J. 1955. Buzzards following the plough. *British Birds* 48: 412.

Panuccio, M., Duchi, A., Lucia, G. & Agostini, N. 2017. Species-Specific Behaviour of Raptors Migrating Across the Turkish Straits in Relation to Weather and Geography. *Ardeola* 64: 305–324.

Papathanasiou, J. & Kenward, R. 2014. Design of a data-driven environmental decision support system and testing of stakeholder data-collection. *Environmental Modelling & Software* 55: 92–106.

Paprocki, N., Heath, J. A. & Novak, S. J. 2014. Regional distribution shifts help explain local changes in wintering raptor abundance: Implications for interpreting population trends. *PLoS ONE* 9.

Paprocki, N., Oleyar, D., Brandes, D., Goodrich, L., Crewe, T. & Hoffman, S. W. 2017. Combining migration and wintering counts to enhance understanding of population change in a generalist raptor species, the North American Red-tailed Hawk. *The Condor* 119: 98–107.

Paradis, E., BailliE, S. R., Sutherland, W. J. & Gregory, R. D. 1998. Patterns of natal and breeding dispersal in birds. *Journal of Animal Ecology* 67: 518–536.

Parejo, D., Avilés, J. M. & Rodríguez, J. 2010. Visual cues and parental favouritism in a nocturnal bird. *Biology letters* 6: 171–173.

Park, K. J., Graham, K. E., Calladine, J. & Wernham, C. W. 2008. Impacts of birds of prey on gemebirds in the UK: a review. *Ibis* 150: 9–26.

Pearce-Higgins, J. W., Stephen, L., Langston, R. H. W., Bainbridge, I. P. & Bullman, R. 2009. The distribution of breeding birds around upland wind farms. *Journal of Applied Ecology* 46: 1323–1331.

Penhallurick, J. & Dickinson, E. C. 2008. The Correct Name of the 'Himalayan Buzzard' Is *Buteo* (*buteo*) *burmanicus*.

Bulletin of The British Ornithologists' Club 128: 131–132.

Pennycuick. 2008. *Modelling the flying bird.* P. *Theoretical Ecology Series.* Academic Press, London.

Pennycuick, C. J. 1972. Soaring behaviour and performance of some east African birds, observed from a motor-glider. *Ibis* 114: 178–218.

Penteriani, V. & Faivre, B. 1997. Breeding density and landscape-level habitat selection of common buzzards (*Buteo buteo*) in a mountain area (Abruzzo apennines, Italy). *Journal of Raptor Research* 31: 208–212.

Percival, S. M. 2005. Birds and windfarms: What are the real issues? *British Birds* 98: 194–204.

Perlman, Y. & Tsurim, I. 2008. Daring, risk assessment and body condition interactions in steppe buzzards *Buteo buteo vulpinus*. *Journal of Avian Biology* 39: 226–228.

Picozzi, N. & Weir, D. 1976. Dispersal and causes of death of Buzzards. *British Birds* 69: 193–201.

Picozzi, N. & Weir, D. N. 1974. Breeding biology of the Buzzard in Speyside. *British Birds* 67: 199–210.

Pinowski, J. & Ryszkowski, L. 1962. The Buzzard's versatility as a predator. *British Birds* 55: 470–475.

Pokrovsky, I., Ehrich, D., Ims, R. A., Kulikova, O., LecomtE, N. & Yoccoz, N. G. 2014. Diet, nesting density, and breeding success of rough-legged buzzards (*Buteo lagopus*) on the Nenetsky Ridge, Arctic Russia. *Polar Biology* 37: 447–457.

Porter, R. F. 1974. *Flight Identification of European Raptors.* A & C Black, London.

Porter, R. F. & Kirwan, G. M. 2010. Studies of Socotran birds VI. The taxonomic status of the Socotra Buzzard. *Bulletin of the British Ornithologists' Club* 130: 116–131.

Potts, G. R. 1998. Global dispersion of nesting Hen Harriers *Circus cyaneus*; implications for grouse moors in the UK. *Ibis* 140: 76–88.

Probst, R. 2002. Greifvogelüberwinterung 1998 bis 2002 im Bleistätter Moos, Kärnten. *Carinthia II* 114: 509–516.

Prop, J. & Quinn, J. L. 2003. Constrained by available raptor hosts and islands: density-dependent reproductive success in red-breasted geese. *Oikos* 102: 571–580.

Prytherch, R. 1997. Buzzards. *BBC Wildlife*: 23–29.

Prytherch, R. 2009. The social behviour of the Common Buzzard. *British Birds* 102: 247–273.

Prytherch, R. 2013. The breeding biology of the Common Buzzard. *British Birds* 106: 239–296.

Prytherch, R. 2016. Common Buzzard nests, nest trees and prey remains in Avon. *British Birds* 109: 256–264.

Prytherch, R. & Roberts, L. 2012. The dispersal of Common Buzzards ringed between 1984 and 2004 in North Somerset. *Bristol Ornithology* 31: 15–27.

Quinn, T. H. & Baumel, J. J. 1990. The digital tendon locking mechanism of the avian foot (Aves). *Zoomoprhology* 109: 281–293.

Raposo do Amaral, F., Sheldon, F. H., Gamauf, A., Haring, E., RiesinG, M., Silveira, L. F. & Wajntal, A. 2009. Patterns and processes of diversification in a widespread and ecologically diverse avian group, the buteonine hawks (Aves, Accipitridae). *Molecular Phylogenetics and Evolution* 53: 703–715.

Ratcliffe, D. A. 1965. Organo-chlorine residues in some raptor and corvid eggs from northern Britain. *British Birds* 58: 65–81.

Ratcliffe, D. A. 1970. Changes attributable to pesticides in egg breakage frequency and eggshell thickness in some British birds. *Journal of Applied Ecology* 7: 67–115.

Ratcliffe, D. A. 1980. *The Peregrine falcon.* T & AD Poyser, Calton.

Reading, C. J. & Davies, J. L. 1996. Predation by grass snakes (*Natrix natrix*) at a site in southern England. *Journal of Zoology* 239: 73–82. Blackwell Publishing, Oxford.

Redpath, S. M. & Thirgood, S. J. 1997. *Birds of Prey and Red Grouse.* Stationery Office, London.

Redpath, S. M., Thirgood, S. J. & Leckie, F. M. 2001. Does supplementary feeding reduce predation of red grouse by hen harriers? *Journal of Applied Ecology* 38: 1157–1168.

Reif, V., Jungell, S., Korpimäki, E., Tornberg, R. & Mykrä, S. 2004a. Numerical response of common buzzards and predation rate of main and alternative prey under fluctuating food conditions. *Annales Zoologici Fennici* 41: 599–607.

Reif, V., Tornberg, R. & Huhtala, K. 2004b. Juvenile grouse in the diet of some raptors. *Journal of Raptor Research* 38: 243–249.

Reif, V., Tornberg, R., Jungell, S. & Korpimäki, E. 2001. Diet variation of common buzzards in Finland supports the alternative prey hypothesis. *Ecography* 24: 267–274.

Riddle, O. 1908. The genesis of fault-bars in feathers and the cause of alternation of light and dark fundamental bars. *The Biological Bulletin* 14: 328–371. Marine Biological Laboratory.

Riesing, M. J., Kruckenhauser, L., Gamauf, A. & Haring, E. 2003. Molecular phylogeny of the genus Buteo (Aves: Accipitridae) based on mitochondrial marker sequences. *Molecular Phylogenetics and Evolution* 27: 328–342.

Robb, M. & Pop, R. 2012. An aberrantly coloured buzzard *Buteo bannermani* on Santo Antão, Cape Verde Islands, in November 2012, with notes on the past and present status of the species. *Zoologia Caboverdiana* 3: 87–90.

Robinson, R. A., Leech, D. & Clark, J. A. 2015. The Online Demography Report: Bird ringing and nest recording in Britain & Ireland in 2015. BTO.

Roche, J. 1977. Un recensement de Buses en plaine de Saone: quelques donnees concemant la nidification. *Jean-le-Blanc* 13: 49–63.

Rodríguez, B., Siverio, F., Rodríguez, A., Siverio, M., Hernández, J. J. & Figuerola, J. 2010. Density, habitat selection and breeding biology of Common Buzzards *Buteo buteo* in an insular environment. *Bird Study* 57: 75–83.

Rooney, E. 2013. Ecology and breeding biology of the common buzzard in Ireland. Queen's University Belfast.

Rooney, E. & Montgomery, W. I. 2013. Diet diversity of the Common Buzzard (*Buteo buteo*) in a vole-less environment. *Bird Study* 60: 147–155.

Rooney, E., Reid, N. & Montgomery, W. I. 2015. Supplementary feeding increases Common Buzzard *Buteo buteo* productivity but only in poor-quality habitat. *Ibis* 157: 181–185.

Roulin, A., Almasi, B., Rossi-PedruzzI, A., Ducrest, A. L., Wakamatsu, K., Miksik, I., BlounT, J. D., Jenni-Eiermann, S. & JennI, L. 2008. Corticosterone mediates the condition-dependent component of melanin-based coloration. *Animal Behaviour* 75: 1351–1358.

Rudebeck, G. 1950. The choice of prey and modes o fhunting of predatory birds with special reference to their selective effect. *Oikos* 2: 67–88.

Rudebeck, G. 1957. *Buteo buteo trizonatus*, a new buzzard from the Union of South Africa. *South African Animal Life* 4: 415–437.

Rudolf, V. H. W. & Antonovics, J. 2007. Disease transmission by cannibalism: rare event or common occurrence? *Proceedings. Biological sciences / The Royal Society* 274: 1205–1210. The Royal Society.

Rutz, C. 2008. The establishment of an urban bird population. *Journal of Animal Ecology* 77: 1008–1019. Blackwell Publishing, Oxford.

Salvati, L., Manganaro, A. & Ranazzi, L. 2001. Occurrence of the Common Buzzard (*Buteo buteo*) in Mediterranean Coastal woodlands: Wood size and vegetation affect patch occupation. *Ornithologischer Anzeiger* 40: 165–171.

Sánchez-Zapata, J. A. & Calvo, J. F. 1999. Raptor distribution in relation to landscape composition in semi-arid Mediterranean habitats. *Journal of Applied Ecology* 36: 254–262.

Sandor, A., Jansen, J. & Vansteelant, W. M. 2017. Understanding hunters' habits and motivations for shooting raptors in the Batumi raptor-migration bottleneck, southwest Georgia. *Sandgrouse* 39: 2–15.

Saurola, P. 1978. Artificial nest construction in Europe. Pp. 72–80 in Geer, T. A. (ed.). *Bird of Prey Management Techniques.* British Falconers Club.

Schindler, S., Hohmann, U., Probst, R., Nemeschkal, H.-L. & Spitzer, G. 2012. Territoriality and Habitat Use of Common Buzzards (*Buteo buteo*) During Late Autumn in Northern Germany. *Journal of Raptor Research* 46: 149–157.

Schmutz, S. & Schmutz, J. K. 1981. Inheritance of Color Phases of Ferruginous Hawks. *Condor* 83: 187–189.

Schnell, J. H. 1958. Nesting behavior and food habits of Goshawks in the Sierra Nevada of California. *Condor* 60: 377–403.

Schwartz, V. 2014. Egerészölyv (*Buteo buteo*) macskabagoly (*Strix aluco*) predációja a Visegrádi-hegységben. *Heliaca* 2014: 104–105.

Selås, V. 2001. Breeding density and brood size of Common Buzzard *Buteo buteo* in relation to snow cover in spring. *Ardea*: 471–479.

Selås, V. 1997. Nest-site selection by four sympatric forest raptors in southern Norway. *Journal of Raptor Research* 31: 16–25.

Selås, V. 2001. Predation on reptiles and birds by the common buzzard, *Buteo buteo*, in relation to changes in its main prey, voles. *Canadian Journal of Zoology* 79: 2086–2093.

Selås, V., Tveiten, R. & Aanonsen, O. M. 2007. Diet of Common Buzzards (*Buteo buteo*) in southern Norway determined from prey remains and video recordings. *Ornis Fennica* 84: 97–104.

Sergio, F., Blas, J., Blanco, G., Tanferna, A., López, L., Lemus, J. A & Hiraldo, F. 2011. Raptor nest decorations are a reliable threat against conspecifics. *Science* 331: 327–330.

Sergio, F., Boto, A., Scandolara, C. & Bogliani, G. 2002. Density, nest sites, diet and productivity of Common Buzzards (Buteo buteo) in the Italian pre-Alps. *Journal of Raptor Research* 36: 24–32.

Sergio, F. & Hiraldo, F. 2008. Intraguild predation in raptor assemblages: A review. *Ibis* 150: 132–145.

Sergio, F., Marchesi, L. & Pedrini, P. 2003. Spatial refugia and the coexistence of a diurnal raptor with its intraguild owl predator. *Journal of Animal Ecology* 72: 232–245.

Sergio, F., Scandolara, C., Marchesi, L., Pedrini, P. & Penteriani, V. 2005. Effect of agro-forestry and landscape changes on common buzzards (*Buteo buteo*) in the Alps: implications for conservation. *Animal Conservation* 7: 17–25.

Shamoun-Baranes, J., Leshem, Y., Yom-Tov, Y. & LiechtI, O. 2003. Differential use of thermal convection by soaring birds over central Israel. *The Condor* 105: 208–218.

Shamoun-Baranes, J., Van Loon, E., Van Gasteren, H., Van Belle, J., Bouten, W. & Buurma, L. 2006. A comparative analysis of the influence of weather on the flight altitudes of birds. *Bulletin of the American Meteorological Society* 87: 47–61.

Shannon, H. D., Young, G. S., Yates, M. A., Fuller, M. R. & Seegar, W. S. 2002. American white pelican soaring flight times and altitudes relative to changes in thermal depth and intensity. *Condor* 104: 679–683.

Sim, I. M. W., Campbell, L., Pain, D. J. & Wilson, J. D. 2000. Correlates of the population increase of Common buzzard *Buteo buteo* in the West Midlands between 1983 and 1996. *Bird Study* 47: 154–164.

Sim, I. M. W., Cross, A. V., Lamacraft, D. L. & Pain, D. J. 2001. Correlates of Common Buzzard *Buteo buteo* density and breeding success in the West Midlands. *Bird Study* 48: 317–329. Taylor & Francis Group, Abingdon.

Simpson, V. R., Walls, S. S., Cooper, J. E. & Kenward, R. E. 1997. Causes of mortality in radio-tracked Eurasian buzzards (*Buteo buteo*) in Dorset. Pp. 188–193 *Proc. 4th Conf. Eur. Comm. Assoc. Avian Vets. Association of Avian Veterinarians, Loughborough,.*

Siverio, F., Rodríguez, A. & Padilla, D. P. 2008. Kleptoparasitism by Eurasian buzzard (*Buteo buteo*) on two Falco species. *Journal Of Raptor Research* 42: 77–78.

Skierczyński, M. 2006. Food niche overlap of three sympatric raptors breeding in agricultural landscape in Western Pomerania region of Poland. *Buteo* 15: 17–22.

Slagsvold, T. & Sonerud, G. 2007. Prey size and ingestion rate in raptors: Importance for sex roles and reversed sexual size dimorphism. *Journal of Avian Biology* 38: 650–661.

Slagsvold, T., Sonerud, G. A., Grønlien, H. E. & Stige, L. C. 2010. Prey handling in raptors in relation to their morphology and feeding niches. *Journal of Avian Biology* 41: 488–497.

Smit, T., Eger, A., Haagsma, J. & Bakhuizen, T. 1987. Tuberculosis in Wild Birds in the Netherlands. *Journal of Wildlife Diseases* 23: 485–487.

Solomon, S., Qin, D., Manning, M., Chen, Z., Marquis, M., Averyt, K. B., M. Tignor, Miller, H. L. & (eds). 2007. IPCC, 2007: Climate Change 2007: The Physical Science Basis. Contribution of Working Group I to the Fourth Assessment Report of the Intergovernmental Panel on Climate Change.

Solonen, T. 1982. Nest-sites of the Common Buzzard Buteo buteo in Finland. *Ornis Fenneca* 59: 191–192.

Sonerud, G. A. 1986. Effect of snow cover on seaonal changes in diet, habitat and regional distribution of raptors that prey on samll mamals in boreal zones of Fennoscandia. *Holarctic Ecology* 9: 33–47.

Sonerud, G. A., Steen, R., Selås, V., Aanonsen, O. M., Aasen, G. H., Fagerland, K. L., Fosså, A., Kristiansen, L., Low, L. M., Ronning, M. E., Skouen, S. K., Asakskogen, E., Johansen, H. M., Johnsen, J. T., Karlsen, L. I., Nyhus, G. C., Roed, L. T., Skar, K., Sveen, B. A., Tveiten, R. & Slagsvold, T. 2014. Evolution of parental roles in provisioning birds: Diet determines role asymmetry in raptors. *Behavioral Ecology* 25: 762–772.

Šotnár, K. & Obuch, J. 2009. Feeding ecology of a nesting population of the Common Buzzard (*Buteo buteo*) in the

Upper Nitra Region, Central Slovakia. *Slovak Raptor Journal* 3: 13–20.

Šotnár, K. & Topercer, J. 2009. Estimating density, population size and dynamics of Common Buzzard (*Buteo buteo*) in the West Carpathian region by a new method. *Slovak Raptor Journal* 3: 1–12.

Spaar, R. 1995. Flight behavior of steppe buzzards (*Buteo buteo vulpinus*) during spring migration in southern Israel: A tracking-radar study. *Israel Journal of Zoology* 41: 489–500.

Spaar, R. & Bruderer, B. 1997. Optimal flight behavior of soaring migrants: a case study of migrating steppe buzzards, *Buteo buteo vulpinus*. *Behavioral Ecology* 8: 288–297.

Spaar, R., Liechti, O. & Bruderer, B. 2000. Forecasting flight altitudes and soaring performance of migrating raptors by the altitudinal profile and atmospheric conditions. *Organisation Scientifique et Technique du Vol a Voile* 24: 49–55.

Spidsø, T. K. & Selås, V. 1988. Prey selection and breeding success in the common buzzard *Buteo buteo* in relation to small rodent cycles in southern Norway. *Fauna Norvegica Serie C Cinclus* 11: 61–66.

Stefanek, P. R., Bowerman, W. W., Grubb, T. G. & HolT, J. B. 1992. Nestling Red-tailed Hawk in occupied Bald Eagle nest. *Journal of Raptor Research* 26: 40–41.

Stewart, A. 2007. *Wildlife Detective: A life fighting wildlife crime.* Argyll Publishing, Edinburgh.

Strandberg, R., Alerstam, T., Hake, M. & Kjellén, N. 2009a. Short-distance migration of the Common Buzzard *Buteo buteo* recorded by satellite tracking. *Ibis* 151: 200–206.

Strandberg, R., Klaassen, R. H. G. & Thorup, K. 2009b. Spatio-temporal distribution of migrating raptors: A comparison of ringing and satellite tracking. *Journal of Avian Biology* 40: 500–510.

Straughan, R. & Rooney, E. 2010. Common Buzzard *Buteo buteo* rearing broods of five and six in successive years in County Armagh. *Irish Birds* 9: 123–125.

Sunde, P., Odderskær, P. & Storgaard, K. 2009. Flight distances of incubating Common Buzzards *Buteo buteo* are independent of human disturbance. *Ardea* 97: 369–372.

Sundev, G., Yosef, R. & Birazana, O. 2009. Brandt's vole density affects nutritional condition of Upland Buzzard *Buteo hemilasius* on the Mongolian grassland steppe. *Ornis Fennica* 86: 131–139.

Svobodová, M., Voříček, P., Votýpka, J. & Weidinger, K. 2004. Heteroxenous coccidia (Apicomplexa: Sarcocystidae) in the populations of their final and intermediate hosts: European buzzard and small mammals. *Acta Protozoologica* 43: 251–260.

Swan, G. 2011. Spatial Variation in the Breeding Success of the Common Buzzard *Buteo buteo* in relation to Habitat Type and Diet. Imperial College, London.

Swann, R. L. & Etheridge, B. 1995. A comparison of breeding success and prey of the Common Buzzard *Buteo buteo* in two areas of northern Scotland. *Bird Study* 42: 37–43.

Sylvén, M. 1978. Interspecific Relations between Sympatrically Wintering Common Buzzards *Buteo buteo* and Rough-Legged Buzzards *Buteo lagopus*. *Ornis Scandinavica* 9: 197.

Taylor, K., Hudson, R. & Horne, G. 1988. Buzzard breeding distribution and abundance in Britain and Northern Ireland in 1983. *Bird Study* 35: 109–118.

Temmink, W. 2004. Common Buzzard *Buteo buteo* scavenges dead fish. *De Takkeling* 12: 156.

Teunissen, W., Schekkerman, H., Willems, F. & Majoor, F. 2008. Identifying predators of eggs and chicks of Lapwing *Vanellus vanellus* and Black-tailed Godwit *Limosa limosa* in the Netherlands and the importance of predation on wader reproductive output. *Ibis* 150: 74–85.

Thiede, S. & Krone, O. 2001. Polygranulomatosis in a common buzzard (*Buteo buteo*) due to Escherichia coli (Hjarre's disease). *Veterinary Record* 149: 774–776.

Thirgood, S. J., Redpath, S. M. & Graham, I. M. 2003. What determines the foraging distribution of raptors on heather moorland? *Oikos* 100: 15–24.

Tornberg, R. & Reif, V. 2007. Assessing the diet of birds of prey: A comparison of prey items found in nests and images. *Ornis Fennica* 84: 21–31.

Tornberg, R., Reif, V. & Korpimäki, E. 2012. What explains forest grouse mortality: Predation impacts of raptors, vole abundance, or weather conditions? *International Journal of Ecology* 2012: 1–10.

Towill, J. 1999. Interlocking of talons between Common Buzzards. *Scottish Birds* 20: 40.

Trivers, R. L. & Willard, D. E. 1973. Natural Selection of Parental Ability to Vary the Sex Ratio of Offspring. *Science* 179: 90–92.

Truszkowski, J. 1976. Role of the Common Buzzard (*Buteo buteo* L.) in agrocenoses of the middle Wielkopolska. *Polish Ecological Studies* 27: 31–60.

Tubbs, C. R. 1967. Population study of Buzzards in the New Forest during 1962-66. *British Birds* 60: 381–395.

Tubbs, C. R. 1972. Analysis of Nest Record Cards for the Buzzard. *Bird Study* 19: 97–104.

Tubbs, C. R. 1974. *The Buzzard*. David & Charles Limited, London.

Tubbs, C. R. & Tubbs, J. M. 1985. Buzzards Buteo buteo and land use in the new forest, Hampshire, England. *Biological Conservation* 31: 41–65.

Tucker, V. A. 2000. The deep fovea, sideways vision and spiral flight paths in raptors. *The Journal of Experimental Biology* 203: 3745–3754.

Turner, C. & SAGE, R. 2003. Fate of released pheasants. *Game Conservancy Annual Review*: 74–75.

Turzański, M. & Czuchnowski, R. 2008. Wybiórczość siedliskowa ptaków szponiastych Falconiformes i kruka *Corvus corax* w Ojcowskim Parku Narodowym: Habitat selection of raptors Falconiformes and the Raven *Corvus corax* in the Ojców National Park Wstęp Materiał i metody. *Prądnik. Prace i Materiały Muzeum im. Prof. Władysława Szafera* 18: 37–52.

Tyack, A. J., Walls, S. S. & KenwarD, R. E. 1998. Behaviour in the post-nestling dependence period of radio-tagged Common Buzzards *Buteo buteo*. *Ibis* 140: 58–63.

Ulfstrand, S. 1970. A procedure for analysing plumage variation and its application to a series of South Swedish Common Buzzards *Buteo buteo* (L.). *Ornis Scandinavica* 1: 107–113.

Ulfstrand, S. 1977. Plumage and Size Variation in Swedish Common Buzzards *Buteo buteo* L. (Aves, Accipitriformes). *Zoologica Scripta* 6: 69–75.

United Nations. 1992. Convention on biological diversity. *Diversity*: 30.

Väli, Ü., Sein, G., Laansalu, A. & Sellis, U. 2015. Milliseid elupaiku eelistavad meie viud? *Eesti Loodus* 2015: 636–640.

Väli, Ü. & Vainu, O. 2015. Short-distance migration of Estonian Common Buzzards *Buteo buteo*. *Ringing & Migration* 30: 81–83.

Vaughan, N., Lucas, E. A., Harris, S. & White, P. C. L. 2003. Habitat associations of European hares *Lepus europaeus* in England and Wales: Implications for farmland management. *Journal of Applied Ecology* 40: 163–175.

Viitala, J., Korplmäki, E., Palokangas, P. & Koivula, M. 1995. Attraction of kestrels to vole scent marks visible in ultraviolet light. *Nature* 373: 425–427. Nature Publishing Group, London.

Voříček, P. 2000. An extremely high population density of Common Buzzard in Biospehere Reserve Palava and its possible causes. *Buteo* 11: 51–56.

Voříček, P., Krištín, A., Obuch, J. & Votypka, J. 1997. Diet of Common Buzzard in the Czech Republic and its importance for game keeping. *Buteo* 9: 57–68.

Voříček, P., Votýpka, J., Zvára, K. & Svobodová, M. 1998. Heteroxenous coccidia increase the predation risk of parasitized rodents. *Parasitology* 117: 521–524.

Walls, S. S. & Kenward, R. E. 1994. The systematic study of radio-tagged raptors: II. Sociality and dispersal. Pp. 317–324 in Meyburg, B.-U. & Chancellor, R. D. (ed.). *Raptor Conservation Today. World Working Group on Birds of Prey.*

Walls, S. S. & Kenward, R. E. 1995. Movements of radio-tagged Common Buzzards Buteo buteo in their first year. *Ibis* 137: 177–182.

Walls, S. S. & Kenward, R. E. 1998. Movements of radio-tagged Buzzards Buteo buteo in early life. *Ibis* 140: 561–568.

Walls, S. S. & Kenward, R. E. 2001. Spatial consequences of relatedness and age in buzzards. *Animal Behaviour* 61: 1069–1078.

Walls, S. S., Kenward, R. E. & Holloway, G. J. 2005. Weather to disperse? Evidence that climatic conditions vertebrate influence dispersal. *Journal of Animal Ecology* 74: 190–197.

Walls, S. S., Mañosa, S., Fuller, R. M., Hodder, K. H. & Kenward, R. E. 1999. Is early dispersal enterprise or exile? Evidence from radio-tagged buzzards. *Journal of Avian Biology* 30: 407–415.

Watson, J. W. & Cunningham, B. 1996. Another occurrence of Bald Eagles rearing a Red-tailed Hawk. *Washington Birds* 5: 51–52.

Watson, J. W., Dawison, M. & Leschner, L. 1993. Bald Eagles rear Red-tailed Hawks. Journal. *Journal of Raptor Research* 27: 126–127.

Watson, M., Aebischer, N. J., Potts, G. R. & Ewald, J. A. 2007. The relative effects of raptor predation and shooting on overwinter mortality of grey partridges in the United Kingdom. *Journal of Applied Ecology* 44: 972–982.

Webb, G. J. W. 2014. *Wildlife Conservation: in the Belly of the Beast.* Charles Darwin University Press, Darwin, Australia.

Weidinger, K. 2009. Nest predators of woodland open-nesting songbirds in central Europe. *Ibis* 151: 352–360.

Weir, D. & Picozzi, N. 1975. Aspects of social behaviour in the Buzzard. *British Birds* 68: 125–141.

Weir, D. & Picozzi, N. 1983. Dispersion of buzzards in Speyside. *British Birds* 76: 66–78.

Weiss, N. & Yosef, R. 2010. Steppe Eagle (*Aquila nipalensis*) Hunts a Eurasian Buzzard (*Buteo buteo vulpinus*) While in Migration Over Eilat, Israel. *Journal of Raptor Research* 44: 77–78.

Widen, P. 1994. Habitat quality for raptors: a field experiment. *Journal of Avian Biology* 25: 219–223.

Wikar, D., Ciach, M., Bylicka, M. & Bylicka, M. 2008. Changes in habitat use by Common Buzzards (Buteo buteo) during non-breeding season in relation to winter conditions. *Polish Journal of Ecology* 56: 119–125.

Wood, H. B. 1950. Growth Bars in Feathers. *The Auk* 67: 486–491.

Woodford, M. 1960. *A Manual of Falconry*. A & C Black, London.

Wuczyński, A. 2003. Abundance of common buzzard (*Buteo buteo*) in the Central European wintering ground in relation to the weather conditions and food supply. *Buteo* 13: 11–20.

Wuczyński, A. 2005. Habitat use and hunting behaviour of Common Buzzards *Buteo buteo* wintering in south-western Poland. *Acta Ornithologica* 40: 147–154.

Yalden, D. W. 1987. The natural history of Domesday Cheshire. *The Naturalist* 112: 125–131.

Yosef, R., Gombobaatar, S. & Bortolotti, G. R. 2013. Sibling Competition Induces Stress Independent of Nutritional Status in Broods of Upland Buzzards. *Journal of Raptor Research* 47: 127–132.

Yosef, R., Tryjanowski, P. & Bildstein, K. 2002. Spring migration of adult and immature buzzards (Buteo buteo) through Elat, Israel: timing and body size. *Journal of Raptor Research* 36: 115–120.

Zahavi, A. 1975. Mate selection-A selection for a handicap. *Journal of Theoretical Biology* 53: 205–214.

Zuberogoitia, I., Martínez, J. A., Zabala, J., Martínez, J. E., Castillo, I., Azkona, A. & Hidalgo, S. 2005. Sexing, ageing and moult of Buzzards Buteo buteo in a southern European area. *Ringing & Migration* 22: 153–158.

Zuberogoitia, I., Martínez, J. E., Martínez, J. A., Zabala, J., Calvo, J. F., Azkona, A. & Pagán, I. 2008. The dho-gaza and mist net with Eurasian Eagle-Owl (*Bubo bubo*) lure: Effectiveness in capturing thirteen species of European raptors. *Journal of Raptor Research* 42: 48–51.

Zuberogoitia, I., Martínez, J. E., Martínez, J. A., Zabala, J., Calvo, J. F., Castillo, I., Azkona, A., Iraeta, A. & Hidalgo, S. 2006. Influence of management practices on nest site habitat selection, breeding and diet of the common buzzard Buteo buteo in two different areas of Spain. *Ardeola* 53: 83–98.

Index

Accipiter
   gentilis 7
   melanoleucus 20
   nisus 48
Accipitridae 15, 17–18
Adder 52
Aldrin 216, 237
Alectoris rufa 48
Alfalfa 93
aliphatic hydrocarbons 215
Alternative Prey Hypothesis (APH) 54, 55
Alytes obstetricans 52
amphibians 52, 204
Anas platyrhynchos 219
Anguila anguila 53
Anguis fragilis 52
Anthus pratensis 46
ants, carpenter 50
Apodemus
   flavicollis 42
   sylvaticus 34
Apus apus 134
Aquila
   adalberti 187
   chrysaetos 134
   nipalensis 210–11
Ardea cinerea 46
artefact collisions 214–15
Arvicola amphibius 42
Asio
   flammeus 53
   otus 53
avian influenza 205, 238
avian tuberculosis 205
Avocet, Pied 47
Aythya fuligula 48

Badger, Eurasian 56, 235
Beech 96, 130, 132, 140, 256–7
beetles, dung 50
Betula pendula 131
bills 20, 64, 71
biodiversity 94, 185, 215, 249, 252–3, 257, 259, 263

Birch, Silver 131
bird flu 205, 238
BirdLife International 229, 249
birds as prey 44–5, 67–8
   gamebirds 48–9, 53–4, 56–7
   juvenile birds 45–8
   songbirds 56
   waders 56
Blackbird, European 45, 50, 56, 71, 159, 178
Blackcap 193
bowerbirds 138
branching 151–2
Branta ruficollis 112
Brassica napus 93
breeding 140
   chicks 145–51
   courtship 125–7
   eggs 141–5
   fledglings 152–3
breeding season habitat 91–6
breeding success 154–6, 165–6
   competition 160–1
   estimating populations 222–9
   food 156–8
   human disturbance 161–2
   individual characteristics 163–5
   lifetime reproductive success (LRS) 164–5
   productivity and geography 162–3
   weather 158–60
British Atmospheric Data Centre (BADC) 186
British Geological Survey 85
British Trust for Ornithology (BTO) 183, 196, 199, 200, 260
Bubo
   bubo 98
   scandiacus 112
Bufo bufo 52
bumblefoot (ulcerative pododermatitis) 206
bustards 211
Buteo 15–19
   augur 13, 18

auguralis 16, 18
brachypterus 13, 16
burmanicus 15
buteo 11, 16
b. arrigonii 15
b. bannermani 14, 19
b. insularum 14, 127
b. menetriesi 15
b. pojana 15
b. refectus 15
b. rothschildi 14
b. vulpinus 14, 15, 16
galapagoensis 17, 154, 229
hemilasius 13, 16
jamaicensis 17, 69
japonicus 13, 15, 17, 18
lagopus 13
oreophilus 13, 16, 17, 18
platypterus 229
regalis 18
ridgwayi 229
rufinus 13, 14, 16, 17, 18
rufofuscus 13
socotraensis 17, 19
swainsoni 20
trizonatus 13, 16, 17
ventralis 229
Butterfly, Large White 50
Buzzard
    Augur 13, 18
    Cape Verde 14, 19
    Eastern 13
    Forest 13, 16, 17
    Himalayan 15
    Jackal 13, 18
    Long-legged 13, 14, 15, 19, 27, 241, 249
    Madagascar 13, 16, 18
    Mountain 13
    Socotra 19, 229
    Upland 13, 14, 34, 150, 241
Buzzard, Rough-legged 15, 18–19, 30, 112, 113, 234, 236, 240–1
    distribution 12–14, 29
    gender differences in territorial behaviour 122–3
    hunting 59–60, 63
    migration 192
    nests 127
    plumage 27–8
    population 229
    prey switching 54
Buzzard, Steppe 14–15, 16, 19, 210–11, 229
    hunting 70
    migration 183, 188, 189, 191, 194, 195, 205, 260
    plumage 24, 27

cainism 149–50
calls 28–9, 114–15
    courtship 125
    fledglings 152–3
Camponotus spp. 50
Canis lupus 263
Capercaillie, Western 48
Caracara cheriway 21
Caracara, Crested 21
Carbofuran 216, 217, 220, 237
Carnid flies (Carnidae) 203
carrion 50–2, 72–4
Carson, Rachel (*Silent Spring*) 216
Cathartes aura 22
ceres 20–1, 35–6, 50, 126, 149
Chaffinch, Common 46
characteristics 20–2
Chicken, Domestic 49
chicks 124–5, 145–51
    branching 151–2
    food requirements 32–3, 148–50
    tagging 140–1, 150–1
    weather 158–9
    wing length growth 146–7
Circaetus gallicus 28
Circus
    cyaneus 55
    pygargus 36
citizen science 27, 196, 260–2
climate change 238–41
Clostridium 136
Columba
    livia 74
    palumbus 33, 49
competition 160–1, 235
conflict, ground 121–2

conservation 249
   current attitudes 249–54
   modern conservation 259–62
   raptors or grouse? 254–9
Convention on Biological Diversity (CBD) 248, 252
copulation 125
Corvus
   corax 73
   cornix 73
   corone 133
   frugilegus 47
   monedula 47
   moneduloides 40
courtship 98, 125–7
Coypu 52
cranes 211
Crataegus spp. 131
crickets, field 50, 85, 182
Crow
   Carrion 133, 255
   Hooded 73, 247
   New Caledonian 40
Cuckoo, Common 68
Cuculus canorus 68
Cygnus oler 205

DDT 216–17, 237–8
Diclofenac 237
Dieldrin 216, 237
diet 31–57
diet studies 36
   cameras at Buzzard nests 37
   cameras at other bird nests 38
   chance observations 36
   combination techniques 40–1
   pellet analysis 38–9
   planned field observations 36–7
   prey remains at nests 37
   stomach content analysis 39
   tagging and tracking 39–40
disease 202–7, 238
dispersal 167–8, 195
   defining dispersal 168–74
   dispersal distance 183–8
   landscapes for settling 181–3
   whether Buzzards disperse 174–7, 179–81
   why Buzzards disperse 178–9

displays 98
   courtship 125–7
   roller-coaster display flight 116
   territorial behaviour 115–18
distribution 12–15
DNA analysis 16–17, 19, 127
Duck, Tufted 48

Eagle
   Bald 150
   Booted 28, 128, 135, 210, 219
   Golden 69, 113, 134, 188
   Harpy 69
   Short-toed Snake 28
   Spanish Imperial 187
   Steppe 210–11
   White-tailed 150
earthworms 31, 33–4, 40, 49–50, 56–7, 70–2, 108–10, 185, 235
   burrows 71–2, 86, 180, 181
   Carbofuran 217
Eel, European 53
eggs 136, 138, 141–5
   clutch sizes 145
   laying dates 143–4
Elaphe quatuorlineata 210–11
Erinaceus europaeus 42
Escherichia coli 205
eyes 21, 64–5, 70–1

Fagus sylvatica 96
Falco
   cherrug 211
   columbarius 256
   naumanni 130
   pelegrinoides 93
   peregrinus 31
   rusticolus 137
   tinnunculus 7
Falcon
   Barbary 93
   Saker 211, 238, 262
falconry 31, 33, 47–8, 50–1, 61, 98, 121, 153, 206, 233, 237–8
falcons 15, 20, 32, 60–5, 68–9, 121, 153, 212, 238
feathers 24–6, 126
   chicks 146–8
   fledglings 152

indicator of pollutants 215–16
  stress bars 34–5, 126
Federation of Associations for Hunting and Conservation in the EU (FACE) 249
feet 20, 22, 61–5, 69
  chicks 151
females 21–2, 150
  courtship 125–7
  female chicks 149
  gender differences in territorial behaviour 122–3
  incubation 142–3
  sex ratio 144–5
Ficedula hypoleuca 46
field survey 85–6
fighting 114, 121–2
fledglings 152–3
flight 22–4, 60–1
  roller-coaster display flight 116
  talon grappling 117–18
  territorial behaviour 115–18
flocking 182
  feeding sites 108–10
  migrating 190, 210
Flycatcher, Pied 46
food 31, 56–7
  breeding success 156–8
  food passing 117–18
  food requirements 31–6
  resources 231–2
  starvation 207–8, 219
Fox
  Arctic 240
  Red 44, 52, 56, 235, 240, 252, 255–6
Fratercula arctica 48
Frenkelia 204
Fringilla coelebs 46
Frog, Common 52

Gallus gallus domesticus 49
gamebirds 48–9, 53–4, 56, 57, 253
  raptors or grouse? 254–9
gamekeepers 245–8, 250–1, 254–5
  prosecutions 251, 260
Garrulus glandarius 38
Geotrupes spp. 50
Geranospiza caerulescens 20
global warming 144, 193, 195, 231, 238–41
Godwit, Black-tailed 46

Goose, Red-breasted 112–13
Goshawk, Northern 21, 31, 44, 53, 104, 121, 123, 135–6, 155, 158, 160, 171, 235, 255, 263
  attacks on Buzzards 208, 220, 236
  breeding success 158, 224–5
  cannibalism 114
  chicks 148–9
  early breeding success 226–7
  eggs 144
  fledglings 153
  grouse 54–5
  home ranges 84
  hunting 61–4, 69, 74
  mortality 198, 201
  nest-switching 111
  nests 93, 111–13, 128–9, 133, 136, 233
  organochlorine contamination 217
  population 229
  territoriality 178
GPS (global positioning system) tags 40, 83, 91, 92, 167, 168, 179, 187, 215
grouse 54–5, 254–9
Grouse
  Black 48, 257
  Hazel 48
Grouse, Red 39
  Joint Raptor Study, Langholm Estate, Scottish Borders 254–8
grouse moors 247
Gryllus campestris 50
Gull
  Great Black-backed 73
  Herring 71
gulls 27, 71, 253
Gyps fulvus 121
Gyrfalcon 137

H5N1 205
habitat use 76–7, 96–7
  breeding season habitat 91–6
  field survey 85–6
  habitat changes 232–4
  habitat use during the non-breeding season 84–91
Haliaeetus
  albicilla 150
  leucocephalus 150

Hare
    Brown 40, 43–4
    Mountain 43
hares 55, 62
Harpia harpyja 69
Harrier
    Hen 59, 65, 201, 254–9
    Marsh 113
    Montagu's 36, 126
harriers 24, 28, 32, 118, 178
Hawk
    Broad-winged 229
    Crane 20
    Ferruginous 17, 18, 20, 30, 234
    Galapagos 154
    Red-tailed 17, 69, 118–19
    Ridgway's 229
    Rufous-tailed 229
    Swainson's 20, 50, 217
hawthorn 131
heavy metals 215
Hedera helix 131
Hedgehog, West European 42
Heptachlor 216
Heron, Grey 46, 74
Hieraaetus pennatus 28
Hippoboscid flies (Hippoboscidae) 207
home ranges 77–80
    home range size 80–4
Honey-buzzard, European 15, 24, 93, 111, 113, 190
horse-flies 50
humans 237–8, 243–4, 263–4
    current attitudes 249–54
    historical human relations 244–9
    human disturbance of nests 161–2
    importance of the Common Buzzard 263
    modern conservation 259–62
    raptors or grouse? 254–9
hunting 58, 75
    adaptations for hunting 58–65
    hunting invertebrates 70–2
    hunting vertebrates 65–70
    scavenging 72–4
hybridisation 15–16, 19
Hydrobatidae 215

identification 22–9
incubation 136, 138, 141–5
indicator of pollutants 215–16
invertebrates 49–50
IUCN (International Union for Conservation of Nature) 211, 229, 241, 262
Ivy 131, 135, 140

Jackdaw, Western 47, 157
Jay 38, 46
Joint Raptor Study 254–8
    Langholm Moor Demonstration Project 255
juveniles 24–6

Kestrel, Common 22, 58, 61, 65, 69, 113, 157, 158, 205, 229
    breeding success 158
    Buzzard attacks 53
    hunting 58, 61, 65, 69
    power lines 212
Kestrel, Lesser 130
Kite
    Black 16, 98, 113, 133, 137, 229, 234
    Red 33, 51, 72, 109–10, 114, 218, 234, 235, 250
kites 24, 28, 32, 136, 178, 217, 244
kleptoparasitism 74, 75, 93

Lacerta vivipara 52
Lagopus lagopus scotica 39
lambs 51, 244, 245
land cover 85–91
    dispersal 180–2, 185
Land Cover Map of Great Britain (LCMGB) 85–6, 88–9, 93
landscapes for settling 181–3
Langholm Estate, Scottish Borders 226, 254–8
    Langholm Moor Demonstration Project 255
Lapwing, Northern 46, 249
Larsen traps 223, 247–8
Larus
    argentatus 71
    marinus 73
lead shot poisoning 218–19
leatherjackets 50

lemmings 112–13
Lemmus lemmus 112
Lepus
    europaeus 40, 43
    timidus 43
Leucocytozoon
    buteonis 204
    toddi 203
lifetime reproductive success (LRS) 164–5
Limosa limosa 46
Linnaeus, Carolus 11
Lizard
    Common 52
    Ocellated 52
longevity 196, 219–20
    disease 202–7
    long-lived Buzzards 196–202
    natural enemies 208–11
    starvation 207–8, 219
    unnatural unintentional deaths 211–19

Magpie 38, 73, 247
Main Prey Hypothesis (MPH) 54
males 21–2, 150
    courtship 125–7
    gender differences in territorial
        behaviour 122–3
    incubation 142, 145
    male chicks 149
    sex ratio 144–5
Mallard 219
mammals as prey 41–4
mantling 69–70, 121
martens 136, 235, 263
Martes spp. 136
Martin, Sand 134
mating 125
Medicago sativa 93
Meles meles 56
Merlin 256
Microtus
    agrestis 32
    arvalis 42
migration 189–95
Milvus
    migrans 16
    milvus 33
Mole 42, 47

molecular analysis 16–17, 20, 59
monogamy 126–7
morphs 20, 26–7, 71, 104, 113
    breeding success 164
    laying dates 143
    mate choice 126
    parasite resistance 202–4
    sex ratio 144–5
mortality 196, 219–20
    disease 202–7
    longevity and survival 196–202
    natural enemies 208–11
    poisoning 215–19, 220, 237–8
    power lines 211–12, 220, 238
    starvation 207–8, 219
    vehicle and artefact collisions 214–15, 220
    wind turbines 212–13
Motacilla alba yarrellii 110
moulting 24–6, 126
Mouse
    Wood 34
    Yellow-necked 42
Mustela
    erminea 235
    nivalis 52
Mycobacterium
    fortuitum 205
    terrae 205
Myocastor coypu 52
Myodes glareolus 42
Myopus schisticolor 112
myxomatosis 43, 44, 55–6, 156–7, 232, 246–8

Natrix natrix 52
natural enemies 208–11
nearest-neighbour distances (NND) 99
nests 124–5, 139, 140–1
    breeding season habitat 91–6
    decoration 136–8
    diet studies 37–8
    how many nests? 134–6
    interspecific nest defence 111–13
    nest-helping 110–11
    nest-sites 127–30
    nest sizes 138
    nest-spacing 98–101

nests on cliffs, buildings and the ground 133–4
tree nests 130–3
North Atlantic Oscillation (NAO) 239

oil spills 215
organochlorines 216–17, 220, 238, 249
Oryctolagus cuniculus 31
Osprey 53, 113
osteodystrophy 206–7
Ostrich 71
Otus scops 36
Ovis aries 51
Owl
    Barn 34–5, 53, 215
    Eurasian Eagle 98, 111, 113, 123, 208, 210, 220, 235–6
    Eurasian Scops 36, 149
    Long-eared 53
    Short-eared 53
    Snowy 112, 122
    Tawny 53, 210, 212, 215

Pandion haliaetus 53
parasites 136, 202–7
    Grouse, Red 254
Partridge
    Grey 48, 253
    Red-legged 48
Parus major 34
Passer domesticus 244
pathogens 136, 203, 238
Pavo cristatus 34
Peafowl, Indian 34, 137–8
pellets 38–9, 51–2
perches 60, 66–7
    territorial behaviour 119–21
Perdix perdix 48
Peregrine 55, 113, 114, 129, 208–10, 214, 220, 233, 255
    attacks on Buzzards 208–10, 220, 263
    DDT 216–17, 237–8
    diet 31, 33, 47–8
    hunting 58, 62, 64, 65, 74
    population 229
Pernis apivorus 15
persecution 244–9
    current attitudes 249–54

persistent organic pollutants (POPs) 215–16
pesticides 161, 181, 215–17, 220, 223, 237–8, 249
Phalarope, Grey 48
Phalaropus fulicarius 48
Phasianus colchicus 40
Pheasant, Ring-necked 40–1, 46, 48, 51, 68, 214, 231
    pheasant pens 51, 201, 250–2
    shooting 218–19, 253, 257
Phylloscopus sibilatrix 46
Pica pica 38
Pieris brassicae 50
Pigeon, Feral 74
pigeons 211
Pine
    Aleppo 135
    Corsican 130, 131
    Scots 91, 133
Pinus
    halepensis 135
    nigra 130
    sylvestris 91
Pipit, Meadow 46, 68, 254
plovers 71
plucking posts 44, 47, 149
plumage 20, 24–6, 111
    morphs 20, 26–7, 71, 104, 113, 126
poisoning 215–19, 220, 237–8
    deliberate poisoning 246, 248
polygamy 126–7
polygranulomatosis 205
populations 221, 241–2
    climate change 238–41
    competition 235
    current Common Buzzard populations 229–31
    disease 238
    estimating populations 222–9
    factors affecting populations 231–41
    poisoning 237–8
    predation 235–7
    resources 231–4
power lines 211–12, 220, 238
predation 235–7
prey 41, 56–7
    birds 44–8
    carrion 50–2

Index

data collection and biases in diet studies 36–41
   gamebirds 48–9
   invertebrates 49–50
   mammals 41–4
   oddities 53
   reptiles and amphibians 52
prey switching 53–4
   dietary switching in the British Isles 55–6
   dietary switching in the north of Europe 54–5
Protection of Birds Act 1954 202, 223
Protection of Birds Act 1967 223
Psammodromus algirus 52
Psammodromus, Large 52
Ptilonorhynchidae 138
Puffin, Atlantic 48

Rabbit 62–3, 65, 69–70, 73, 84–5, 97, 108, 110, 133, 149, 185, 214, 216–17, 231–2, 235, 253
   breeding success in Buzzards 155–7, 159, 163
   burrows 85, 93–4, 157, 180
   Buzzard diets 31, 32–4, 36–8, 40–4, 46–8, 51, 55–7
   myxomatosis 43, 44, 55–6, 156–7, 232, 246–8
radio-tagging 39–40, 76–7
   chicks 140–1, 150–1
   estimating populations 224–8
   home range size 80–4
   longevity and survival 196–202, 219, 224
   tracking dispersal 167–81
Rana temporaria 52
Rape, Oilseed 93, 232
Ratcliffe, Derek (*The Peregrine Falcon*) 209, 216
Raven, Northern 73, 133, 209, 211, 218, 244, 247
Recurvirostra avosetta 47
reptiles 52, 54, 66, 158
ringing 154, 175–6, 184, 189, 194, 224–5, 246
   longevity and survival 196–202, 219, 224
Riparia riparia 134

rodenticides 217–18, 220
Rook 47
roosting 38, 40, 64, 82, 84, 86, 88, 96, 105, 108, 121, 123, 255, 256
Rowan 131

Sarcocystis dispersa 204
scavenging 50–2, 72–4
   indirect poisoning 218–19
Sciurus
   carolinensis 32
   vulgaris 42
Scolopax rusticola 49
sexual dimorphism 21–2, 143
Shared Prey Hypothesis (SPH) 54–5
sheep 51–2, 72, 73, 95, 216, 245, 247, 254–6
shooting 50, 196, 198, 245, 249
   gamebirds 48–9, 53–4, 162, 218–19, 245, 250, 252–61, 263
shrews 24, 40, 42, 55, 148
skulls 21
Slow Worm 52
Snake, Four-lined 210–11
   Grass 52
songbirds 50, 56, 253
Sorbus aucuparia 131
Sparrow, House 244
Sparrowhawk, Black 20
Sparrowhawk, Eurasian 48, 49, 58, 136, 164, 191, 233, 253, 260
   breeding success 164
   nest-helping 111
   nests 128, 136
   population 229
   territoriality 178
   territory-switching 114
Squirrel
   Grey 32, 133, 256–7
   Red 42
squirrels 38, 62, 86
Starling, European 45
starvation 207–8, 219
Stoat 235, 255
storm-petrels 215
Streptococcus aureus 206
Strigea falconispalumbi 204
Strix aluco 53

Struthio camelus 71
Sturnus vulgaris 45
Swan, Mute 205
Swift, Common 134
Sylvia atricapilla 193

Tabanidae 50
tails 28, 34–5
    peafowl 34, 137–8
talon grappling 117–18
Talpa europaea 42
taxonomy 15–19
    molecular analysis 16–17, 20, 59
territorial behaviour 113–14, 123
    calls 114–15
    flights 115–18
    gender differences 122–3
    ground conflicts 121–2
    perches 119–21
    roller-coaster display flight 116
    territorial ceilings 118–19
territoriality 98–105
    dispersal after fledging 178
    interspecific nest defence 111–13
    single birds 104–5
    territorial boundaries 101–2
    territory sizes 102–3
    tolerance 105–11
Tetrao
    tetrix 48
    urogallus 48
Tetrastes bonasia 48
thermals 22, 23, 102, 115, 125, 169, 207, 213
    dispersal 186, 188, 195
    migration 189–91
Thrush, Song 45, 56
ticks (Ixodida) 207
Timon lepidus 52
Tipula spp. 50
Tit, Great 34, 154
Toad
    Common 52
    Midwife 52
Toxoplasma 204
tracking 39–40
tree nesting 130–3
trematodosis 204

Tubbs, Colin (*The Buzzard*) 39, 244
Turdus
    merula 45
    philomelos 45
Typhaeus typhaeus 50
Tyto alba 34–5, 53, 215

Vanellus vanellus 46
vehicle collisions 214–15, 220
Vipera berus 52
viral haemorrhagic enteritis 55–6
Vole
    Bank 42
    Common 42, 232
    Short-tailed 32, 42
    Water 42
voles 32–4, 37, 40–5, 47–8, 63–5, 70, 148–9, 163, 182, 247, 254–6
    Frenkelia 204
    population fluctuations 54–7, 85, 113, 156–7, 158–9, 225, 232, 235
Vulpes
    lagopus 240
    vulpes 44, 240
Vulture
    Griffon 121, 169
    Turkey 22, 51, 188
vultures 15, 20, 51, 169, 122, 213, 234, 238
    Diclofenac 237

waders 56, 68, 71, 257
Wagtail, Pied 110
Warbler, Wood 46
Weasel, Least 42, 52
weather 158–60, 242
    climate change 238–41
    dispersal 186–8
White Nile fever 238
Wildlife and Countryside Act 1981 223, 247
wind turbines 212–13
Wolf, Grey 263
Woodcock, Eurasian 49
woodlands 36, 78, 83–9, 91–7, 232–3
    nest-sites 127–9
    tree nests 130–3
Woodpigeon 33, 34, 40, 44, 46, 49, 67, 157, 163
worms *see* earthworms